THE PENDULUM

The Pendulum
A Case Study in Physics

GREGORY L. BAKER

Bryn Athyn College of the New Church,
Pennsylvania, USA

and

JAMES A. BLACKBURN

Wilfrid Laurier University,
Ontario, Canada

OXFORD
UNIVERSITY PRESS

Great Clarendon Street, Oxford OX2 6DP

Oxford University Press is a department of the University of Oxford.
It furthers the University's objective of excellence in research, scholarship,
and education by publishing worldwide in

Oxford New York

Auckland Cape Town Dar es Salaam Hong Kong Karachi
Kuala Lumpur Madrid Melbourne Mexico City Nairobi
New Delhi Shanghai Taipei Toronto

With offices in

Argentina Austria Brazil Chile Czech Republic France Greece
Guatemala Hungary Italy Japan Poland Portugal Singapore
South Korea Switzerland Thailand Turkey Ukraine Vietnam

Oxford is a registered trade mark of Oxford University Press
in the UK and in certain other countries

Published in the United States
by Oxford University Press Inc., New York

First published 2005
First published in paperback 2009

British Library Cataloguing in Publication Data
(Data available)

Library of Congress Cataloguing in Publication Data
(Data available)

Typeset by Newgen Imaging Systems (P) Ltd., Chennai, India
Printed in Great Britain
on acid-free paper by Antony Rowe, Chippenham

ISBN 978-0-19-856754-7 (Hbk.) 978-0-19-955768-4 (Pbk.)

10 9 8 7 6 5 4 3 2 1

Preface

To look at a thing is quite different from seeing a thing

(Oscar Wilde, from *An Ideal Husband*)

The pendulum: a case study in physics is an unusual book in several ways. Most distinctively, it is organized around a single physical system, the pendulum, in contrast to conventional texts that remain confined to single fields such as electromagnetism or classical mechanics. In other words, the pendulum is the central focus, but from this main path we branch to many important areas of physics, technology, and the history of science.

Everyone is familiar with the basic behavior of a simple pendulum—a pivoted rod with a mass attached to the free end. The grandfather clock comes to mind. It might seem that there is not much to be said about such an elemental system, or that its dynamical possibilities would be limited. But, in reality, this is a very complex system masquerading as a simple one.

On closer examination, the pendulum exhibits a remarkable variety of motions. By considering pendulum dynamics, with and without external forcing, we are drawn to the essential ideas of linearity and nonlinearity in driven systems, including chaos. Coupled pendulums can become synchronized, a behavior noted by Christiaan Huygens in the seventeenth century. Even quantum mechanics can be brought to bear on this simple type of oscillator. The pendulum has intriguing connections to superconducting devices. Looking at applications of pendulums we are led to measurements of the gravitational constant, viscosity, the attraction of charged particles, the equivalence principle, and time.

While the study of physics is typically motivated by the wish to understand physical laws, to understand how the physical world works, and, through research, to explore the details of those laws, this science continues to be enormously important in the human economy and polity. The pendulum, in its own way, is also part of this development. Not just a device of pure physics, the pendulum is fascinating because of its intriguing history and the range of its technical applications spanning many fields and several centuries. Thus we encounter, in this book, Galileo, Cavendish, Coulomb, Foucault, Kamerlingh Onnes, Josephson, and others.

We contemplated a range of possibilities for the structure and flavor of our book. The wide coverage and historical connections suggested a broad approach suited to a fairly general audience. However, a book without equations would mean using words to try to convey the beauty of the theoretical (mathematical) basis for the physics of the pendulum. Graphs and equations give physics its predictive power and preeminent place in our understanding of the physical world. With this in mind, we opted instead for a thorough technical treatment. In places we have supplied background material for the nonexpert reader; for example, in the chapter on the quantum pendulum, we include a short introduction to the main ideas of quantum physics.

There is another significant difference between this book and standard physics texts. As noted, this work focuses on a single topic, the pendulum. Yet, in conventional physics books, the pendulum usually appears only as an illustration of a particular theory or phenomenon. A classical mechanics text might treat the pendulum within a certain context, whereas a book on chaotic dynamics might describe the pendulum with a very different emphasis. In the event that a book on quantum mechanics were to consider the pendulum, it would do so from yet another point of view. In contrast, here we have gathered together these many threads and made the pendulum the unifying concept.

Finally, we believe that *The Pendulum: A Case Study in Physics* may well serve as a model for a new kind of course in physics, one that would take a thematic approach, thereby conveying something of the interrelation of disciplines in the real progress of science. To gain a full measure of understanding, the requisite mathematics would include calculus up to ordinary differential equations. Exposure to an introductory physics course would also be helpful. A number of exercises are included for those who do wish to use this as a text. For the more casual reader, a natural curiosity and some ability to understand graphs are probably sufficient to gain a sense of the richness of the science associated with this complex device.

We began this project thinking to create a book that would be something of an encyclopedia on the topic, one volume holding all the facts about pendulums. But the list of potential topics proved to be astonishingly extensive and varied—too long, as it turned out, for this text. So from many possibilities, we have made the choices found in these pages.

The book, then, is a theme and variations. We hope the reader will find it a rich and satisfying discourse.

Acknowledgments

We are indebted to many individuals for helping us bring this project to completion. In the summer of 2003, we visited several scientific museums in order to get a first hand look at some of the famous pendulums to which we refer in the book. In the course of these visits various curators and other staff members were very generous in allowing us access to the museum collections. We wish to acknowledge the hospitality of Adrian Whicher, Assistant Curator, Classical Physics, the Science Museum of London, Jonathan Betts, Senior Curator of Horology, Matthew Read, Assistant Curator of Horology, and Janet Small, all of the National Maritime Museum, Greenwich, G. A. C. Veeneman, Director, Hans Hooijmaijers, Curator, and Robert de Bruin all of the Boerhaave Museum of Leyden, and Laurence Bobis, Directrice de la Bibliothèque de L'Observatoire de Paris. We wish also to thank William Tobin for helpful information on Léon Foucault and his famous pendulum, James Yorke for some historical information, Juan Sanmartin for further articles on O Botafumeiro, Bernie Nickel for useful discussions about the Foucault pendulum at the University of Guelph, Susan Henley and William Underwood of the Society of Exploration Geophysicists, and Rajarshi Roy and Steven Strogatz for useful discussions. Others to whom we owe thanks are Margaret Walker, Bob Whitaker, Philip Hannah, Bob Holström, editor of the Horological Science Newsletter, and Danny Hillis and David Munro, both associated with the Long Now clock project. For clarifying some matters of Latin grammar, JAB thanks Professors Joann Freed and Judy Fletcher of Wilfrid Laurier University. Finally, both of us would like to express gratitude to our colleague and friend, John Smith, who has made significant contributions to the experimental work described in the chapters on the chaotic pendulum and synchronized pendulums.

Library and other media resources are important for this work. We would like to thank Rachel Longstaff, Nancy Mitzen, and Carroll Odhner of the Swedenborg Library of Bryn Athyn College, Amy Gillingham of the Library, University of Guelph, for providing copies of correspondence between Christiaan Huygens and his father, Nancy Shader, Charles Greene, and the staff of the Princeton Manuscript Library. GLB wishes to thank Charles Lindsay, Dean of Bryn Athyn College for helping to arrange sabbaticals that expedited this work, Jennifer Beiswenger and Charles Ebert for computer help, and the Research committee of the Academy of the New Church for ongoing financial support.

Financial support for JAB was provided through a Discovery Grant from the Natural Sciences and Engineering Research Council of Canada.

Acknowledgments

The nature of this book provided a strong incentive to use figures from a wide variety of sources. We have made every effort to determine original sources and obtain permissions for the use of these illustrations. A large number, especially of historical figures or pictures of experimental apparatus, were taken from books, scientific journals, and from museum sources. Credit for individual figures is found in the respective captions. Many researchers generously gave us permission to use figures from their publications. In this connection we thank G. D'Anna, John Bird, Beryl Clotfelter, Richard Crane, Jens Gundlach, John Lindner, Gabriel Luther, Riley Neuman, Juan Sanmartin, Donald Sullivan, and James Yorke. The book contains a few figures created by parties whom we were unable to locate. We thank those publishers who either waived or reduced fees for use of figures from books.

It has been a pleasure working with OUP on this project and we wish to express our special thanks to Sonke Adlung, physical science editor, Tamsin Langrishe, assistant commissioning editor, and Anita Petrie, production editor.

Finally, we wish to express profound gratitude to our wives, Margaret Baker and Helena Stone, for their support and encouragement through the course of this work.

Contents

Introduction

1

The pendulum is a familiar object. Its most common appearance is in old-fashioned clocks that, even in this day of quartz timepieces and atomic clocks, remain quite popular. Much of the pendulum's fascination comes from the well known regularity of its swing and thus its link to the fundamental natural force of gravity. Older students of music are very familiar with the adjustable regularity of that inverted ticking pendulum known as a metronome. The pendulum's influence has extended even to the arts where it appears as the title of at least one work of fiction—Umberto Eco's *Fourcault's Pendulum*, in the title of an award winning Belgian film *Mrs. Foucault's Pendulum*, and as the object of terror in Edgar Allen Poe's 1842 short story *The Pit and the Pendulum*.

The history of the physics of the pendulum stretches back to the early moments of modern science itself. We might begin with the story, perhaps apocryphal, of Galileo's observation of the swinging chandeliers in the cathedral at Pisa. By using his own heart rate as a clock, Galileo presumably made the *quantitative* observation that, for a given pendulum, the time or period of a swing was independent of the amplitude of the pendulum's displacement. Like many other seminal observations in science, this one was only an approximation of reality. Yet it had the main ingredients of the scientific enterprise; observation, analysis, and conclusion. Galileo was one of the first of the modern scientists, and the pendulum was among the first objects of scientific enquiry.

Chapters 2 and 3 describe much of the basic physics of the pendulum, introducing the pendulum's equation of motion and exploring the implications of its solution. We describe the concepts of period, frequency, resonance, conservation of energy as well as some basic tools in dynamics, including phase space and Fourier spectra. Much of the initial treatment—Chapter 2—approximates the motion of the pendulum to the case of small amplitude oscillation; the so-called *linearization* of the pendulum's gravitational restoring force. Linearization allows for a simpler mathematical treatment and readily connects the pendulum to other simple oscillators such as the idealized spring or the oscillations of certain simple electrical circuits.

For almost two centuries geoscientists used the small amplitude, linearized pendulum, in many forms to determine the acceleration due to gravity, g, at diverse geographical locations. More refined studies led to a better understanding of the earth's density near geological formations. The variations in the local gravitational field imply, among other things, that the earth has a slightly nonspherical shape. As early as 1672, the French astronomer Jean Richer observed that a pendulum clock at the equator would only keep correct time if the pendulum were shortened as compared to its length in Paris. From this empirical fact, the Dutch physicist Huygens made some early (but incorrect) deductions about the earth's shape. On the other hand, the nineteenth century Russian scientist, Sawitch timed a pendulum at twelve different stations and computed the earth's shape distortion from spherical as one part in about 300—a number close to the presently accepted value. During the period from the early 1800s up into the early twentieth century, many local measurements of the acceleration due to gravity were made with pendulum-like devices. The challenge of making these difficult measurements and drawing appropriate conclusions captured the interest of many workers such as Sir George Airy and Oliver Heaviside, who are more often known for their scientific achievements in other areas.

Chapter 3 continues the discussion first by adding the physical effects of damping and forcing to the *linearized* pendulum and then by a consideration of the full *nonlinear* pendulum, which is important for large amplitude motion. Furthermore, real pendulums do not just keep going forever, because in this world of increasing entropy, motion is dissipated. These dissipative effects must be included as must the compensating addition of energy that keeps the pendulum moving in spite of dissipation. The playground swing is a common yet surprisingly interesting example. A child can pump the swing herself using either sitting or standing techniques. Alternatively, she can prevail upon a friend to push the swing with a periodic pulse. Generally the pulse resonates with the natural motion of the swing, but interesting phenomena occur when forcing is done at an off-resonant frequency. Analysis of these possibilities involves a variety of mechanical considerations including, changing center of mass, parametric pumping, conservation of angular momentum, and so forth. Another, more exotic, example is provided by the large amplitude motion of the huge incense pendulum in the cathedral of Santiago de Compostela, Spain. For almost a thousand years, centuries before Galileo's pendular experiments, pilgrims have worshiped there to the accompanying swishing sound of the incense pendulum as it traverses a path across the transept with an angular amplitude of over eighty degrees. Finally, the chapter ends with a consideration of the most famous literary use of the pendulum; Edgar Allan Poe's nightmarish story *The Pit and the Pendulum*. Does Poe, the non-scientist, provide enough details for a physical analysis? Chapter 3 suggests some answers.

Chapter 4 connects the pendulum to the rotational motion of the earth. From the early nineteenth century, it was supposed that the earth's rotation on its axis should be amenable to observation. By that time, classical mechanics was a developed and mature mathematical science. Mechanics predicted that additional noninertial forces, centrifugal and coriolis forces, would arise in the description of motion as it appeared from an accelerating (rotating) frame of reference such as the earth. Coriolis force—causing an apparent sideways displacement in the motion of an object—as seen by an earthbound observer, would be a dramatic demonstration of the earth's rotation. Yet the calculated effect was small.

In 1851, Léon Foucault demonstrated Coriolis force with a very large pendulum hung from the dome of the Pantheon in Paris (Tobin and Pippard 1994). His pendulum oscillated very slowly and with each oscillation the plane of oscillation rotated very slightly. With the pendulum, the coriolis force was demonstrated in a cumulative fashion. While the pendulum gradually ran down and needed to be restarted every 5 or 6 hours, its plane of oscillation rotated by about 60 or 70 degrees in that time. The plane rotated through a full circle in about 30 hours. In actuality, the plane of oscillation did *not* rotate; *the earth rotated under the pendulum*. If the pendulum had been located at the North or South poles, the full rotation would occur in 24 hours, whereas a pendulum located at the equator would not appear to rotate at all. Foucault's demonstration was very dramatic and immediately captured the popular imagination. Even Louis Napoleon, the president of France, used his influence to hasten the construction of the Pantheon version. Foucault's work was also immediately and widely discussed in the scientific literature (Wheatstone 1851).

The large size of the original Foucault demonstration pendulum masked some important secondary effects that became the subject of much experimental and theoretical work. As late as the 1990s the scientific literature shows that efforts are still being made to devise apparatus that controls these spurious effects (Crane 1995).

Foucault's pendulum demonstrates the rotation of the earth. But more than this, its behavior also has implications for the nature of gravity in the universe, and it has been suggested that a very good pendulum might provide a test of Einstein's general theory of relativity (Braginsky et al. 1984).

Chapter 5 focuses on the torsion pendulum, where an extended rigid mass is suspended from a flexible fiber or cable that allows the mass to oscillate in a horizontal plane. The restoring force is now provided by the elastic properties of the suspending fiber rather than gravity. While the torsion pendulum is intrinsically interesting, its importance in the history of physics lies in its repeated use in various forms to determine the universal gravitational constant, G. The torsion pendulum acquired this role when Cavendish, in 1789, measured the effect on a torsion pendulum of large masses placed near the pendulum bob. Since that time a whole stream of measurements with similar devices have provided improved estimates of this universal constant. In fact, the search for an accurate value of G continues into the third millennium. New results were described at the American Physical Society meeting in April, 2000 held in Long Beach, California, that reduce the error in G to about 0.0014%. This new result was obtained with apparatus based upon the torsion pendulum, not unlike the original Cavendish device. The value of the universal gravitational constant and possible variations in that constant over time and space are fundamental to the understanding of cosmology—our global view of the universe.

The next part of our story has its origins in a quiet revolution that occurred in the field of mathematics toward the end of the nineteenth century, a revolution whose implications would not be widely appreciated for another 80 years. It arose from asking an apparently simple question: "Is the solar system stable?" That is, will the planets of the solar system continue to orbit the sun in predictable, regular orbits for the calculable future? With others, the French mathematician and astronomer, Henri Poincaré tried to answer the question definitively. Prizes were offered and panels of judges poured over the lengthy treatises (Barrow 1997). Yet the important point here is not the answer, but that in the search for the answer, Poincaré discovered a new type of mathematics. He developed a *qualitative* theory of differential equations, and found a pictorial or geometric way to view the solutions in cases *for which there were no analytic solutions*. What makes this theory revolutionary is that Poincaré found certain solutions or *orbits* for some nonlinear equations that were quite irregular. The universe was not a simple periodic or even quasi-periodic (several frequencies) place as had been assumed previously. The oft-quoted words of Poincaré tell the story,

> it may happen that small differences in the initial conditions produce very great ones in the final phenomena. A small error in the former will produce an enormous error in the latter. Prediction becomes impossible, and we have the fortuitous phenomenon.[1]

"Fortuitous" or random-appearing behavior was not expected and, if it did occur, it was typically ignored as anomalous or too complex to be modeled. Thus was born the science that eventually came to be known as *chaos*, the name much later coined by Yorke and Li of the University of Maryland.

The field of *chaos* would have never emerged without another, much later revolution—the computer revolution. The birth of a full-scale science of chaos coincided with the application of computers to these special types of equations. In 1963 Edward Lorenz of the Massachusetts Institute of Technology was the first to observe (Lorenz 1963) the chaotic power of nonlinear effects in a simple model of meteorological convection—flow of an air mass due to heating from below. With the publication of Lorenz's work a flood of scientific activity in chaos ensued. Thousands of scientific articles appeared in the existing physics and mathematics journals, and new, often multidisciplinary, journals appeared that were especially devoted to nonlinear dynamics and chaos. Chaos was found to be ubiquitous. Chaos became a new paradigm, a new world view.

Many of the original and archetypical systems of equations or models found in the literature of chaos are valued more for their mathematical properties than for their obvious correspondence with physically realizable systems. However, as one of the simplest physical *nonlinear* systems, the

[1] See (Poincaré 1913, p. 397).

pendulum is a natural and rare candidate for practical study. It is modeled quite accurately with relatively simple equations, and a variety of actual physical pendulums have been constructed that correspond very well to their model equations. Therefore, the chaotic classical pendulum has become an object of much interest, and quantitative analysis is feasible with the aid of computers. Many configurations of the chaotic pendulum have been studied. Examples include the torsion pendulum, the inverted pendulum, and the parametric pendulum. Special electronic circuits have been developed whose behavior exactly mimics pendular motion.

Intrinsic to the study of chaotic dynamics is the intriguing mathematical connection with the unusual geometry of fractals. Fractal structure seems to be ubiquitous in nature and one wonders if the underlying mechanisms are universally chaotic, in some sense—unstable but nevertheless constrained in ways that are productive of the rich complexity that we observe in, for example, biology and astronomy. The pendulum is a wonderful example of chaotic behavior as it exhibits all the complex properties of chaos while being itself a fully realizable physical system. Chapter 6 describes many aspects of the chaotic pendulum.

Chapter 7 explores the effects of coupling pendulums together. As with the single pendulum, the origins of coupled pendulums reach back to the golden age of physics. Three hundred years ago, Christiaan Huygens observed the phenomenon of synchronization of two clocks attached to a common beam. The slight coupling of their motions through the medium of the beam was sufficient to cause synchronization. That is, after an initial period in which the pendulums were randomly out of phase, they gradually arrived at a state of perfectly matched (but opposite) motions. In another venue, synchronization of the flashes of swarms of certain fireflies has been documented. While that phenomenon is not explicitly physical in origin, some very interesting mathematical analysis and experiments have been done in this context (Strogatz 1994). Similarly chaotic pendulums, both in numerical simulation and in reality, have been shown to exhibit synchronization. As is true with many synchronized chaotic pairs, one pendulum can be made to dominate over another pendulum. Surprisingly, such a "master" and "slave" relationship can form the basis for a system of somewhat secure communications. Again, the pendulum is an obvious choice for study because of its simplicity. Real pendulums can be coupled together with springs or magnets (Baker et al. 1998). This story continues today as scientists consider the fundamental notion of what it means for physical systems to be synchronized and ask the question, "How synchronized is synchronized?"

During its long history the pendulum has been an important exemplar through several paradigm shifts in physical theory. Possibly the most profound of these scientific discontinuities is the quantum revolution of Planck, Einstein, Bohr, Schrodinger, Heisenberg, and Born in the first quarter of the twentieth century. It led scientists to see that a whole new mechanics must be applied to the world of the very small; atoms, electrons, and so forth. Much has been written on the quantum revolution, but its effect on that simple device, the pendulum, is not perhaps widely known. Many classical mechanical systems have interesting and fascinatingly different behaviors when considered as quantum systems. We might inquire as to what happens when a pendulum is scaled down to atomic dimensions. What are the consequences of pendulum "quantization"? For the pendulum with no damping and no forcing, the process of quantization is relatively straightforward and proceeds according to standard rules as shown in Chapter 7. One of the pioneering researchers in quantum mechanics, Frank Condon, produced the seminal paper on the quantum pendulum in 1928 just a couple of years after the new physics was made broadly available in the physics literature. We learn that the pendulum, like other confined systems, is only allowed to exist with certain fixed energies. Just as the discovery of discrete frequency lines in atomic spectra ultimately vindicated the quantum mechanical prediction of discrete atomic energies, so also does the quantum simple pendulum exhibit a similar discrete energy spectrum.

Does the notion of a quantum pendulum have a basis in physical reality? We find it difficult to imagine that matter is composed of tiny pendulums. Yet surprisingly, there are interactions at the molecular level that have the same mathematical form as the pendulum. One example is motion of molecular complexes in the form of "hindered rotations". We will describe the temperature dependence of hindered rotations and show that the room temperature dynamics of such complexes depends heavily on the particular atomic arrangement.

As a further complication, many researchers have asked if quantum mechanics, with its inherent uncertainties, washes away many of the effects of classical chaotic dynamics—described in the previous chapter. The classical unstable orbits of chaotic systems diverge rapidly from each other as Poincaré first predicted, and yet this "kiss and run" quality could be smeared out by the fact that specific orbits are not well defined in quantum physics. In classical physics, we presume to know the locations and speed of the pendulum bob at all times. In quantum physics, our knowledge of the pendulum's state is only probabilistic. The quantized, but macroscopic, gravity driven pendulum provides further material for this debate.

In one of nature's surprising coincidences, quantum physics does present us with one very clear analogy with the classical, forced pendulum; namely, the Josephson junction. The Josephson junction is a superconducting quantum mechanical device for which the classical pendulum is an exact mathematical analogue. The junction consists of a pair of superconductors separated by an extremely thin insulator (a sort of superconducting diode). Josephson junctions are very useful as ultra-fast switching devices and in high sensitivity magnetometers. Because of their analogy with the pendulum, all the work done with the pendulum in the realm of control and synchronization of chaos can be usefully applied to the Josephson junction. And so ends the ninth chapter.

For the tenth chapter, we return to the sixteenth century and Galileo to consider the role of the pendulum in time keeping. Galileo was the first to design (although never build) a working pendulum clock. He is also reputed to have built, in 1602, a special medical pendulum whose length (and therefore period) could be adjusted to match the heart rate of a patient. This measurement of heart rate would then aid in the diagnostic process. The medical practitioner would find various diseases listed at appropriate locations along the length of the pendulum.

Galileo was keenly aware of the need for an accurate chronometer for the measurement of longitude at sea. Portugal, Spain, Holland, and England had substantial investments in accurate ocean navigation. Realizing the economic benefits of accurate navigation, governments and scientific societies offered financial prizes for a workable solution. Latitude was relatively easy to measure, but the determination of longitude required either an accurate clock or the use of very precise astronomical measurements and calculations. In Galileo's time neither method was feasible. While the mechanical clock was invented in the early fourteenth century and the pendulum was conceived as a possible regulator by Leonardo da Vinci and the Florentine clockmaker Lorenzo della Volpaia, these ideas were not combined successfully. Galileo's contribution to clock design was an improved method of linking the pendulum to the clock.

The earliest practicable version of a clock based upon Galileo's design was constructed by his son, but in the meantime in 1657, Christiaan Huygens became the first to build and patent a successful pendulum clock (Huygens 1986). Although much controversy developed over how much Huygens knew of Galileo's design, Huygens is generally credited with developing the clock independently. There is also a felicitous connection between Huygens invention of a method to keep the regulating pendulum's period independent of amplitude, and the mathematics of the cycloid, a connection that we discuss analytically. The longitude problem was ultimately solved (Sobel 1996) by the Harrison chronometer, built with a spring regulator, but the pendulum clock survives today as a beautiful and accurate timekeeper.

Pendulum clocks exemplify important physical concepts. The clock needs to have some method of transferring energy to the pendulum to maintain its oscillation. There also needs to be a method whereby the pendulum regulates the motion of the clock. These two requirements are encompassed in one remarkable mechanism called the *escapement*. The escapement is a marvelous invention in that it makes the pendulum clock one of the first examples of an automaton with self-regulating feedback. Chapter 10 concludes with a brief look at some of the world's most interesting pendulum clocks.

Finally, there are interesting configurations and applications of the pendulum that do not fit neatly into the book's structure. Therefore we include descriptions of some of these pendulums as separate notes in Appendices A–F.

2 Pendulums somewhat simple

Fig. 2.1
Portrait of Galileo. ©Bettmann/Corbis/Magma.

Fig. 2.2
Cathedral at Pisa. The thin vertical wire indicates a hanging chandelier.

There are many kinds of pendulums. In this chapter, however, we introduce a simplified model; the small amplitude, linearized pendulum. For the present, we ignore friction and in doing so obviate the need for energizing the pendulum through some forcing mechanism. Our initial discussion will therefore assume that the pendulum's swing is relatively small; and this approximation allows us to linearize the equations and readily determine the motion through solution of simplified model equations. We begin with a little history.

2.1 The beginning

Probably no one knows when pendulums first impinged upon the human consciousness. Undoubtedly they were objects of interest and decoration after humankind learnt to attend routinely to more basic needs. We often associate the first scientific observations of the pendulum with Galileo Galilei (1564–1642; Fig. 2.1).

According to the usual story (perhaps apocryphal), Galileo, in the cathedral at Pisa, Fig. 2.2 observed a lamplighter push one of the swaying pendular chandeliers. His earliest biographer Viviani suggests that Galileo then timed the swings with his pulse and concluded that, even as the amplitude of the swings diminished, the time of each swing was constant. This is the origin of Galileo's apparent discovery of the approximate isochronism of the pendulum's motion. According to Viviani these observations were made in 1583, but the Galileo scholar Stillman Drake (Drake 1978) tells us that guides at the cathedral refer visitors to a certain lamp which they describe as "Galileo's lamp," a lamp that was not actually installed until late in 1587. However, there were undoubtedly earlier swaying lamps. Drake surmises that Galileo actually came to the insight about isochronism in connection with his father's musical instruments and then later, perhaps 1588, associated isochronism with his earlier pendulum observations in the cathedral. However, Galileo did make systematic observations of pendulums in 1602. These observations confirmed only approximately his earlier conclusion of isochronism of swings of differing amplitude. Erlichson (1999) has argued that, despite the nontrivial empirical evidence to the contrary, Galileo clung to his earlier conclusion, in part, because he believed that the universe had been ordered so that motion would be simple and that there was "no reason" for the longer path to take a longer time than the shorter path. While Galileo's most famous conclusion about the pendulum has only partial legitimacy, its importance resides (a) in it being the first known scientific deduction about the pendulum, and (b) in the fact that the insight of approximate isochronism is part of the opus of a very famous seminal character in the history of physical science. In these circumstances, the pendulum begins its history as a significant model in physical science and, as we will see, continues to justify its importance in science and technology during the succeeding centuries.

2.2 The simple pendulum

The simple pendulum is an idealization of a real pendulum. It consists of a point mass, m, attached to an infinitely light rigid rod of length l that is itself attached to a frictionless pivot point. See Fig. 2.3. If displaced from its vertical equilibrium position, this idealized pendulum will oscillate with a constant amplitude forever. There is no damping of the motion from friction at the pivot or from air molecules impinging on the rod. Newton's second law, mass times acceleration equals force, provides the equation of motion:

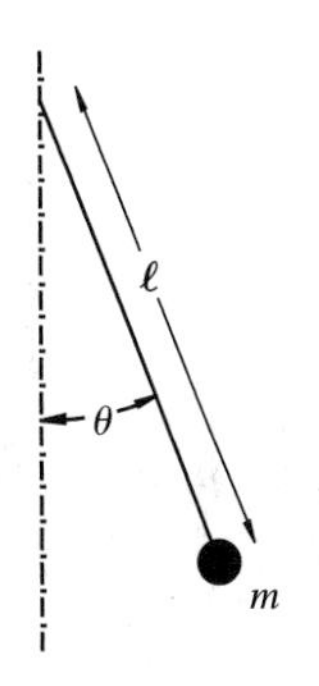

Fig. 2.3
The simple pendulum with a point mass bob.

$$ml\frac{d^2\theta}{dt^2} = -mg\sin\theta, \tag{2.1}$$

where θ is the angular displacement of the pendulum from the vertical position and g is the acceleration due to gravity. Equation (2.1) may be simplified if we assume that amplitude of oscillation is small and that $\sin\theta \approx \theta$. We use this *linearization* approximation throughout this chapter. The modified equation of motion is

$$\frac{d^2\theta}{dt^2} + \frac{g}{l}\theta = 0. \tag{2.2}$$

The solution to Eq. (2.2) may be written as

$$\theta = \theta_0 \sin(\omega t + \phi_0), \tag{2.3}$$

where θ_0 is the angular amplitude of the swing,

$$\omega = \sqrt{\frac{g}{l}} \tag{2.4}$$

is the angular frequency, and ϕ_0 is the initial phase angle whose value depends on how the pendulum was started—its initial conditions. The period of the motion, *in this linearized approximation*, is given by

$$T = 2\pi\sqrt{\frac{l}{g}}, \tag{2.5}$$

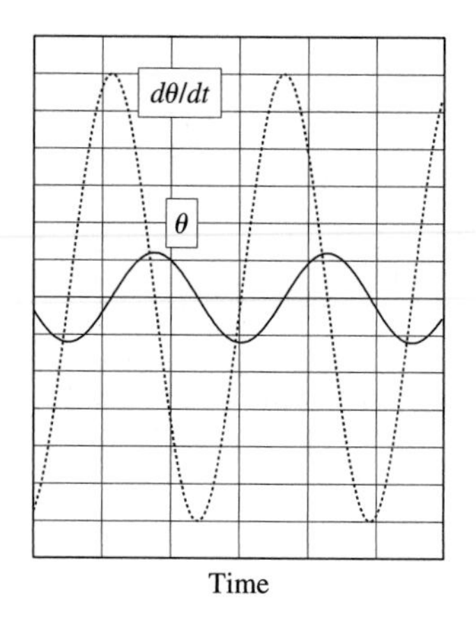

Fig. 2.4
Time series for the angular displacement θ and the angular velocity, $\dot{\theta}$.

which is a constant for a given pendulum, and therefore lends support to Galileo's conclusion of isochronism.

The dependence of the period on the geometry of the pendulum and the strength of gravity has very interesting consequences which we will explore. But for the moment we consider further some of the mathematical relationships. Figure 2.4 shows the angular displacement $\theta = \theta_0 \sin(\omega t + \phi_0)$ and the angular velocity $\dot{\theta} = \theta_0 \omega \cos(\omega t + \phi_0)$, respectively, as functions of time. We refer to such graphs as time series. The displacement and velocity are 90 degrees out of phase with each other and therefore when one quantity has a maximum absolute value the other quantity is zero. For example, at the bottom of its motion the pendulum has no angular displacement yet its velocity is greatest.

The relationship between angle and velocity may be represented graphically with a *phase plane diagram.* In Fig. 2.5 angle is plotted on the horizontal axis and angular velocity is plotted on the vertical axis. As time goes on, a point on the graph travels around the elliptically shaped curve. In effect, the equations for angle and angular velocity are considered to be parametric equations for which the parameter is proportional to time. Then the *orbit* of the *phase trajectory* is the ellipse

$$\frac{\theta^2}{\theta_0^2} + \frac{\dot{\theta}^2}{(\omega\theta_0)^2} = 1. \tag{2.6}$$

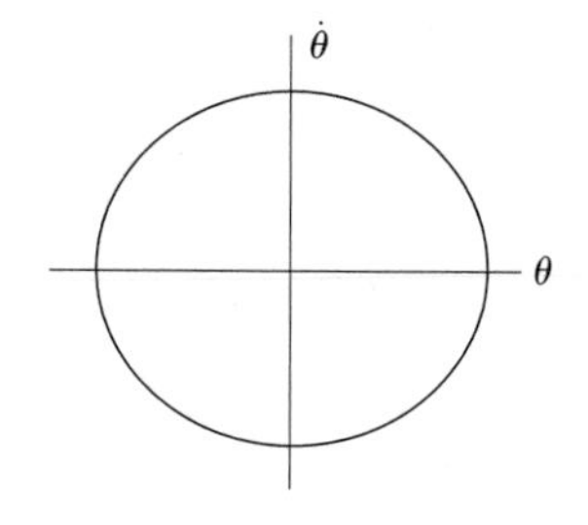

Fig. 2.5
Phase plane diagram. As time increases the phase point travels around the ellipse.

Since the motion has no friction nor any forcing, energy is conserved on this phase trajectory. Therefore the sum of the kinetic and potential energies at any time can be shown to be constant as follows. In the linearized approximation,

$$E = \frac{1}{2} m l^2 \dot{\theta}^2 + \frac{1}{2} m g l \theta^2 \tag{2.7}$$

and, using Eqs. (2.3) and (2.4), we find that

$$E = \frac{1}{2} m g l \theta_0^2, \tag{2.8}$$

which is the energy at maximum displacement.

The *phase plane* is a useful tool for the display of the dynamical properties of many physical systems. The linearized pendulum is probably one of the simplest such systems but even here the phase plane graphic is helpful. For example, Eq. (2.6) shows that the axes of the ellipse in Fig. 2.5 are determined by the amplitude and therefore the energy of the motion. A pendulum of smaller energy than that shown would exhibit an ellipse that sits inside the ellipse of the pendulum of higher energy. See Fig. 2.6. Furthermore the two ellipses would never intersect because such intersection implies that a pendulum can jump from one energy to another without the agency of additional energy input. This result leads to a more general conclusion called the *no-crossing theorem*; namely, that orbits in phase space never cross. See Fig. 2.7.

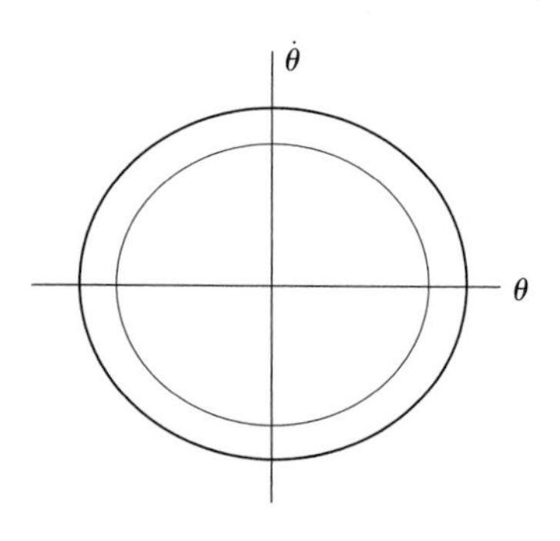

Fig. 2.6
Phase orbits for pendulums with different energies, E_1 and E_2.

Why should this be so? Every orbit is the result of a deterministic equation of motion. Determinism implies that the orbit is well defined and that there would be no circumstance in which a well determined particle would arrive at some sort of ambiguous junction point where its path would be in doubt. (Later in the book we will see *apparent* crossing points but these false crossings are the result of the system arriving at the same phase coordinates at *different* times.)

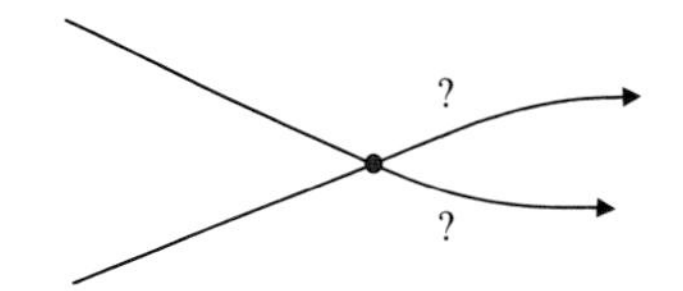

Fig. 2.7
If two orbits in phase space intersect, then it is uncertain which orbit takes which path from the intersection. This uncertainty violates the deterministic basis of classical mechanics.

We introduce one last result about orbits in the phase plane. In Fig. 2.6 there are phase trajectories for two pendulums of different energy. Now think of a large collection of pendulums with energies that are between the two trajectories such that they have very similar, but not identical, angles and velocities. This cluster of pendulums is represented by a set of many *phase points* such that they appear in the diagram as an approximately solid block between the original two trajectories. As the group of pendulums executes their individual motions the set of phase points will move between the two ellipses in such a way that the area defined by the boundaries of the set of points is preserved. This preservation of *phase area*, known as Liouville's theorem (after Joseph Liouville (1809–1882)) is a consequence of the conservation of energy property for each pendulum. In the next chapter we will demonstrate how such areas decrease when energy is lost in the pendulums. But for now let us show how phase area conservation is true for the very simple case when $\theta_0 = 1, \phi = 0$, and $\omega = 1$. In this special case, the ellipses becomes circles since the axes are now equal. See Fig. 2.8. A block of points between the circles is bounded by a small polar angle interval $\Delta\alpha$, in the phase space, that is proportional to time. Each point in this block rotates at the same rate as the motion of its corresponding pendulum progresses. Therefore, after a certain time, all points in the original block have rotated, by the same polar angle, to new positions again bounded by the two circles. Clearly, the size of the block has not changed, as we predicted.

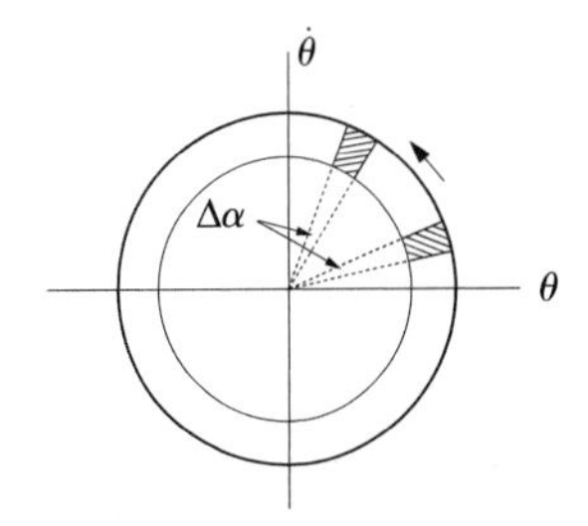

Fig. 2.8
Preservation of area for conservative systems. A block of phase points keeps its same area as time advances.

The motion of the pendulum is an obvious demonstration of the alternating transformation of kinetic energy into potential energy and the reverse. This phenomenon is ubiquitous in physical systems and is known as *resonance*. The pendulum resonates between the two states (Miles 1988*b*). Electrical circuits in televisions and other electronic devices resonate. The terms *resonate* and *resonance* may also refer to a sympathy between two or more physical systems, but for now we simply think of resonance as the periodic swapping of energy between two possible formats.

We conclude this section with the introduction of one more mathematical device. Its use for the simple pendulum is hardly necessary but it will be increasingly important for other parts of the book. Almost two hundred years ago, the French mathematician Jean Baptiste Fourier (1768–1830) showed that periodic motion, whether that of a simple sine wave like our pendulum, or more complex forms such as the triangular wave that characterizes the horizontal sweep on a television tube, are simple linear sums of sine and cosine waves now known as *Fourier Series*. That is, let $f(t)$

be a periodic function such that $f(t) = f(t + (2\pi)/\omega_0)$, where $T = (2\pi)/\omega_0$ is the basic periodicity of the motion. Then Fourier's theorem says that this function can be expanded as

$$f(t) = \sum_{n=1}^{\infty} b_n \cos n\omega_0 t + \sum_{n=1}^{\infty} c_n \sin n\omega_0 t + d, \tag{2.9}$$

where the coefficients b_n and c_n give the strength of the respective cosine and sine components of the function and d is constant. The coefficients are determined by integrating $f(t)$ over the fundamental period, T. The appropriate formulas are

$$d = \frac{1}{T}\int_{-T/2}^{T/2} f(t)\,dt, \quad b_n = \frac{1}{T}\int_{-T/2}^{T/2} f(t)\cos n\omega_0 t\,dt,$$
$$c_n = \frac{1}{T}\int_{-T/2}^{T/2} f(t)\sin n\omega_0 t\,dt. \tag{2.10}$$

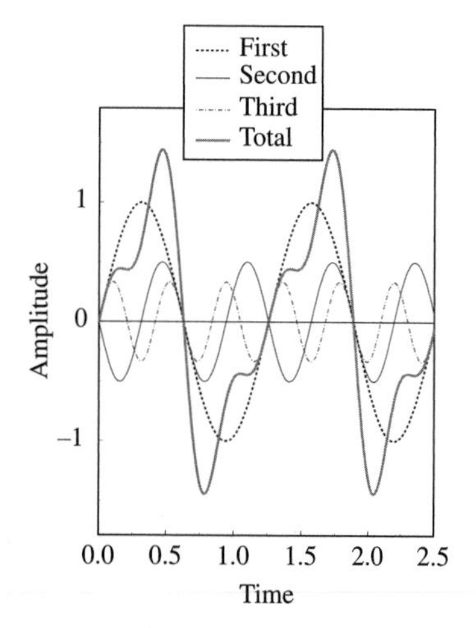

Fig. 2.9
The first three Fourier components of the sawtooth wave. The sum of these three components gives an approximation to the sawtooth shape.

These Fourier coefficients are sometimes portrayed crudely on stereo equipment as dancing bars in a dynamic bar chart that is meant to portray the strength of the music in various frequency bands.

The use of complex numbers allows Fourier series to be represented more compactly. Then Eqs. (2.9) and (2.10) become

$$f(t) = \sum_{n=-\infty}^{n=\infty} a_n e^{in\omega_0 t}, \quad \text{where } a_n = \frac{\omega_0}{2\pi}\int_{-\pi/\omega_0}^{\pi/\omega_0} f(t)e^{-in\omega_0 t}\,dt. \tag{2.11}$$

Example 1 *Consider the time series known as the "sawtooth," $f(t) = t$ when $-\frac{T}{2} < t < \frac{T}{2}$, with the pattern repeated every period, T. Using Eq. (2.11) it can be shown that*

$$a_n = 0 \ \textit{for}\ n = 0,$$

$$a_n = \tfrac{1}{in\omega_0}\ \textit{for}\ n = \textit{odd integer, and}$$

$$a_n = \tfrac{-1}{in\omega_0}\ \textit{for}\ n = \textit{even integer}.$$

Through substitution and appropriate algebraic manipulation we obtain the final result:

$$f(t) = \frac{2}{\omega_0}\left[\sin\omega_o t - \frac{1}{2}\sin 2\omega_0 t + \frac{1}{3}\sin 3\omega_0 t + \cdots\right]. \tag{2.12}$$

The original function and the first three frequency components are shown in Figs. 2.9 and 2.10.

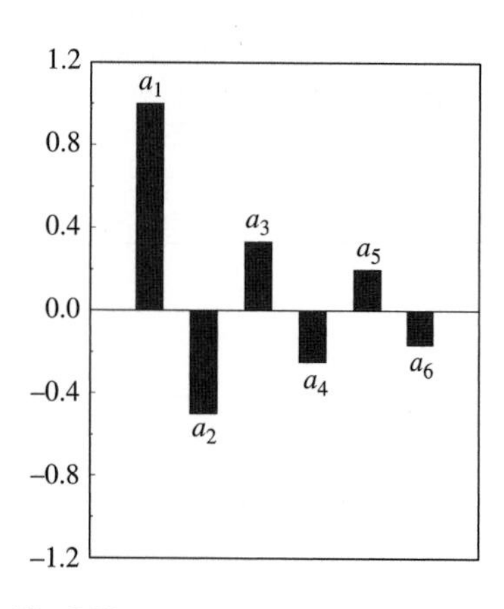

Fig. 2.10
The amplitudes of several Fourier components for the sawtooth waveform.

The time variation of the motion of the linearized version of the simple pendulum is just that of a single sine or cosine wave and therefore one frequency, the resonant frequency ω_0 is present in that motion. Obviously, the machinery of the Fourier series is unnecessary to deduce that result.

However, we now have it available as a tool for more complex periodic phenomena.

Fourier, like other contemporary French mathematicians, made his contribution to mathematics during a turbulent period of French history. He was active in politics and as a student during the "Terror" was arrested although soon released. Later when Napoleon went to Egypt, Fourier accompanied the expedition and coauthored a massive work on every possible detail of Egyptian life, *Description de l'Egypt*. This is multivolume work included nine volumes of text and twelve volumes of illustrations. During that same campaign, one of Napoleon's engineers uncovered the Rosetta Stone, so-named for being found near the Rosetta branch of the Nile river in 1799. The significance of this find was that it led to an understanding of ancient Egyptian Hieroglyphics. The stone, was inscribed with the same text in three different languages, Greek, demotic Egyptian, and Hieroglyphics. Only Greek was understood, but the size and the juxtaposition of the texts allowed for the eventual understanding of Hieroglyphics and the ability to learn much about ancient Egypt. In 1801, the victorious British, realizing the significance of the Rosetta stone, took it to the British Museum in London where it remains on display and is a popular artifact. Much later, the writings from the Rosetta Stone become the basis for translating the hieroglyphics on the Rhind Papyrus and the Golenischev Papyrus; these two papyri provide much of our knowledge of early Egyptian mathematics. The French Egyptologist Jean Champollion (1790–1832) who did much of the work in the translation of Hieroglyphics is said to have actually met Fourier when the former was only 11 years old, in 1801. Fourier had returned from Egypt with some papyri and tablets which he showed to the boy. Fourier explained that no one could read them. Apparently Champollion replied that he would read them when he was older—a prediction that he later fulfilled during his brilliant career of scholarship (Burton 1999). After his Egyptian adventures, Fourier concentrated on his mathematical researches. His 1807 paper on the idea that functions could be expanded in trigonometric series was not well received by the Academy of Sciences of Paris because his presentation was not considered sufficiently rigorous and because of some professional jealousy on the part of other Academicians. But eventually Fourier was accepted as a first rate mathematician and, in later life, acted a friend and mentor to a new generation of mathematicians (Boyer and Merzbach 1991).

We have now developed the basic equations for the linearized, undamped, undriven, very simple harmonic pendulum. There are an amazing number of applications of even this simple model. Let us review some of them.

2.3 Some analogs of the linearized pendulum

2.3.1 The spring

The linearized pendulum belongs to a class of systems known as harmonic oscillators. Probably the most well known realization of a harmonic oscillator is that of a mass suspended from a spring whose restoring force is proportional to its stretch. That is

$$F_{\text{restoring}} = -kx, \tag{2.13}$$

where k is the spring constant and rate at which the spring's response increases with stretch, x. This force law was discovered by Robert Hooke in 1660. The equation of motion

$$m\frac{d^2x}{dt^2} + kx = 0 \tag{2.14}$$

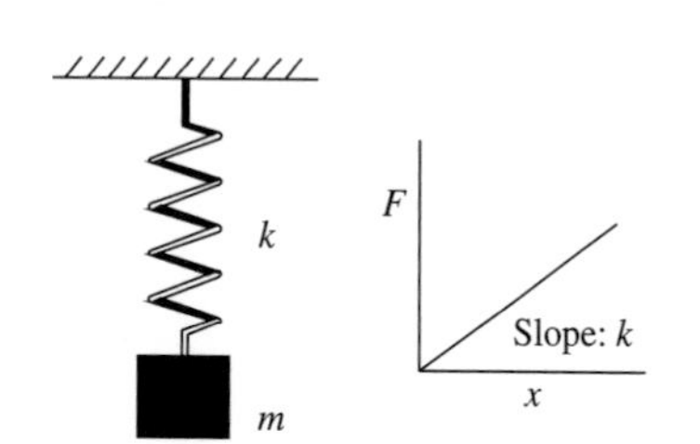

Fig. 2.11
A mass hanging from a spring. The graph shows the dependence of the extension of the spring on the force (weight). The linear relationship is known as "Hooke's law."

is identical in form to that of the linearized pendulum and therefore its solution has corresponding properties: single frequency periodic motion, resonance, energy conservation and so forth. A schematic drawing of the spring and a graph of its force law are shown in Fig. 2.11.

The functional dependence of the spring force (Eq. (2.13)) can be viewed more generally. Consider any force law that is derived from a smooth potential energy $V(x)$; that is $F(x) = -dV/dx$. The potential energy may be expanded in a power series about some arbitrary position x_0 which, for simplicity, we will take as $x_0 = 0$. Then the series becomes

$$V(x) = V(0) + V'(0)x + \frac{1}{2}V''(0)x^2 + \frac{1}{6}V'''(0)x^3 + \cdots . \tag{2.15}$$

The first term on the right side is constant and, as the reference point of a potential energy, is typically arbitrary and may be set equal to zero. The second, linear, term contains $V'(0)$ which is the negative of the force at the reference point. Since this reference point is, again typically, chosen to be a point of stable equilibrium where the forces are zero, this second term also vanishes. For the spring, this would be the point where the mass attached to the spring hangs when it is not in motion. Thus, the first nonvanishing term in the series is the quadratic term $\frac{1}{2}V''(0)x^2$ and comparison of it with the spring's restoring force (Eq. 2.13) leads to the identification

$$k = V''(0). \tag{2.16}$$

The spring constant is the second derivative of any smooth potential.

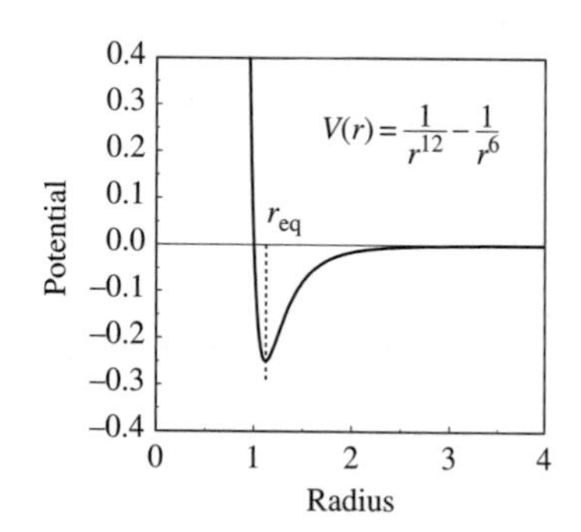

Fig. 2.12
A typical Lennard–Jones potential curve that can effectively model, for example, intermolecular interactions. For this illustration, $a = b = 1$.

Example 2 *The Lennard–Jones potential energy is often used to describe the electrostatic potential energy between two atoms in a molecule or between two molecules. Its functional form is displayed in Fig. 2.12 and is given by the equation*

$$V(r) = \frac{a}{r^{12}} - \frac{b}{r^6}, \tag{2.17}$$

where a and b are constants appropriate to the particular molecule. The positive term describes the repulsion of the atoms when they are too close and the negative term describes the attraction if the atoms stray too far from each other. Hence, the two terms balance at a stable equilibrium point as shown in the figure, $r_{\text{eq}} = \left(\frac{2a}{b}\right)^{1/6}$. *The second derivative of the potential energy may* *be evaluated at r_{eq} to yield the spring constant of the equivalent harmonic oscillator,*

$$V''(r_{eq}) = \frac{18b^2}{a}\left(\frac{b}{2a}\right)^{1/3} = k. \tag{2.18}$$

Knowledge of the molecular bond length provides r_{eq} and observation of the vibrational spectrum of the molecule will yield a value for the spring constant, k. With just these two pieces of information, the parameters, a and b of the Lennard–Jones potential energy may be determined.

The linearized pendulum is therefore equivalent to the spring in that they both are simple harmonic oscillators each with a single frequency and therefore a single spectral component. Occasionally we will refer to a pendulum's equivalent oscillator or equivalent spring, and by this terminology we will mean the linearized version of that pendulum.

2.3.2 Resonant electrical circuit

We say that a function $f(t)$ or operator $L(x)$ is linear if

$$\begin{aligned} L(x+y) &= L(x) + L(y) \\ L(\alpha x) &= \alpha L(x). \end{aligned} \tag{2.19}$$

Examples of linear operators include the derivative and the integral. But functions such as $\sin x$ or x^2 are nonlinear. Because linear models are relatively simple, physics and engineering often employ linear mathematics, usually with great effectiveness. Passive electrical circuits, consisting of resistors, capacitors, and inductors are realistically modeled with linear differential equations. A circuit with a single inductor L and capacitor C, is shown in Fig. 2.13. The sum of the voltages measured across each element of a circuit is equal to the voltage provided to a circuit from some external source. In this case, the external voltage is zero and therefore the sum of the voltages across the elements in the circuit is described by the linear differential equation

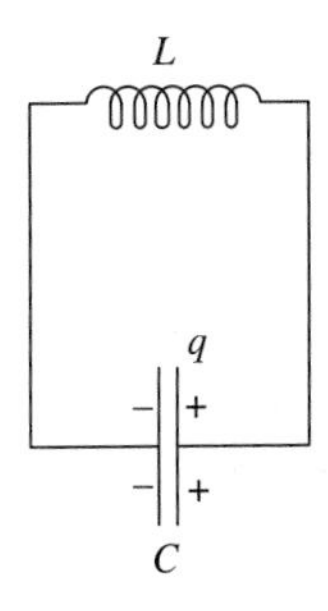

Fig. 2.13
A simple LC (inductor and capacitor) circuit.

$$L\frac{d^2q}{dt^2} + \frac{1}{C}q = 0, \tag{2.20}$$

where q is the electrical charge on the capacitor. The form of Eq. (2.20) is exactly that of the linearized pendulum and therefore a typical solution is

$$q = q_0 \sin(\omega t + \phi), \tag{2.21}$$

where the resonant frequency depends on the circuit elements:

$$\omega = \sqrt{\frac{1}{LC}}. \tag{2.22}$$

The charge q plays a role analogous to the pendulum's angular displacement θ and the current $i = dq/dt$ in the circuit is analogous to the pendulum's angular velocity, $d\theta/dt$. All the same considerations, about the motion in phase space, resonance, and energy conservation, that previously held for the linearized pendulum, also apply for this simple electrical circuit. In a (q, i) phase plane, the point moves in an elliptical curve around the origin. The charge and current oscillate out of phase with each other. The capacitor alternately fills with positive and negative charge. The voltage across the inductor is always balanced by the voltage across the capacitor such that the total voltage across the circuit always adds to zero as expressed by Eq. (2.20). As with the spring, we will return to this electrical analog with additional complexity. For now, we turn to some applications and complexities of the linearized pendulum.

2.3.3 The pendulum and the earth

From ancient times thinkers have speculated about, theorized upon, calculated, and measured the physical properties of the earth (Bullen 1975). About 900BC, the Greek poet Homer suggested that the earth was a convex dish surrounded by the Oceanus stream. The notion that the earth was spherical seems to have made its first appearance in Greece at the time of Anaximander (610–547BC). Aristotle, the universalist thinker, quoted contemporary mathematicians in suggesting that the circumference of the earth was about 400,000 stadia—one stadium being about 600 Greek feet. Mensuration was not a precise science at the time and the unit of the stadium has been variously estimated as 178.6 meters (olympic stadium), 198.4 m (Babylonian–Persian), 186 m (Italian) or 212.6 m (Phoenician–Egyptian). Using any of these conversion factors gives an estimate that is about twice the present measurement of the earth's circumference, 4.0086×10^4 km. Later Greek thinkers somewhat refined the earlier values. Eratosthenes (276–194BC), Hipparchus (190–125BC), Posidonius (135–51BC), and Claudius Ptolemy (AD100–161) all worked on the problem. However the Ptolemaic result was too low. It is rumored that a low estimate of the distance to India, based on the Ptolemy's result, gave undue encouragement to Christopher Columbus 1500 years later.

In China the astronomer monk Yi-Hsing (AD683–727) had a large group of assistants measure the lengths of shadows cast by the sun and the altitudes of the pole star on the solstice and equinox days at thirteen different locations in China. He then calculated the length L of a degree of meridian arc (earth's circumference/360) as 351.27 li (a unit of the Tang Dynasty) which, with present day conversion, is about 132 km, an estimate that is almost 20% too high.

The pendulum clock, invented by the Dutch physicist and astronomer Christiaan Huygens (1629–1695) and presented on Christmas day, 1657, provided a powerful tool for measurement of the earth's gravitational field, shape, and density. The daily rotation of the earth was, by then, an accepted fact and Huygens, in 1673, provided a theory of centrifugal motion that required the effective gravitational field at the equator to be less than that at the poles. Furthermore, the centrifugal effect should also have the effect of fattening the earth at the equator, thereby further weakening gravity at the surface near the equator. In 1687 Newton published his universal law of gravity in the *Principia*. It is the existence of the relationship between gravity and the length of the pendulum (Eq. (2.5)), established through the work of Galileo and Huygens, that makes the pendulum a useful tool for the measurement of the gravitational field and therefore a tool to infer the earth's shape and density. The first recorded use of the pendulum in this context is usually attributed to the measurements of Jean Richer, the French astronomer, made in 1672 . Richer (1630–1696) found that a pendulum clock beating out seconds in Paris at latitude 49° North lost about 2(1/2) minutes per day near the equator in Cayenne at 5° North and concluded that Cayenne was further from the center of the earth than was Paris. Newton, on hearing of this result ten years later by accident at a meeting of the Royal Society, used it to refine his theory of the earth's oblateness (Bullen 1975). However, Richer's result also helped lead to the eventual demise of the idea of using a pendulum clock as a reliable timing standard for the measurement of longitude (Matthews 2000).

A clever bit of theory by Pierre Bougeur (1698–1758), a French professor of hydrography and mathematics, allowed the pendulum to be an instrument for estimating the earth's density, σ (Bouguer 1749). In 1735 Bouguer was sent, by the French Academy of Sciences, to Peru, to measure the length of a meridian arc, L near the equator. (A variety of such measurements at different latitudes would help to determine the earth's oblateness.) But while in Peru he made measurements of the oscillations of a pendulum, which in Paris beat out seconds whereas in Quito, (latitude 0.25° South) the period was different. His original memoir is a little confusing as to whether he maintained a constant length pendulum or whether, as his data suggests, he modified the length of the pendulum to keep time with his pendulum clock that he adjusted daily. At any rate, he used the pendulum to measure the gravitational field. But more than this he made measurements of the gravitational field close to sea level and then on top of the Cordilleras mountain range. In this way Bouguer was able to estimate the relative size of the mean density of the earth.

In order to appreciate the cleverness of Bouguer's method, we derive his result. Consider the schematic diagram of the earth with the height increment (the mountain range) shown in Fig. 2.14. The acceleration due to gravity at the surface of the earth is readily shown to be

$$g_0 = \frac{GM_E}{a^2} = \frac{4}{3}\pi G\sigma a, \tag{2.23}$$

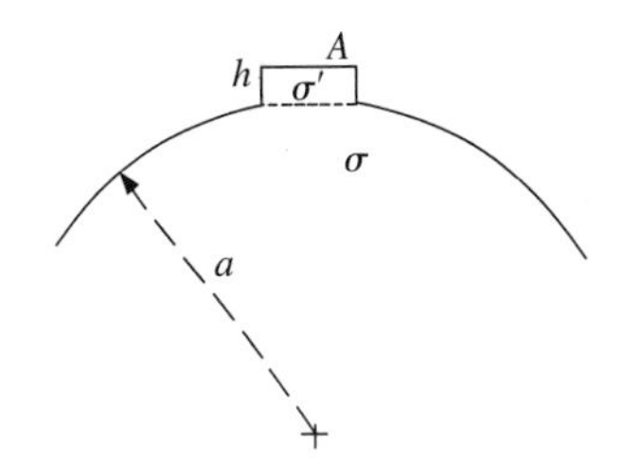

Fig. 2.14
The little "bump" on the earth's surface represents a whole mountain range.

where a is the earth's radius, σ is the mean density, and G is the universal gravitational constant. Now consider the acceleration due to gravity on the mountain range. There are two effects. First, the gravitational field is reduced by the fact that the field point is further from the center of the earth, and second, the field is enhanced by the gravitational pull of the mountain range. The first effect is found through a simple ratio using Newton's law of gravity, but the second effect is a little more involved and requires the use of a fundamental relation in the theory of gravitational fields; Gauss' law. It is expressed mathematically as

$$\int \mathbf{g} \cdot \mathbf{n}\, dA = 4\pi G \int \sigma\, dV, \tag{2.24}$$

where **g** is the acceleration due to gravity (expressed as a vector) and **n** is the outward unit normal vector. The integral on the left side is calculated over a closed surface and the integral on the right side is a volume integral throughout the inside of the closed surface boundary. Essentially the flux or "amount" of gravitational field coming out of the surface is proportional to the mass contained inside the surface. In the diagram, the mountain range is approximated by a "pill box" with height h and top and bottom areas A. We suppose that h is much less than any lateral dimension and therefore assume that the gravitational field is directed only out of the top and the bottom of the pill box. Then Eq. (2.24) becomes

$$2gA = 4\pi G\sigma' Ah \tag{2.25}$$

or

$$g = 2\pi G\sigma' h, \tag{2.26}$$

where σ' is the density of the mountain range as determined by sampling the local soil materials. With these equations we can now write an expression for the acceleration due to gravity as measured on top of the mountain range:

$$g = \frac{4}{3} G\pi\sigma \frac{a^3}{(a+h)^2} + 2\pi G\sigma' h. \tag{2.27}$$

Since $h \ll a$, the first term on the right can be approximated using the binomial expansion and then the ratio of the two measurements of the gravitational field is found to be

$$\frac{g}{g_0} \approx 1 - \frac{2h}{a} + \frac{3h}{2a}\frac{\sigma'}{\sigma}. \tag{2.28}$$

The two corrections terms on the right side of the equation are the first of several corrections that were eventually incorporated into experiments of this or similar types. The first term $2h/a$ is the so-called *free air* term and the other term is referred to as the *Bouguer* term. The point of Eq. (2.28) is that, with data on the relative accelerations due to gravity, it should be possible to calculate the ratio of the density of the mountainous material to that of the rest of the earth. Bouguer's pendulum measurements convinced him that the earth's mean density was about four times that of the mountains, a ratio not too different from a modern value of 4.7. In Bouguer's own words

> Thus it is necessary to admit that the earth is much more compact below than above, and in the interior than at the surface... Those physicists who imagined a great void in the middle of the earth, and who would have us walk on a kind of very thin crust, can think so no longer. We can make nearly the same objections to

Woodward's theory of great masses of water in the interior. (page 33 of (Bouguer 1749)).

Bouguer's experiment was the first of many of this type. A common variant on the mountain range experiment was to measure the difference in gravitational field at the top and the bottom of a mine shaft. In this case, the extra structure was not just a mountain range but a spherical shell above the radius of the earth to the bottom of the shaft. See Fig. 2.15. An equation similar to Eq. (2.28) holds although the *Bouguer* term must be modified as

$$\frac{g_{\text{top}}}{g_{\text{bottom}}} \approx 1 - 2\frac{h}{a} + 3\frac{h}{a}\frac{\sigma'}{\sigma} \tag{2.29}$$

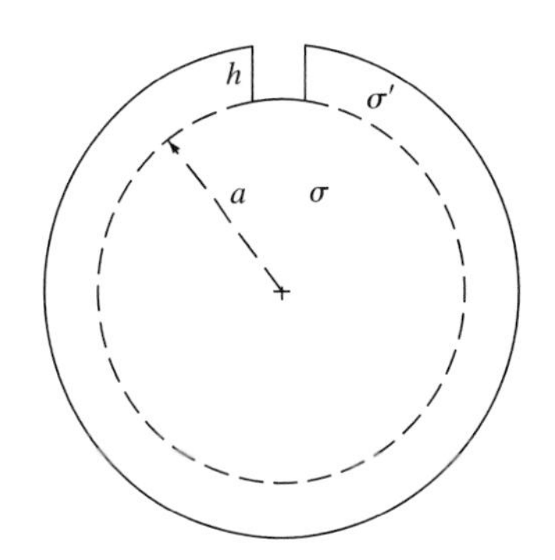

Fig. 2.15
A schematic of the earth with a mine shaft of depth, h.

because of the shape of the spherical shell of density σ', and the radius a is measured from the center of the earth to *bottom* of the mine shaft. Coal mines were widely available in England and the seventh astronomer royal and Lucasian professor of mathematics at Cambridge, George Airy (1801–1892) was one of many to attempt this type of experiment. His early efforts in 1826 and 1828 in Cornwall were frustrated by floods and fire. But much later in 1854 he successfully applied his techniques at a Harton coal-pit in Sunderland and obtained a value for the earth's density of $\sigma = 6.6\,\text{gm}/\text{cm}^3$ (Bullen 1975, p. 16).

Pendulum experiments continued to be improved. Von Sterneck explored gravitational fields at various depths inside silver mines in Bohemia and in 1887 invented a four pendulum device. Two pairs of 1/2 second pendulums were placed at right angles. Each pendulum in a given pair oscillated out of phase with its partner, thereby reducing flexure in the support structure that ordinarily contributed a surprising amount of error to measurements. The two mutually perpendicular pairs provided a check on each other. Von Sterneck's values for the mean density of the earth ranged from 5.0 to $6.3\,\text{gm/cm}^3$. The swing of the pendulums in a pair is compared with a calibrated 1/2 second pendulum clock by means of an arrangement of lights and mirrors as observed through a telescope. Because they are slightly out of phase, the gravity pendulum and the clock pendulum eventually get out of phase by a whole period. The number of counts between such "coincidences" is observed and used in calculating the precision of the gravity pendulum period. Accuracies as high as 2×10^{-7} were claimed for the apparatus.

Other types of pendulums have also been used in geological exploration, but they are based upon pendulums that are more involved than the simple pendulum that is the fundamental ingredient of the experiments and equipment described above.

2.3.4 The military pendulum

Since the mid-twentieth century physics has had a strong relationship with the engineering of military hardware. Yet there are precursors to this modern connection. Benjamin Robin (1707–1751), a British mathematician

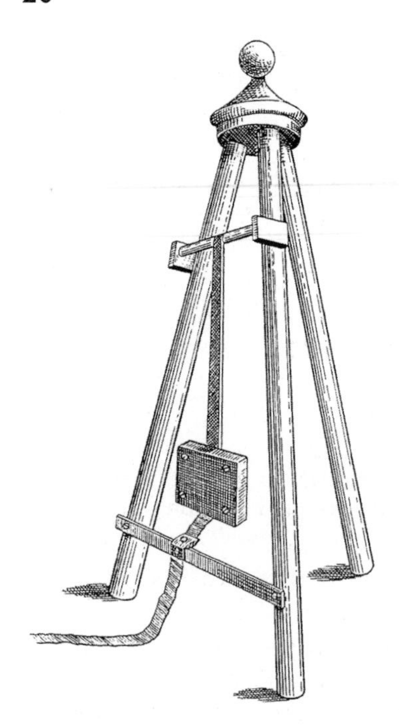

Fig. 2.16
Robin's 1742 ballistic pendulum. (From Taylor (1941) with permission from Dover).

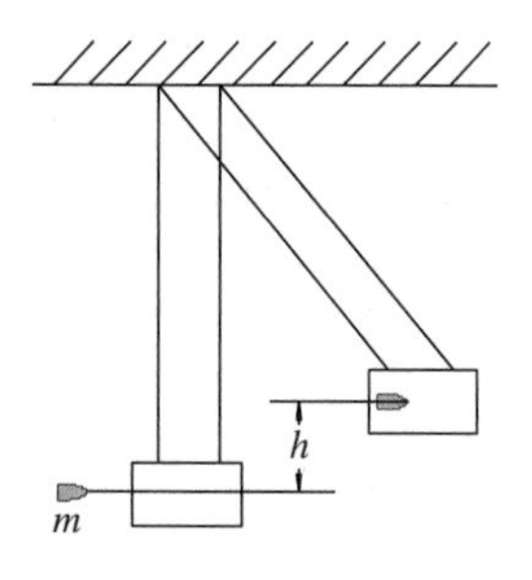

Fig. 2.17
Schematic diagram of the Blackwood ballistic pendulum used in undergraduate laboratories.

and military engineer gave a giant boost to the "modern" science of artillery with the 1742 publication of his book, "New Principles of Gunnery." One of his contributions was a method for determining the muzzle velocity of a projectile; the apparatus is illustrated in Fig. 2.16. (Even today, undergraduate physics majors do an experiment with a version of this method using an apparatus known as the Blackwood ballistic pendulum—Blackwood was a professor of physics in the early twentieth century at the University of Pittsburgh.)

With a relatively modern apparatus a "bullet" is fired into a pendulum consisting of a large wooden bob suspended by several ropes. The projectile is trapped in the bob, causing the bob to pull laterally against the ropes and therefore rise to some measurable height. See Fig. 2.17. Application of the elementary laws of conservation of energy and momentum produce the required value of projectile muzzle velocity.

Here is the simple analysis. Prior to the moment of collision between the projectile of mass m and the pendulum bob of mass M, the projectile has a velocity v. After the collision, the projectile quickly embeds in the bob and imparts a velocity V to the bob. Momentum before and after the collision is preserved so that

$$mv = (M + m)V. \tag{2.30}$$

After the particle is embedded in the bob, the kinetic energy of the combination of projectile and bob thrusts the pendulum outward and upward to a height h. All the kinetic energy is transformed to potential energy and therefore

$$\frac{1}{2}(M + m)V^2 = (M + m)gh. \tag{2.31}$$

Mutual solution of Eqs. (2.30) and (2.31) yields the muzzle velocity of the projectile,

$$v = \frac{M + m}{m}\sqrt{2gh}. \tag{2.32}$$

The beauty of this result is that it bypasses the need to have any sort of measure of the energy lost as the projectile is trapped by the pendulum bob. That lost kinetic energy simply produces heat in the pendulum.

One wonders if the many students who perform this laboratory experiment each year are aware that they are replicating early military research.

2.3.5 Compound pendulum

The model of a simple pendulum requires that all mass be concentrated at a single point. Yet a real pendulum will have some extended mass distribution as indicated in Fig. 2.18. Such a pendulum is called a *compound* pendulum. If I_p is the moment of inertia about the pivot point, l is the distance from the pivot to the center of mass, and m is the mass of the pendulum, then Newton's second law prescribes the following equation of motion:

$$I_p \frac{d^2\theta}{dt^2} + mgl \sin\theta = 0, \tag{2.33}$$

and for small angular displacements we again substitute θ for $\sin\theta$ The linearized equation of motion is

$$I_p \frac{d^2\theta}{dt^2} + mgl\theta = 0 \tag{2.34}$$

with period equal to

$$T = \frac{1}{2\pi}\sqrt{\frac{I_p}{mgl}}. \tag{2.35}$$

This expression reverts to that for the simple pendulum when all the mass is concentrated at the lowest point.

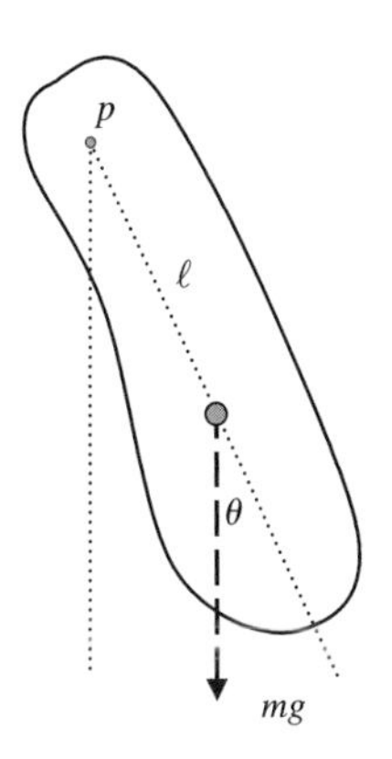

Fig. 2.18
A compound pendulum with an arbitrary distribution of mass.

2.3.6 Kater's pendulum

The formulas for the period of the simple pendulum and the compound pendulum both contain a term for g, the acceleration due to gravity, and therefore one should be able to time the oscillations of the small amplitude pendulum and arrive at an estimate of the local gravitational field. Yet without special effort the results obtained tend to be inaccurate. For example, it is often difficult to determine the appropriate length of the pendulum as there is ambiguity in the measurement at the pivot or at the bob. At the suggestion of the German astronomer F. W. Bessel (1784–1847), Captain Henry Kater (1777–1835) of the British Army invented a reversible pendulum in 1817 that significantly increased the accuracy of the measurement of g Kater's pendulum, shown schematically in Fig. 2.19, consists of a rod with two pivot points whose positions along the rod are adjustable. In principle, the determination of g is made by adjusting the pivot points until the periods of small oscillation about both positions are equal. In practice, it is difficult to adjust the pivot points—usually knife edges—and instead counterweights are attached to the rod and are easily positioned along the rod until the periods are equal. In this way, the pivot positions are defined by fixed knife edges that provide the possibility of accurate measurement. Once the periods are found to be equal and measured, the acceleration due to gravity is calculated from the formula

$$T = 2\pi\sqrt{\frac{h_1 + h_2}{g}}, \tag{2.36}$$

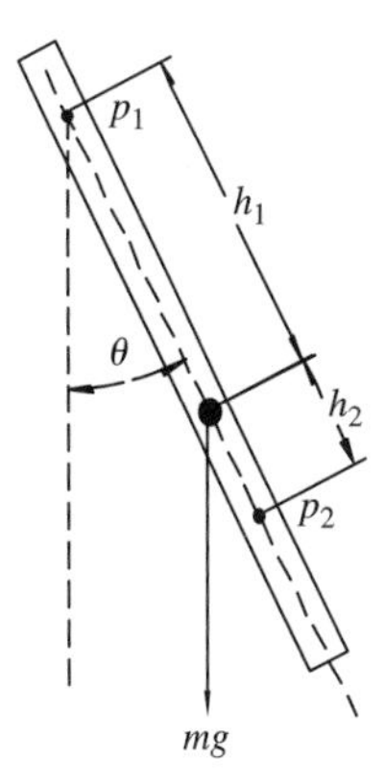

Fig. 2.19
The Kater reversing pendulum.

where h_1 and h_2 are the respective distances from the pivots to the center of mass of the pendulum. But more importantly their *sum* $(h_1 + h_2)$ is easily measurable as the distance between the two knife edge pivot points.

Equation (2.36) is not obvious and its derivation is of some interest. Referring to Fig. 2.19, the pendulum, of mass m, may be suspended about either point P_1 or point P_2. The distances of these suspension points from the center of mass are h_1 and h_2, respectively. The moments of inertia of the pendulum about each of the pivots are denoted as I_1 and I_2. Therefore the linearized equations of motion corresponding to the two pivot points are

$$I_1 \frac{d^2\theta}{dt^2} + mgh_1\theta = 0 \quad I_2 \frac{d^2\theta}{dt^2} + mgh_2\theta = 0. \tag{2.37}$$

The moments of inertia may be expanded using the parallel axis theorem such that

$$I_1 = mk^2 + mh_1^2 \qquad I_2 = mk^2 + mh_2^2, \tag{2.38}$$

where k is the *radius of gyration*, an effective radius of the system about the center of mass such that mk^2 is equal to the moment of inertia about the center of mass. Solution of Eq. (2.37) leads to periods of

$$T_1 = 2\pi\sqrt{\frac{k^2 + h_1^2}{gh_1}} \qquad T_2 = 2\pi\sqrt{\frac{k^2 + h_2^2}{gh_2}}. \tag{2.39}$$

With a little algebra we see that the periods are equal if

$$h_1 h_2 (h_1 - h_2) = k^2 (h_1 - h_2). \tag{2.40}$$

While it would seem easiest to set h_1 equal to h_2, realization of this condition is difficult to achieve with good accuracy in a physical configuration. Instead, the counterweights are used to establish the other algebraic condition;

$$k^2 = h_1 h_2 \tag{2.41}$$

and therefore

$$T_1 = T_2 = 2\pi\sqrt{\frac{h_1 + h_2}{g}}. \tag{2.42}$$

Example 3 *Consider a Kater pendulum in the form of a rod of length L and mass m. Suppose, not very realistically, that we can arbitrarily position the pivot points rather than achieve equality of swing periods with counter weights. Now let us suppose that the pendulum is swung about a pivot located at one end of the rod. We ask ourselves where the other pivot (on the other half of the rod) could be located that would give an equal swing period. (This example is due to Peters (1999).) The moment of inertia of the rod is* $(1/12)\ mL^2$ *about its center. By the parallel axis theorem the moment of inertia about one end is* $(1/12)\ mL^2 + m(L/2)^2 = (1/3)\ mL^2$. *Referring to Eq. (2.35), the period of an oscillation for the pivot located at one end becomes* $T_A = (1/2\pi)\sqrt{(I/mg\,L/2)} = (1/2\pi)\sqrt{(2L/3g)}$. *Let x be the distance from the center along the other half of the rod where the other pivot point is located. By the parallel axis theorem, the moment of inertia about this point is* $(1/12)\ mL^2 + mx^2$ *so that the period is* $T_B = (1/2\pi)\times \sqrt{((1/12)\,m\,L^2 + mx^2)/mgx}$. *Setting* $T_A = T_B$ *leads to a quadratic expression for x with the two roots L/2 and L/6. The root at L/2 is obvious and uninteresting and therefore we choose x = L/6. Substitution of this root* *into the equation leads to* $T_B = (1/2\pi)\sqrt{((1/12)mL^2 + mx^2)/mgx} = (1/2\pi)\sqrt{(2L/3g)} = T_A$ *as expected. As noted previously, the pivot points can not be set exactly and some adjustments are required, using small counter weights, in order to obtain equality of periods.*

In practice, the lengths in Eq. (2.41) are difficult to predict accurately and the experimenter uses a convergence process to arrive at equality of periods. The counterweights are moved systematically until equality is achieved. With this type of pendulum the National Bureau of Standards, in 1936, determined the acceleration due to gravity at Washington, DC as $g = 980.080 \pm 0.003$ cm/s^2 (Daedalon 2000).

After its invention, many of the pendulum gravity experiments were done with the Kater "reversing" pendulum. One of the original pendulums, number 10, constructed by a certain Thomas Jones, rests in the Imperial Science Museum in London. The display card reads as follows.

This pendulum was taken together with No. 11, which was identical, . . . on a voyage lasting from 1828–1831. During this time Captain Henry Foster swung it at twelve locations on the coasts and islands of the South Atlantic. Subsequently it was used in the Euphrates Expedition, of 1835–6, then taken to Antarctic by James Ross in 1840.

2.4 Some connections

One of the fascinating aspects of the history of the pendulum is the remarkable number of famous and not-so-famous physical scientists that have some connection to the pendulum. This phenomenon will come into sharper relief as our story unfolds. We have mentioned a few of these people; here are some others. Marin Mersenne (1588–1648) , a friar of the order of Minims in Paris, proposed the use of the pendulum as a timing device to Christiaan Huygens thereby inspiring the creation of Huygen's pendulum clock. Mersenne is perhaps better known as the inventor of Mersenne numbers. These numbers are generated by the formula

$$2^p - 1, \tag{2.43}$$

where p is prime. Most, but not all, of the numbers generated by this formula are also prime. Jean Picard (1620–1682), a professor of astronomy at the College de France in Paris, introduced the use of pendulum clocks into observational astronomy and thereby enhanced the precision of astronomical data. Picard is perhaps better known for being the first to accurately measure the meridian distance L and his observations, like Richer's observations were used by Newton in calculating the earth's shape. Robert Hooke (1635–1703) well known for the linear law of elasticity, Eq. (2.13), for his invention of the microscope, a host of other inventions, and his controversies with Newton, was one of the first to suggest, in 1666, that the pendulum could be used to measure the acceleration due to gravity. Edmond Halley (1656–1742) , astronomer royal, of Halley's comet fame, was another user of the pendulum. In 1676 Halley sailed to St. Helena's island, the southernmost British possession, located in the south Atlantic, in order to make a star catalog for the southern hemisphere. As a friend of Hooke, he was aware of Hooke's suggested use of the pendulum to measure gravity and did make such measurements while on St. Helena. (While Halley is famous for having his name applied to the comet, he probably rendered a significantly more important service to mankind by pressing for and financially supporting the publication of Newton's *Principia.*) In the next century, Sir Edward Sabine (1788–1883) , an astronomer with Sir William Parry in the search for the northwest passage (through the Arctic ocean across the north of Canada) spent the years from 1821 to 1825 determining measurements of the gravitational field along the coasts of North America and Africa, and, of course, in the Arctic, with the pendulum.

The American philosopher Charles Saunders Peirce (1839–1914) makes a surprising appearance in this context. Known for his contributions to logic and philosophy, Peirce rarely held academic position in these branches of learning, but made his living with the US Coast and Geodetic Survey. Between 1873 and 1886, Pierce conducted pendulum experiments at a score of stations in Europe and North America in order to improve the determination the earth's ellipticity. However, his relationship with the Survey administration was fractious, and he resigned in 1891. And finally, in the twentieth century, we note the work of Felix Andries Vening Meinesz (1887–1966), a Dutch geophysicist who, as part of his Ph.D. (1915) dissertation, devised a pendulum apparatus which, somewhat like Von Sterneck's device, used the concept of pairs of perpendicularly oriented pendulums swinging out of phase with each other. (See Fig. 2.20).

Fig. 2.20
Vening Meinesz pendulum. Four pendulums arranged in mutually perpendicular pairs are visible. (Courtesy of the Society of Exploration Geophysicists Geoscience center. Photo ©2004 by Bill Underwood.)

In this way Vening Meinesz eliminated a horizontal acceleration term due to the vibration of peaty subsoil that seemed to occur in many places where gravity was measured. Vening Meinesz' apparatus was also especially fitted for measurements on or under water and contained machinery that compensated for the motion of the sea. Aside from the interruption caused by the Second World War, some version of this device was used on submarines from 1923 until the late 1950s (Vening 1929).

In the next chapter we add some complexity to the pendulum. We include friction and then compensate for the energy loss with an external source of energy. Eventually, we also relax the condition of small amplitude motion and therefore the equations of motion become nonlinear, a significant complication in our discussion. However the small amplitude motion of the linearized pendulum will predominate in three of the chapters; those on the Foucault pendulum, the torsion pendulum (which is well modeled as linear), and the pendulum clock. Obviously, the linearized pendulum is the basis of important applications.

2.5 Exercises

1. In a later chapter we discuss the Foucault pendulum that was the first explicit demonstration of the rotation of the earth. The original Foucault pendulum was 67 meters in length. Calculate the frequency and period of its motion. The plane of oscillation of the pendulum rotated through a full 360 degrees in 31.88 hours. How many oscillations does the pendulum make in that time?

2. In the early days of gravity measurement by pendulum oscillation, a "seconds" pendulum had a length of about 1 m. This connection between the meter and the second was thought to have some special significance. What was the actual period of the "seconds" pendulum? From your result how do you think the period of the pendulum was initially defined?
3. A particle undergoing uniform acceleration from a standing start at the position $x = 0$ has the following parametric equations (or time series) for position and velocity:

$$v = at$$
$$x = \frac{1}{2}at^2.$$

Determine the equation for its orbit in an x, v phase space and sketch the orbit.
4. Consider the phase orbit given by Eq. (2.6). Form the phase space diagram such that the x-axis is θ and the y-axis is $\dot{\theta}/\omega$ Then the phase orbit becomes a circle of radius θ_0. Note also that $\theta = \theta_0 \cos \omega t$. Therefore the phase point traces out a circular orbit with a polar angle $\alpha = \omega t$. We are now ready to easily prove that areas in phase space are preserved in time. Proceed as follows. Consider two boundary orbits in phase space defined by two pendulums of different amplitudes (energies), $\theta_0(1)$ and $\theta_0(2)$. These orbits are two concentric circles. Now imagine a region between these two orbits bounded on the other sides by angles $\alpha_1 = \omega t_1$ and $\alpha_2 = \omega t_2$ Using polar coordinates calculate the area of this region and show that for some later times $t_1 + \Delta t$ and $t_2 + \Delta t$, the area still only depends upon the difference, $t_2 - t_1$. That is, the area is preserved in time, and the system is conservative. See Fig. 2.21.
5. Find the Fourier series for the periodic function,

$$f(t) = 1 : 0 < t < T/2$$
$$f(t) = -1 : -T/2 < t < T.$$

6. The complete restoring force of the pendulum is $F = -mg \sin \theta$. Various approximations may be obtained using a Taylor series expansion in which the expansion variable is the length along the arc of the pendulum's swing, $s = l\theta$. That is

$$F(s) = F(s_0) + F'(s_0)(s - s_0) + F''(s_0)(s - s_0)^2/2! + F'''(s_0)(s - s_0)^3/3! + \cdots$$

where $F' = dF/ds$. Express F in terms of s. Let $s_0 = 0$ and show that the first nonvanishing term in the expansion is the usual small angle linear approximation, $F \approx -mg\theta$. Now let $s_0 = l\pi/4$, and show that the linear approximation, in the region of $\theta = \pi/4$, is

$$F \approx \frac{mg}{\sqrt{2}}\left[\left(\frac{\pi}{4} - \frac{1}{\sqrt{2}}\right) - \theta\right].$$

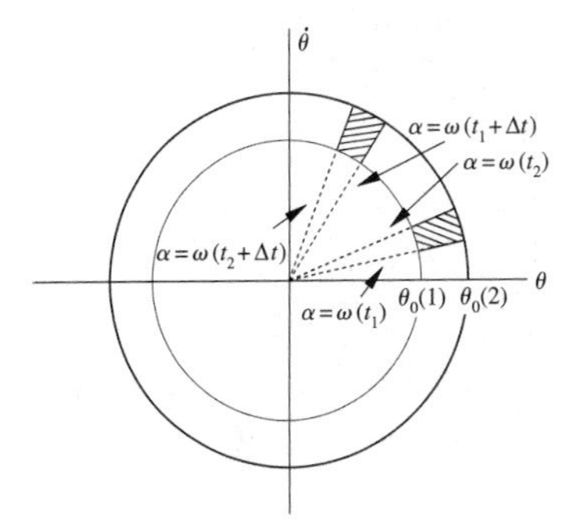

Fig. 2.21
Figure for problem 4.

7. Determine equations for the constants a and b in the Lennard–Jones potential, in terms of given values of the molecular spring constant, k and the equilibrium bond length, r_{eq} Note that the force is zero at $r = r_{eq}$.
8. Derive Eq. (2.29) for the ratio of densities σ'/σ where σ' is the density near the surface of the earth (above the mine shaft), and σ is the average density of the earth. For this derivation try the following sequence of calculations. First calculate g at the bottom of the mine shaft using Gauss' law, and remember that the earth at a radius above that of the bottom of the shaft contributes nothing to the gravitational field. Then use Gauss' law to calculate the gravitational field on top of the earth by dividing the earth into two parts: one at a depth below the shaft with density σ, and the shell above the bottom of the shaft with density σ'. Finally, examine the ratio of g_{top}/g_{bottom} and use the binomial expansion in terms of h/a where needed. Neglect any terms that are more than first degree in the ratio h/a.

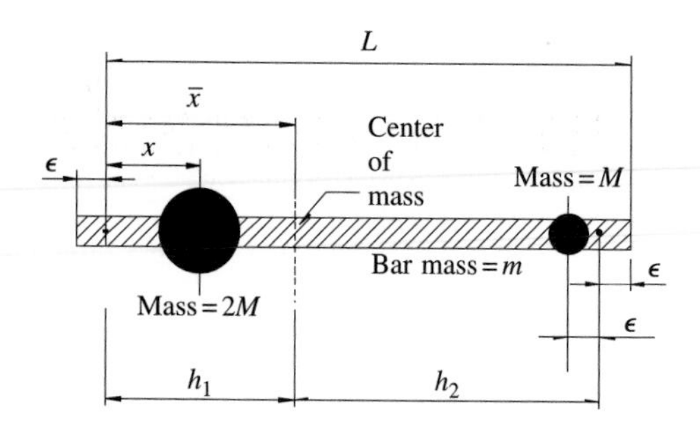

Fig. 2.22
Figure for problem 9.

9. Figure 2.22 shows a Kater pendulum with two attached masses, M and $2M$ The pivot points are just inside the ends of the bar (mass m) at a distance ϵ from the ends. The smaller mass is fixed at a distance of ϵ from the right pivot point. The larger mass is located a variable distance x from the left point. The point of this exercise is to find the location of the mass $2M$ such that the pendulum will oscillate with equal period from either pivot point.

(a) Find the center of mass $\bar{x}$ of the system in terms of the quantities shown in Fig. 2.22.
(b) Find h_1 and h_2.
(c) Check that $h_1 + h_2 = L - 2\epsilon$.
(d) Use the condition that $h_1 = h_2$ to find the appropriate value of x.

10. For the example in the text, $h_1 = L/2$ and $h_2 = L/6$. Using Eq. (2.42) show that these values lead to the correct result for the period.

11. Repeat the analysis for the Kater pendulum example in the text by putting one pivot point half-way between the center and the end of the rod; that is, at $L/4$ from the center. One position for the other pivot is, trivially, a distance $L/4$ from the center on the opposite side of the center line. (a) Using the analysis in the example, show that there is another location for the second pivot point at a distance $L/3$ from the center on the opposite side from the first pivot point. Show that the periods of oscillation for the pendulum from each pivot point are equal.

12. Consider a pendulum that consists of a uniform rod of length L and mass M that hangs from a frictionless peg that passes through a small hole drilled in the rod. The rod is free to oscillate (without friction) and assume that the oscillations are of small amplitude and therefore the equation of motion may be written as

$$I\frac{d^2\theta}{dt^2} + MgD\theta = 0,$$

where I is the moment of inertia and D is the distance between the center of mass of the rod and the pivot point.

(a) What is the frequency of oscillation of this pendulum?
(b) If the pivot point is located very near the top of the rod ($D = L/2$), find the frequency in terms of L and g.
(c) If the pivot point is located 1/3 of the way from the end of the rod, find the frequency of oscillation.
(d) If, in general, the pivot point is located a distance $D = L/k$ from the center of mass where $k \in [2,\infty)$, find a general expression for the frequency in terms of L, g, and k.
(e) For what value of k is the frequency a maximum?
(f) For what value of k is the frequency a minimum?

13. Find the Mersenne primes for $p = 3, 5, 7, 11, 13, 17, 19, 31$.

Pendulums less simple

3

3.1 O Botafumeiro

In the northwest corner of Spain, in the province of Galicia, lies the mist shrouded town of Santiago de Compostela, the birthplace of the cult of Santiago (St. James, the major apostle), and the home of the magnificent cathedral that is presumably built upon the bones of that martyred apostle (Adams 1999) (see Fig. 3.1). For a thousand years, pilgrims have sought out this cathedral as a shrine to Saint James where they might worship and receive salvation. The most famous and unique feature of the celebration of the mass at this cathedral, at least since the fourteenth century, is *O Botafumeiro*, a very large incense burner suspended by a heavy rope from a point seventy feet above the floor of the nave, and swung periodically through a huge arc of about eighty degrees (Sanmartin 1984). The rapid motion through the air fans the hot incense coals, making copious amounts of blue smoke, and the censer itself generates a frightening swooshing sound as it passes through the bottom of its arc. Some of the physics in this chapter is manifested by the remarkable motion of *O Botafumeiro* and therefore we provide some details of its structure and dynamics (Sanmartin 1984).

Fig. 3.1
The cathedral of Santiago de Compostela in northern Spain. Photo by Margaret Walker.

The censer, or incense burner itself, stands more than a meter high and is suspended by a thick rope whose diameter is 4.5 cm. (One can imagine something about the size of a backyard barbecue grill.) Over a period of seven hundred years, a variety of censers have been used. The original censer seems to have been silver, which was later replaced by another silver one, donated by the French king Louis XI. A papal bull from Pope Nicholas V, in 1447, threatened excommunication to anyone who stole it. It was probably this latter censer that was destroyed when it suffered a violent fall in 1499. The censer is sporadically mentioned in records over the next couple of centuries with yet another silver replacement being made as late as 1615. There is some evidence—but not conclusive—that French troops took a silver censer during Napoleon's 1809 campaign. At some point prior to 1852, the censer was made of iron, but at that date it was replaced by a censer of silvered brass, which is the one in use today (see Fig. 3.2). The current censer has a mass of 53 kg and is about 1.5 m in overall height. Its center of mass is about 55 cm above the base. Approximately three meters of rope are used just in tying the heavy knot that secures the rope to the censer.

The other end of the rope is connected, high above the floor of the nave, to rollers on a frame that is supported by an iron structure. The iron structure consists of four large struts, joined at the central frame, that stretch to four posts near the upper part of the tower. The rollers provide the mechanism used for pumping the censer (see Fig. 3.3). In the early days, the complete support structure was wooden; the iron structure was built later, in 1602, in part because the cumbersome wooden support obscured light coming into the tower. The length of rope between the rollers and the top of the knot is 19.4 m. Therefore the length from the rollers to the center of mass of the censer is 20.6 m and the length to the floor is 21.8 m. This means that the bottom of the censer comes within about half a meter of the floor of the nave as it passes through the bottom of its arc.

The censer is pumped by periodically shortening and lengthening the rope as it is wound up and then down around the rollers. This kind of pumping is now known as *parametric* forcing. The pumping action is carried out by a squad of priests, called *tiraboleiros* or ball swingers, grouped to the side and each holding a rope that is a strand of the main rope that goes from the pendulum to the rollers and back down to near the floor. The tiraboleiros periodically pull on their respective ropes in response to orders from the chief verger of the cathedral. Unlike the child on a swing who senses the effect of her pumping, the team must simply act as a unit and follow the timing prescribed by the verger. One of the more terrifying aspects of the pendulum's motion is the fact that the amplitude of its swing is very large—about 82°—and therefore it has a high velocity—about 68 km/h—as it pass through the bottom of its arc, spewing smoke and flames.

How does *O Botafumeiro* differ from the simple pendulum described in Chapter 2? First, the censer is not a point mass or even a mass with spherical symmetry, and therefore it is a *compound* pendulum, a pendulum

Fig. 3.2
O Botafumeiro, the giant censer that hangs in the transept of the cathedral of Santigo de Compostela. (Reproducido con autorización, copyright © 1990. Prensa Científica, S.A. Reservados todos los derechos.)

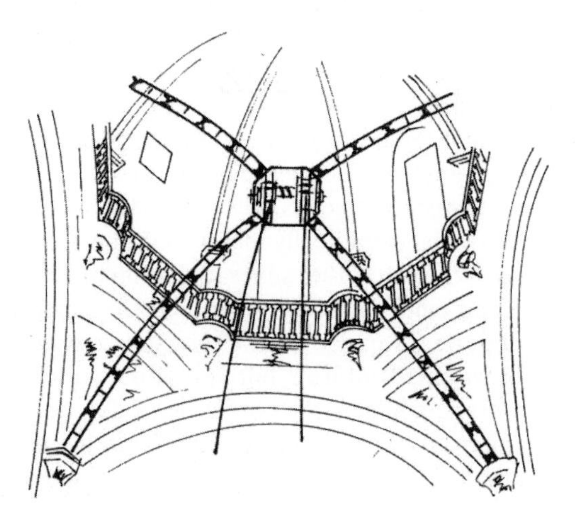

Fig. 3.3
Support structure for the rope holding O Botafumeiro. (Reprinted with permission from Sanmartin (1984, p. 939). ©1984, American Association of Physics Teachers.)

whose bob is an extended nonspherically symmetric mass. Second, the fact that the censer needs to be pumped implies that there is energy *dissipation* in the system and therefore damping of the motion. Third, pumping or *forcing* of the system is a fruitful topic in itself and will be seen to lead to rich dynamical behavior. Fourth, the swing of the pendulum is quite large, and, contrary to Galileo's conclusion of isochronism based upon his crude observations, the period of the motion is *not* independent of the amplitude of the motion. Finally there are some subtle effects associated with the rope itself in the particular case of *O Botafumeiro*. That is, the mass of the rope is not insubstantial and should be included in analysis of the censer's motion as should the fact that the inertia of the system leads to the rope not being stretched straight throughout the oscillation. This latter phenomenon suggests that the motion might be treated as that of a *double pendulum*. (Although we will not do so in this book. See (Sanmartin 1984).) The subject of this chapter is, then, some of the complications to pendulum dynamics suggested by the motion of *O Botafumeiro*.

As a postscript to this introduction, Sanmartin notes that the history of *O Botafumeiro* has not been without mishap and even violence. On July 25, 1499, there was an accident involving the four chains that attach the censer to the rope. The chains broke, thereby sending the censer crashing into one of the side doors. On May 23, 1622, the rope broke. This time, the censer fell approximately vertically since the amplitude of the motion was small at the moment. In a more sinister vein, it has been suggested that the suspending rope was used in the murder of the Archbishop Suero Gomez de Toledo at the entrance to the cathedral on June 29, 1366. Two noblemen, F. Perez Churruchao and G. Gomez Gallinato were involved. Apparently they also killed the Dean of the cathedral, Pedro Alverez, who, when he discovered the fate of the archbishop, ran back into the cathedral. The killers and a third person, Pedro el Cruel, chased him around the triforium before catching him. Pedro later denied any involvement.

Let us now leave the exotic world of *O Botafumeiro*, and return to the generic pendulum. We first enhance our simple model of the linearized pendulum, of the previous chapter, with the addition of various physical factors. With each new factor the mathematics becomes more complex and, eventually, reaches the point of intractability. At that point one must resort to numerical solution of the equations on a computer. In order to keep the analysis manageable, we will first treat many of the factors separately. In the second half of the chapter, we consider the full large amplitude, nonlinear, case.

3.2 The linearized pendulum with complications

3.2.1 Energy loss—friction

A common observation is that the periodic motion of a pendulum gradually diminishes and that, eventually, a pendulum will come to rest. In fact all large systems that are not continually energized lose energy . There is no such thing as perpetual motion. In a mechanical system, the causes of energy loss are multitudinous, and include friction between solid surfaces and drag on the system from fluids and gases. Objects that travel through the atmosphere or other "fluids" at relatively high speeds experience a drag force that is proportional to the square of the velocity. For slower objects, the drag force may be realistically modeled as proportional to the first power of the velocity, a relationship named after the nineteenth century Cambridge University mathematician and physicist, Sir George Stokes (1819–1903). (Stokes, like Newton, Airy and the present day cosmologist Stephen Hawking, was, in his time, the Lucasian Professor. Like many other prominent scientists, Stokes was drawn to the study of the pendulum. He was particularly interested in the motion of pendulums within a fluid where friction would be important. His work on pendulums led to the study of geodesy and by the 1850s he had become the foremost British authority on the subject.) What constitutes a relatively high or low speed depends on the physical properties of the medium. But for a slow pendulum travelling in air, the linear dependence of drag on velocity is a reasonable approximation. Furthermore, other frictional effects for real pendulums seem also to be accounted for with the same velocity dependence. (However, O Botafumeiro travels fast enough that a velocity squared dependence is more realistic.) For now, we approximate the aggregate effect of drag or friction for typical pendulum as

$$F_D = -2\gamma \frac{d\theta}{dt}, \tag{3.1}$$

where γ is a constant. The equation of motion for the linearized pendulum now has an added term and becomes

$$\frac{d^2\theta}{dt^2} + 2\gamma \frac{d\theta}{dt} + \frac{g}{l}\theta = 0. \tag{3.2}$$

The general solution of this second-order differential equation has the form

$$\theta(t) = Ae^{r_1 t} + Be^{r_2 t},$$

where A and B are constants to be determined from initial conditions, and r_1 and r_2 are the two roots of the quadratic equation created by substitution of a trial solution, e^{rt}, into Eq. (3.2):

$$r^2 + (2\gamma)r + (g/l) = 0$$

$$r_1 = -\gamma + \sqrt{\gamma^2 - \omega_0^2} \tag{3.3}$$

$$r_2 = -\gamma - \sqrt{\gamma^2 - \omega_0^2}. \tag{3.4}$$

The angular frequency of undamped oscillations is $\omega_0 = \sqrt{g/l}$. The strength of the damping now determines the time-dependent pendulum behavior. Three distinct regimes exist.

Underdamped $\gamma^2 < \omega_0^2$
For this case, define a new frequency

$$\omega^2 = \omega_0^2 - \gamma^2. \tag{3.5}$$

Then the two roots are complex quantities which can be written

$$r_1 = -\gamma + i\omega$$
$$r_2 = -\gamma - i\omega.$$

The negative real part of the roots implies exponential decay, while the imaginary parts ($\pm i\omega$) yield oscillations at a new frequency that is slightly smaller than the undamped value ω_0. Combining these components, the angular motion can be expressed in the form

$$\theta = Ce^{-\gamma t}\cos(\omega t + \phi). \tag{3.6}$$

The canonical solution given earlier contained two constants A and B. An illustration of this form of damped oscillation is given in Fig. 3.4

Two constants remain in the new form of solution as expressed in Eq. (3.6)—one (C) defines the amplitude; the other (ϕ) appears as a phase shift. As noted before, specification of initial conditions for the pendulum will explicitly determine C and ϕ. For example, suppose it is known that at $t=0$, $\theta=\theta_0$ and the pendulum is at rest. Then

$$\theta_0 = C\cos(\phi)$$

and

$$\omega\sin(\phi) + \gamma\cos(\phi) = 0.$$

From these two equations C and ϕ are easily calculated.

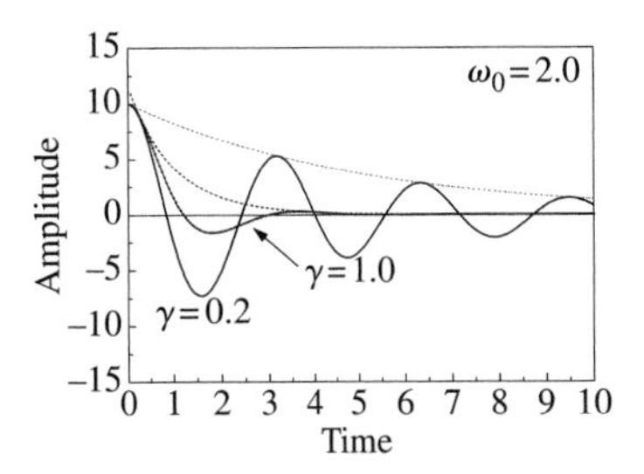

Fig. 3.4
Illustation of an underdamped oscillation in which the amplitude slowly decays under control of the exponential prefactor. The pendulum is released from rest at an angle of 10°. Two different damping values are shown. The dotted lines indicate the decay of the respective amplitudes.

Critically Damped $\gamma^2 = \omega_0^2$
In this special case, the two roots become equal and the method described above cannot be applied . The solution to the equation of motion is instead

$$\theta(t) = e^{-\gamma t}[A + Bt]. \tag{3.7}$$

The two constants can be determined as before from given initial conditions. If at $t=0$ the pendulum is at rest at a displacement θ_0, then

$$\theta_0 = A$$

and

$$B - A\gamma = 0.$$

A critically damped response using the previous choice $\omega_0=2$ is included in Fig. 3.5.

As can be seen in the figure, for critical damping the decay is optimum in the sense that the angle asymptotically approaches zero without overshoot (underdamping) or undershoot (overdamping).

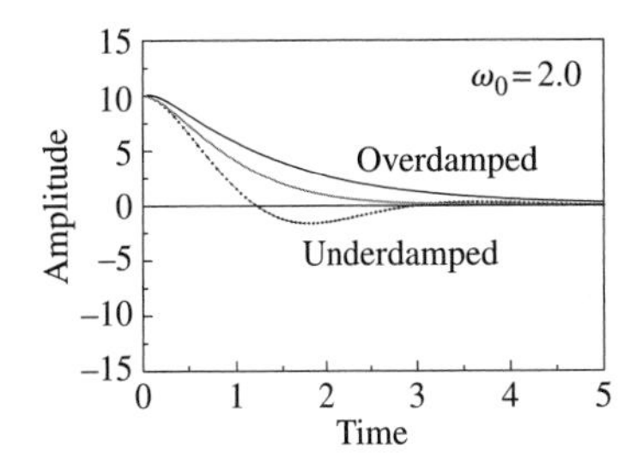

Fig. 3.5
Decay of the pendulum for three regimes: underdamped (lower curve: $\gamma=1$); critically damped (middle curve: $\gamma=2$); overdamped (upper curve: $\gamma=3$).

Overdamped $\gamma^2 > \omega_0^2$
For the overdamped case, both roots are real, so the decay is entirely exponential. With

$$k = \sqrt{\gamma^2 - \omega_0^2}, \tag{3.8}$$

which is of course a real quantity in this situation, the solution is

$$\theta(t) = e^{-\gamma t}\left[Ae^{kt} + Be^{-kt}\right]. \tag{3.9}$$

As in the previous two cases, the constants A and B will be fixed by the initial conditions of the problem.

The time dependence of the pendulum's angular displacement for this case is also included in Fig. 3.5. Furthermore, each of these motions may also be represented in phase space. The underdamped case is illustrated in Fig. 3.6.

In the ideal case of the completely undamped simple pendulum, the total energy of the system is conserved. The energy flows alternatively between kinetic and potential energy. But for the damped pendulum the energy dissipates exponentially. We calculate the energy of the pendulum as follows. Starting with Eq. (2.7);

$$E = \frac{1}{2}ml^2\dot{\theta}^2 + \frac{1}{2}mgl\theta^2, \tag{3.10}$$

we obtain the total energy of the linearized pendulum. The θ and $\dot{\theta}$ terms are replaced by Eq. (3.6) and its derivative. The final result is cumbersome and contains several terms,

$$E = \frac{1}{2}ml^2\theta_0^2 e^{-2\gamma t}[(\gamma^2 + \omega_0^2)\sin^2\alpha + \omega^2\cos^2\alpha - \gamma\omega\sin 2\alpha], \tag{3.11}$$

where $\alpha = \omega t + \phi$, but, unlike the nondissipative case, each term contains a decreasing exponential factor

$$f(t) = e^{-2\gamma t} \tag{3.12}$$

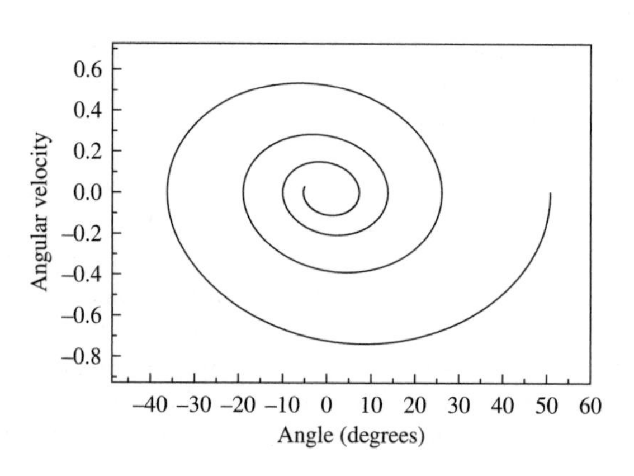

Fig. 3.6
Phase portrait of the underdamped linearized pendulum.

that decays at twice the rate of either the angular displacement or velocity time series. The remaining parts of each term in the expression are periodic with either the fundamental frequency ω or double that frequency.

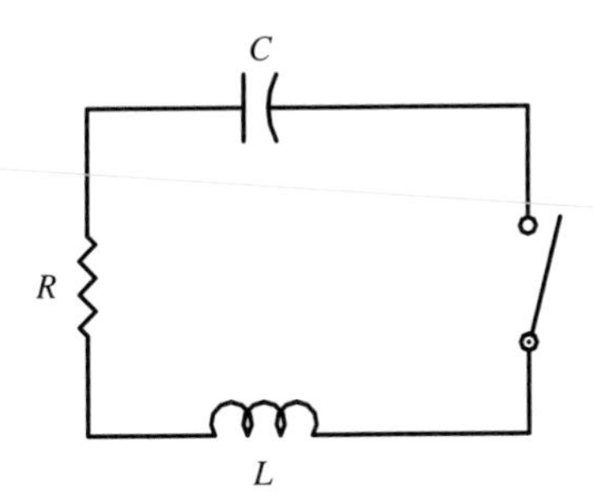

Fig. 3.7
Tuned circuit with resistance.

Example 4 *The electrical analog of pendulum damping is the presence of resistance R now added to the tuned LC circuit of Fig. 2.13. Figure 3.7 shows the modified schematic diagram for the tuned circuit. Eq. (2.20) is now augmented with a term proportional to the current, the simple Ohm's law term , as follows:*

$$L\frac{d^2q}{dt^2} + R\frac{dq}{dt} + \frac{q}{C} = 0. \tag{3.13}$$

Substitution of the trial solution $q = q_0 e^{-\gamma t}\sin(\omega t + \phi)$ yields the frequency and the damping as

$$\omega = \sqrt{\omega_0^2 - \gamma^2}, \tag{3.14}$$

where

$$\omega_0 = \sqrt{\frac{1}{LC}} \quad \textit{and} \quad \gamma = \frac{R}{2L}. \tag{3.15}$$

We know from electrical theory that the power dissipation occurs only in the resistance R, and is given by the formula

$$\textit{Power} = i^2 R. \tag{3.16}$$

Again the time dependent expression is cumbersome, but it does contain the exponential decay factor

$$f(t) = e^{-(R/L)t} \tag{3.17}$$

as expected. The quantity L/R is sometimes referred to as the decay constant or time constant or ring time of the circuit. It is the time during which the energy decays by a factor of 1/e from the original amount.

Example 5 *There are many examples of the decay of oscillations found in nature. For example, in magnetic resonance imaging, large numbers of tiny nuclei are momentarily energized by radio frequency electromagnetic radiation. Once the radiation is turned off, the nuclei "relax" to their equilibrium state. The relaxation time (equivalent to the decay time) tells the observer something about the environment and, in medical applications, the relaxation time can help with diagnosis.*

Example 6 *A more obvious example is found in the "ring" time of a large room, such as a concert hall. Each hall has its own damping constant (the inverse of the ring time) which will depend on many properties of the hall, including the size of the audience and the materials on the inside surfaces of the hall. A conductor can sometimes make dramatic use of the ring time. Suppose a large choir and orchestra are performing at a high volume and then have a sudden rest in the music. The musicians all stop playing very abruptly. The conductor can then draw out the pause—the time when the musicians are* *silent—such that the music will still 'linger' in the air even while the musicians are quiet. This phenomenon is sometimes used with powerful effect just prior to a grand finale. Conductors of the famous choral work "Messiah" by George Handel exploit this bit of physics just before the last few chords of the popular "Hallelujah" chorus.*

3.2.2 Energy gain—forcing

The addition of dissipation to the pendulum causes energy decay and, strictly speaking, the motion continues at increasingly reduced energy for an infinite time. However, except for very long, massive pendulums with heavy masses, the oscillations do not actually last more than a few moments. Therefore continuous operation of most pendulums necessitates that some kind of forcing be applied to provide an energy infusion that compensates for the energy loss caused by the damping. In this section, we discuss several types of forcing: sinusoidal forcing, pulsed forcing, and parametric forcing.

3.2.2.1 Sinusoidal forcing

Sinusoidal forcing is simply a force with a constant amplitude and a sinusoidal time variation and it is inserted into the equation of motion as

$$\frac{d^2\theta}{dt^2} + 2\gamma\frac{d\theta}{dt} + \frac{g}{l}\theta = F \sin \Omega t. \tag{3.18}$$

(Note that in Eq. (3.18) F does not have the units of a force, but that its dimensions are $(\text{time})^{-2}$). This differential equation may be solved by standard methods (See, for example, chap. 5 of (Zill and Cullen 1993)). The complete solution consists of a general solution to the corresponding homogeneous equation and a particular solution appropriate to the full inhomogeneous equation. Assuming that the damping is relatively light, the solution is a linear combination of (a) an exponentially decaying, periodic solution of the corresponding homogeneous differential equation and (b) a periodic, particular solution due to the forcing term. It has the form

$$\theta(t) = \sqrt{c_1^2 + c_2^2}e^{-\gamma t} \sin(\omega t + \phi_1) + \frac{F}{\sqrt{4\gamma^2\Omega^2 + (\omega_0^2 - \Omega^2)^2}} \sin(\Omega t + \phi_2), \tag{3.19}$$

where

$$\phi_1 = \tan^{-1}\left(\frac{c_1}{c_2}\right) \quad \text{and} \quad \phi_2 = \tan^{-1}\left(\frac{2\gamma\Omega}{\omega_0^2 - \Omega^2}\right) \tag{3.20}$$

and

$$\omega_0 = \sqrt{\frac{g}{l}} \quad \text{and} \quad \omega = \sqrt{\omega_0^2 - \gamma^2}. \tag{3.21}$$

The constants c_1 and c_2 depend on the initial conditions for the pendulum. If we assume that the pendulum is a rest when the clock is started, then $\theta(0)=0$ and $d\theta(0)/dt=0$. These conditions lead to the following expressions for the constants:

$$c_1 = \frac{-2F\gamma\Omega}{\left[4\gamma^2\Omega^2 + (\omega_0^2 - \Omega^2)^2\right]}$$
$$c_2 = \frac{-F\Omega}{\omega}\left[\frac{(\omega_0^2 - \Omega^2) + 2\gamma^2}{4\gamma^2\Omega^2 + (\omega_0^2 - \Omega^2)^2}\right]. \tag{3.22}$$

The general solution, Eq. (3.19), has (a) a term that decays—a *transient solution*— and (b) a *steady state solution* that predicts long term periodic motion of the pendulum at the forcing frequency, Ω. It is interesting to note that while the frequency of the sinusoidal term of the decay part is ω, which depends on the resonant frequency ω_0 and damping γ, the frequency of the steady state sinusoidal term depends only on the frequency of forcing, Ω.

The *amplitude* of the steady state periodic motion depends on the size of the damping term. If the damping is quite small and if the forcing frequency Ω is close to ω_0 then the amplitude of the motion is relatively large. In fact, the expression goes to infinity if there is no damping at all. This case is described below. But for finite damping the shape of the amplitude curve as a function of Ω for several values of damping coefficient, γ, is given in Fig. 3.8. Each curve has a finite maximum that indicates the frequency value for maximum absorption of energy. In the previous chapter we discussed the concept of *resonance* as a periodic exchange of kinds of energy. We can also talk about resonance as the state of a system of maximum energy absorption. When Ω equals ω_0 the system is in the *resonant* state: and absorbs energy most easily. Furthermore, the resonant state is characterized by a certain width in frequency units that depends on the amount of damping. If the system is lightly damped then the resonant state is narrow but very absorbent with an extremely efficient transfer of energy from the forcing mechanism to the pendulum. On the other hand, if the damping is relatively strong, then the resonance is broad band and not very sensitive to frequency changes near ω_0, but neither is the transfer of energy very efficient because much energy is dissipated as heat. Additional details on resonance and the *quality* of the resonance as measured by the "Q" factor are discussed in Appendix A.

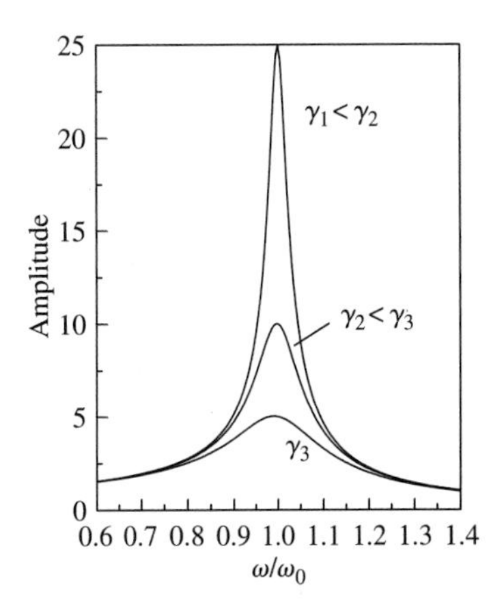

Fig. 3.8
Amplitude as a function of forcing frequency for various amounts of damping.

The study of periodic motion is at least partially motived by the fact that its mathematical form is relatively simple and the resulting differential equation of motion may be readily solved. To actually build a mechanical model of a pendulum (even a linearized pendulum) that is driven by the simple sinusoidal forcing function is not trivial. We discuss a similar (but nonlinear) pendulum later in this book but for now simply note that in most cases, simple mechanical pendular forcing is not well modeled by this function. However, there is one widely used application that does fit the linearized model of sinusoidal forcing very well; that is, the tuned electrical circuit that is subject to a sinusoidal electromagnetic field.

We have already analyzed the LRC electrical circuit and found that, like the unforced, damped, linearized pendulum, the motion of charge among the elements gradually dies away as the energy is dissipated in the heating of the resistor. However, a periodic electromagnetic energy field impinging on the circuit will induce a time varying voltage across the circuit elements. This voltage will sustain the flow of charge through the circuit. In more exotic applications, such as radio and television, the electromagnetic field in the circuit may then be amplified to reproduce messages that were transmitted as a modulation of the original periodic electromagnetic field. This chain of events is the essence of radio and television reception. Returning to the basic mathematics of sinusoidal forcing in the circuit we find that, with the addition of the periodic electromagnetic field, the balance equation for the voltage across the circuit is

$$L\frac{d^2q}{dt^2} + R\frac{dq}{dt} + \frac{q}{C} = V\sin\Omega t. \tag{3.23}$$

This equation is identical in form to Eq. (3.18) and therefore all the mathematics derived from that equation of motion apply to this differential equation for the charge. We observe transient and steady state solutions, and we observe resonance. In a radio or television, we tune the circuit elements so that they will resonate with the carrier field of the desired transmitting station. In this way the radio or television receiver discriminates among transmitting stations and is able to select a particular station with its particular carrier frequency, ω_0

Let us conclude this section on sinusoidal forcing by looking at what happens if the system is forced *without* damping. Intuition suggests that the effect of an energy input from forcing without a corresponding dissipation of energy through damping will result in an oscillatory pendulum whose amplitude of motion will grow without limit. But this result is not obvious from simply setting the damping equal to zero in the solution of the general case. Therefore we begin again with the equation of motion from which the damping term is removed:

$$\frac{d^2\theta}{dt^2} + \frac{g}{l}\theta = F\sin\Omega t. \tag{3.24}$$

Using standard techniques, we find the general solution to be

$$\theta(t) = b_1\sin\omega_0 t + b_2\cos\omega_0 t + \frac{F}{\omega_0^2 - \Omega^2}\sin\Omega t. \tag{3.25}$$

As before we note that two sinusoidal terms contain the natural frequency ω_0, while the other sinusoidal term has the forcing frequency, Ω. Again, we assume that the pendulum is initially at rest; $\theta(0) = 0 = \theta'(0)$, and thereby determine the $b_{1,2}$ coefficients. This calculation results in the solution

$$\theta(t) = \frac{(F/\omega_0)[\omega_0\sin\Omega t - \Omega\sin\omega_0 t]}{(\omega_0^2 - \Omega^2)}. \tag{3.26}$$

This solution is only defined when the forcing frequency is *not equal* to the resonant frequency. In this case, and perhaps somewhat surprisingly, the nonresonant forcing does not lead to an unlimited growth in displacement.

For resonant forcing the case is different. We can determine the behavior when the frequencies *are equal* by a limit process whereby Ω is allowed to approach ω_0 Application of L'Hopital's rule results in

$$\lim_{\Omega \to \omega_0} \theta(t) = \frac{F}{2\omega_0^2}[\sin \omega_0 t - \omega_0 t \sin \omega_0 t]. \tag{3.27}$$

The second term of this solution grows without limit, as our intuition suggested. A sketch of the solution is shown in Fig. 3.9.

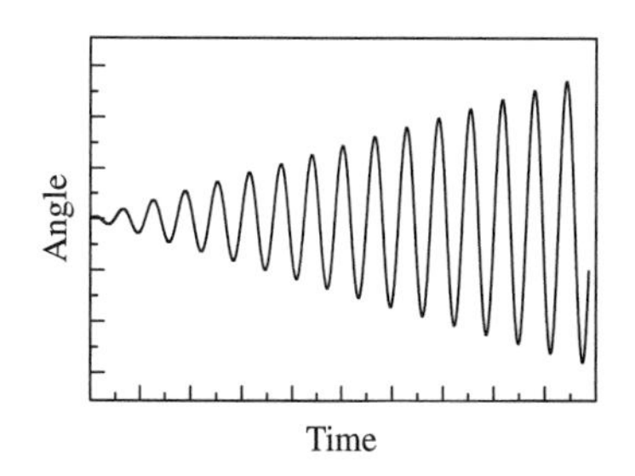

Fig. 3.9
Angular displacement as a function of time due to forcing at the resonant frequency.

3.2.2.2 Pulsed forcing

The common playground swing is one of the most ubiquitous examples of the driven pendulum. One pictures a small rider on the swing being pushed by an older child or adult. We might be tempted to model the periodic pushing by the second person as sinusoidal forcing. But, in reality, the rider receives a push or *impulse*, as the swing moves forward through the bottom of its arc, rather than sinusoidal forcing. Thus, a realistic model of the pushed swing will incorporate pulsed forcing. The profile of the push may be simply modeled as a rectangular pulse. Of course, any real pulse will have "soft" edges, but in either case the periodicity of the forcing and its nonsinusoidal form will guarantee that any realistic shape will have many Fourier components. We therefore expect complexity beyond that observed for sinusoidal forcing. While the mathematics is more complex, the final results are similar to the case of sinusoidal forcing. The fundamental frequency component of the forcing is a dominant factor in the resultant motion.

Let us assume that the swing starts from rest and that the periodic pulsing occurs at the beginning of each cycle at the bottom of the arc. In this first attempt at a model we ignore damping and let the amplitude of the motion be small in order that the swing may be treated as a linearized pendulum. The push function is illustrated in Fig. 3.10 and is written as

$$\begin{aligned} F(t) &= F, \quad 0 < t < \tau \\ &= 0, \quad \tau < t < T = n\tau. \end{aligned} \tag{3.28}$$

For simplicity of solution we choose the push time to be an integer fraction of the period of the pendulum. Furthermore, the period T is made to coincide with the resonant motion of the pendulum, as is the common practice in pushing a swing. That is

$$T = 2\pi\sqrt{\frac{l}{g}} \text{ with an angular frequency of } \omega_0 = \sqrt{\frac{g}{l}}. \tag{3.29}$$

With the pulsed forcing term , the equation of motion is now

$$\frac{d^2\theta}{dt^2} + \omega_0^2\theta = F(t) \text{ with initial conditions } \theta(0) = 0, \ \frac{d\theta(0)}{dt} = 0. \tag{3.30}$$

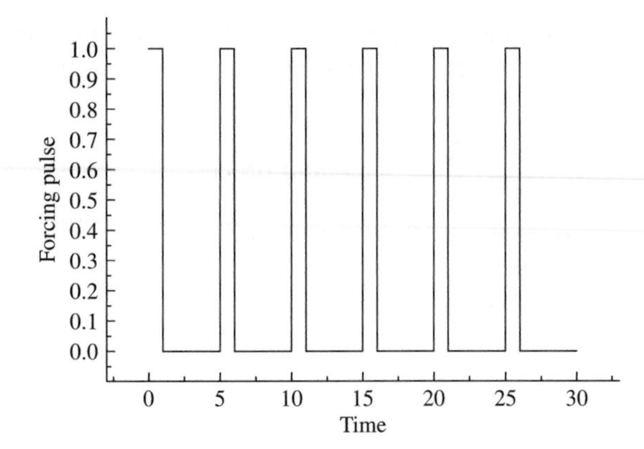

Fig. 3.10
Pulsed drive as a function of time.

The nature of the forcing function is such that the differential equation is readily solved by Laplace transform techniques. (For the uninitiated, we cryptically note that the Laplace transform technique consists of (a) converting a differential equation to an algebraic equation, (b) solving the algebraic equation, and (c) by the appropriate inverse transform operation, converting that algebraic solution into the desired solution of the differential equation. Its name comes from its originator, the French mathematician and astronomer, Pierre Simon Laplace (1749–1827). For a standard treatment see, for example, chapter 7 of (Zill and Cullen 1993).) The following definition of the transform provides the notation:

$$\mathcal{L}\{\theta(t)\} = \Theta(s) = \int_0^{\infty} e^{-st}\theta(t)\,dt. \tag{3.31}$$

The transformed version of the equation of motion is then

$$s^2\Theta(s) + \omega_0^2\Theta(s) = \frac{F(1-e^{-s\tau})}{s(1-e^{-sT})}, \tag{3.32}$$

with the transformed solution being

$$\begin{aligned}\Theta(s) &= \frac{F(1-e^{-s\tau})}{s(s^2+\omega_0^2)(1-e^{-sT})} \\ &= \frac{F}{s(s^2+\omega_0^2)}\begin{bmatrix} 1+e^{-sn\tau}+e^{-s2n\tau}\dots \\ -e^{-s\tau}-e^{-s(n+1)\tau}\dots \end{bmatrix}.\end{aligned} \tag{3.33}$$

The inverse transform may be found with the help of the following identities:

$$L^{-1}\left[\frac{1}{s(s^2+\omega_0^2)}\right] = \frac{1}{\omega_0^2}(1-\cos\omega_0 t) \tag{3.34}$$

and

$$L^{-1}[e^{-as}L\{\theta(t)\}] = \theta(t-a)U(t-a), \tag{3.35}$$

where $U(t-a)$ is the unit step function equal to zero for $t<a$ and equal to one for $t>a$. The full solution to the equation of motion is then

$$\theta(t)=\frac{F}{\omega_0^2}\Big[1-\cos\omega_0 t+U(t-n\tau)-U(t-n\tau)\cos\omega_0(t-n\tau)$$
$$+U(t-2n\tau)-U(t-2n\tau)\cos\omega_0(t-2n\tau)\cdots\Big]$$
$$+\frac{F}{\omega_0^2}\Big[-U(t-\tau)+U(t-\tau)\cos\omega_0(t-\tau)-U(t-\{n+1\}\tau$$
$$+U(t-\{n+1\}\tau)\cos\omega_0(t-\{n+1\}\tau)\cdots\Big], \tag{3.36}$$

where we have utilized our original assumption that $n\tau=T$, and the periodic properties of trigonometric functions. We can also write the solution for each time interval, either during a pulse, or between pulses. For example (in units of F/ω_0^2),

$$\begin{aligned}
\theta(t) &= [1-\cos\omega_0 t] && 0<t<\tau\\
&= [\cos\omega_0(t-\tau)-\cos\omega_0 t] && \tau<t<T\\
&= [1+\cos\omega_0(t-\tau)-2\cos\omega_0 t] && T<t<T+\tau\\
&= [2\cos\omega_0(t-\tau)-2\cos\omega_0 t] && T+\tau<t<2T\\
&= [1+2\cos\omega_0(t-\tau)-3\cos\omega_0 t] && 2T<t<2T+\tau\\
&= [3\cos\omega_0(t-\tau)-3\cos\omega_0 t] && 2T+\tau<t<3T\\
&= [1+3\cos\omega_0(t-\tau)-4\cos\omega_0 t] && 3T<t<3T+\tau
\end{aligned} \tag{3.37}$$

and so on in the same pattern. Every second piece of the solution can be written as a simple sine or cosine function with a suitable amplitude and fixed phase angle. The alternative pieces are similar except that each has the additional factor of $(1-\cos\omega_0 t)$. The overall result is a sinusoidal function with a linearly increasing amplitude as seen in Fig. 3.11. For that illustration the forcing pulse occupies ten percent of the period of the motion. Note that Fig. 3.11 is very similar to Fig. 3.9, which illustrates sinusoidal forcing. Our intuition that the two types of forcing would lead to similar results has been vindicated. As usual, the model becomes unrealistic when the angular displacement becomes large.

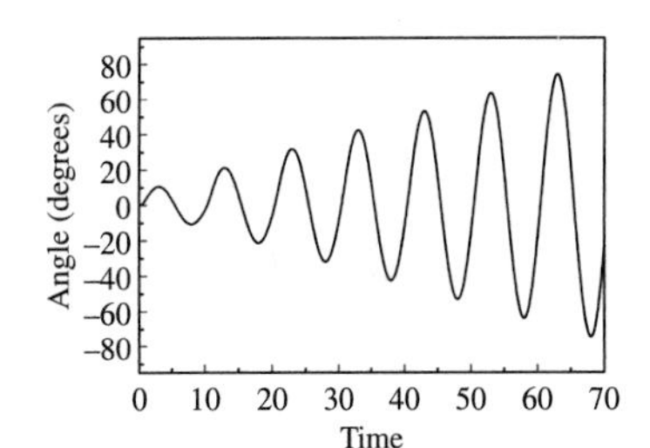

Fig. 3.11
Increasing angular displacement due to pulsed resonant forcing.

We have ignored two effects: damping and the actual sinusoidal dependence of the gravitational restoring force. For the moment we defer treatment of the latter effect and just add damping to our linearized, pulsed pendulum. With damping, the motion of the pendulum no longer diverges but is bounded, and depending on the damping factor, can be made to have a relatively small amplitude. The equation of motion now becomes

$$\frac{d^2\theta}{dt^2}+2\gamma\frac{d\theta}{dt}+\omega_0^2\theta=F(t) \tag{3.38}$$

with the same initial conditions. Following the methodology of the undamped case we find the Laplace transform of the solution to be

$$\Theta(s) = \frac{F}{s((s+\gamma)^2+\omega^2)}\left[1 - e^{-s\tau} + e^{-ns\tau} - e^{-(n+1)s\tau} + e^{-2ns\tau} - e^{-(2n+1)s\tau}\right], \quad (3.39)$$

where

$$\omega^2 = \omega_0^2 - \gamma^2.$$

The solution to this equation is proportionately more complex than that of the undamped equation of motion. Computation of the inverse transform is left as an exercise for the persistent reader. It is illuminating to view the graphical representation of the solution shown in Fig. 3.12. Note that the time series is asymmetric and that the pulse causes the pendulum to be strongly displaced during and after the pulse, but that the pendulum loses energy, due to damping, between pulses.

The rather cumbersome form of the solutions to the pulsed pendulum equations of motion can motivate the search for a model that will produce less complex solutions. It turns out that some simplification can be achieved by letting the pulse become very strong during a very short time period. We consider the limit for which the pulse strength approaches infinity while the pulse duration approaches zero. This mathematical wizardry is achieved with the Dirac delta function, defined (Hundhausen 1998) as

$$\delta(t-t_0) = 0 \ \text{for} \ \ t \neq t_0, \qquad \text{and} \quad \int_{-\infty}^{\infty} \delta(t-t_0)\,dt = 1. \quad (3.40)$$

This function was invented by the theoretical physicist Paul Adrien Maurice Dirac (1902–1984) in order to deal with certain mathematical difficulties in the formulations of quantum mechanics. It has proved useful in a variety of contexts. Sometimes it is referred to as the "needle" function. We may think of the shape as being that of an infinitely thin vertical spike based at the coordinate t_0. Although this sort of infinitely quick, infinitely

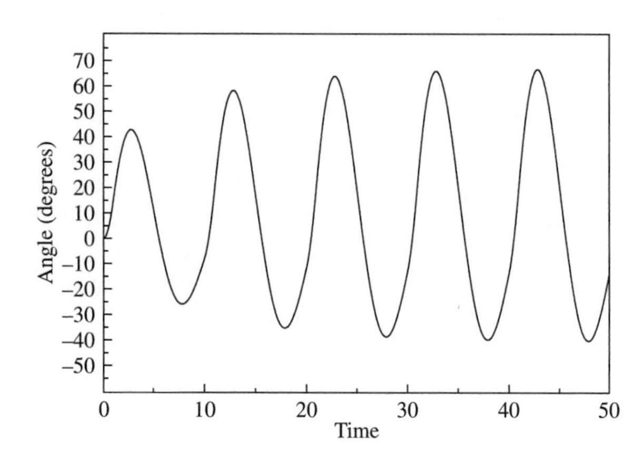

Fig. 3.12
Angular displacement due to pulsed forcing, but with added damping.

hard push on a pendulum is a little unrealistic, the use of the delta function leads to some simplification of the analysis. We rewrite the equation of motion as

$$\frac{d^2\theta}{dt^2} + 2\gamma\frac{d\theta}{dt} + \omega_0^2\theta = F[\delta(t) + \delta(t-T) + \delta(t-2T) + \delta(t-3T)\cdots], \tag{3.41}$$

where F is now a constant and $T = 2\pi/\omega_0$. The Laplace transform of this equation is

$$\Theta(s) = \frac{F}{[(s+\gamma)^2 + \omega^2]}\left[1 + e^{-sT} + e^{-2sT} + e^{-3sT} + \cdots\right], \tag{3.42}$$

where $\omega^2 = (\omega_0^2 - \gamma^2)$. Using various inverse transform relations, the solution to the equation of motion becomes

$$\theta(t) = \frac{F}{\omega}\left[e^{-\gamma t}\sin\omega t + U(t-T)e^{-\gamma(t-T)}\sin\omega(t-T)\cdots\right]. \tag{3.43}$$

Since $\omega \neq 2\pi/T$, it is cumbersome to write this solution as a set of piecewise solutions. The general effect is that while the amplitude decays slightly during the interval between each pulse, the next pulse provides a strong jump in the amplitude. The sharpness of the push causes a slight discontinuity in the slope of the time series at the moment of the pulse. Eventually the decay caused by the dissipative term, and the energy input caused by the Dirac function pulse balance each other and the system arrives at a steady periodic state. For a lightly damped swing, the motion is periodic. A simulation of the swing's motion for the first few pushes is shown in Fig. 3.13.

If the equation of motion does *not* include damping, then the solution in each interval is almost trivial:

$$\theta(t) = \frac{Fn}{\omega_0}\sin\omega_0 t \qquad (n-1)T < t < nT. \tag{3.44}$$

The amplitude grows linearly with n, as it does for the finite pulse width case, shown in Fig. 3.11.

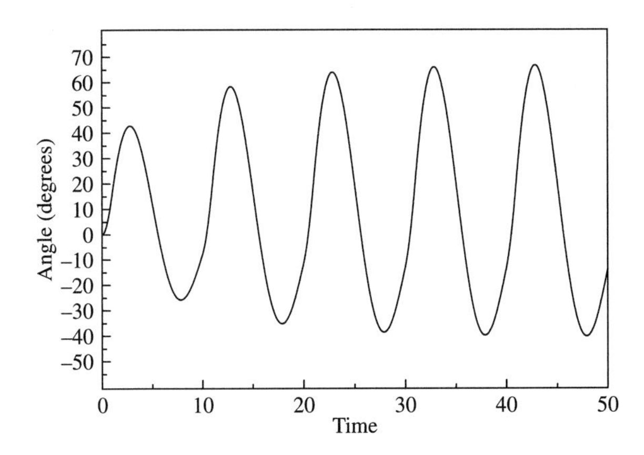

Fig. 3.13
Angular displacement due to Dirac delta function forcing (and damping). Note the slight discontinuity in the slope of the angle at the point of forcing.

Delta-function forcing is, of course not entirely realistic, but the motion it promotes is quite similar to that provided by forcing from finite width pulses. And the delta function has the huge advantage of providing simpler mathematics for the motion of the pendulum. Therefore, as a model, delta function forcing is a viable alternative to the more complex problem of finite width forcing.

3.2.3 Parametric forcing

Parametric forcing is any method of forcing that varies some parameter of the pendulum's configuration in a periodic manner such that the forcing energy becomes at least partially converted to motional energy for the pendulum. An ubiquitous example of periodic parametric forcing is the pumping of a playground swing performed by the rider in order to build and then maintain the swing's motion. In the case of the playground swing the rider's pumping action involves movement of his or her body so that the effective length of the swing is periodically altered. A more exotic example is found with the pumping of *O Botafumeiro* by the team of priests. The actual length of the pendulum is varied as the rope is periodically wound up and wound down around the rollers high above the transept. In both cases, the energy input periodically changes the *effective* length (parameter) of the pendulum. These changes also provide a change in the effective gravitational field on the pendulum bob due to the momentary forcing. The energy of motion along the length converts to angular motion of the pendulum. Both of these applications of parametric forcing will be discussed more fully after the nonlinear pendulum has been introduced. For now, we develop a simple model of a linearized parametric pendulum.

In complex systems, determination of the correct equations of motion may be a daunting task. It is not always easy to specify, with precision or confidence, the applicable forces in systems with several degrees of freedom. Another approach, commonly termed the Lagrangian approach after Joseph Louis Lagrange (1736–1813) , may often be less perilous even though the initial equations seem abstract.[1] The justification for this method is based upon an optimization and may be found in texts on advanced mechanics. See, for example, (Goldstein 1950) or (Chow 1995). Here we simply provide the prescription for its use. Let us define the excess of kinetic energy over potential energy as

$$L = T - V, \tag{3.45}$$

where L is called the Lagrangian , T is the kinetic energy, and V is the potential energy of the system. The system position is specified by generic coordinates that form the set (q_i) and therefore the Lagrangian, and particularly the potential energy are both functions of the q_i coordinates. The kinetic energy will depend upon velocity coordinates $(\dot{q}_i)$. Because only

[1] Lagrange's most famous treatise "Mécanique Analytique" contained not a single diagram. Lagrange took pride in employing only "algebraic operations." He went on to say that, "Those who love Analysis will, with joy, . . . be grateful to me for thus having extended its field." (Quoted on p. 333 in Dugas (1955)).

energies are involved in the creation of the Lagrangian, it is relatively easy to write the appropriate Lagrangian. The required equations of motion are found by taking certain derivatives of the Lagrangian such that

$$\frac{d}{dt}\left(\frac{\partial L}{\partial \dot{q}_i}\right) - \left(\frac{\partial L}{\partial q_i}\right) = 0. \tag{3.46}$$

This expression gives one equation of motion for each of the coordinates. We now demonstrate this approach with the linearized parametric pendulum.

A typical parametric pendulum, whether a swing or a pumped incense burner has a time varying length. Polar coordinates relative to the pivot point are a good choice. The length of the pendulum may be represented by $r(t)$ and the angular displacement of the bob by $\theta(t)$. The Lagrangian for the linearized pendulum becomes

$$L = \frac{1}{2}m\dot{r}^2 + \frac{1}{2}mr^2\dot{\theta}^2 - mgr\left(\frac{\theta^2}{2}\right), \tag{3.47}$$

which, through use of Eq. (3.46), gives the equations of motion:

$$\begin{aligned} \frac{d}{dt}\left(\frac{\partial L}{\partial \dot{r}}\right) - \left(\frac{\partial L}{\partial r}\right) &\Rightarrow \ddot{r} - r\dot{\theta}^2 + g\theta^2 = 0 \\ \frac{d}{dt}\left(\frac{\partial L}{\partial \dot{\theta}}\right) - \left(\frac{\partial L}{\partial \theta}\right) &\Rightarrow r^2\ddot{\theta} + 2r\dot{r}\dot{\theta} + gr\theta = 0. \end{aligned} \tag{3.48}$$

The advantage of the Lagrangian approach becomes clear with the appearance of the nonintuitive middle terms in each differential equation. Our primary concern is the swinging motion as described by the second equation. Division by r^2 yields

$$\ddot{\theta} + 2\frac{\dot{r}}{r}\dot{\theta} + \frac{g}{r}\theta = 0, \tag{3.49}$$

an equation very similar to that of the simple linearized pendulum. We anticipate that the middle term will provide the forcing and thereby create the increasing amplitude of the motion. The natural angular frequency of the pendulum is given by $\omega_0 = \sqrt{g_0/r_0}$ where r_0 is the unperturbed length of the pendulum and g_0 is the acceleration due to gravity.

For parametric pumping the optimum effect is achieved by pumping at twice the natural frequency of the pendulum as shown by the path of the bob of O Botafumeiro in Fig. 3.14.Therefore, a possible time variation of the pendulum length may be expressed as

$$r(t) = \left(r_0 - \frac{\Delta r}{2}\right) + \frac{\Delta r}{2}\cos 2\omega_0 t, \tag{3.50}$$

where Δr is twice the amplitude of the rope length variation which, in turn, changes the effective field of gravity on the bob. Thus, the effective gravitational field becomes

$$g(t) = g_0 - \ddot{r} = g_0 + \frac{\Delta r}{2}4\omega_0^2\cos 2\omega_0 t. \tag{3.51}$$

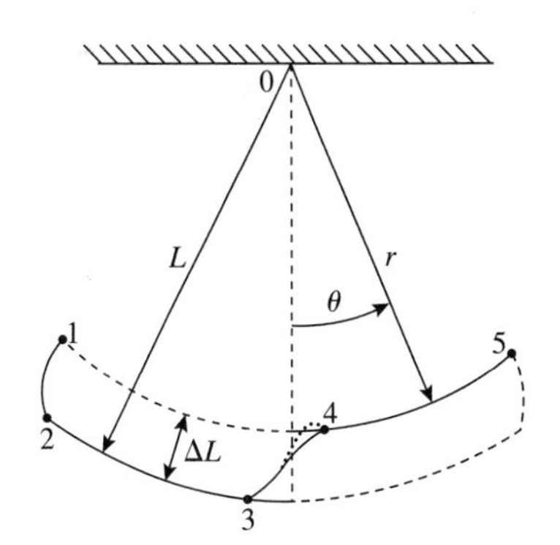

Fig. 3.14
Schematic diagram of the path of O Botafumeiro during a cycle. (Reprinted with permission from Sanmartin (1984, p. 940). ©1984, American Association of Physics Teachers.)

Numerical solution of the equation of motion subject to the time variations of r and g gives rise to the time series shown in Fig. 3.15.

As we will show, the frequency of the *nonlinear* pendulum is amplitude dependent. Optimum forcing requires that the forcing frequency track the pendulum's natural frequency, and therefore optimum forcing of a *nonlinear* parametric pendulum requires that the forcing frequency be adjusted downward as the amplitude increases. For O Botafumeiro, the adjustment is made by the *tiraboleiros* as the censer swings toward its maximum amplitude of about 80 degrees. Fig. 3.16 shows a monotonic increase in O Botafumeiro's amplitude.

We now temporarily leave the question of forcing and turn to the difficult problem of the nonlinear pendulum with its sinusoidal restoring force. Its mathematics becomes significantly more complex than that of the linear case. For the general case, with damping and even simple periodic forcing, there is no analytic solution and therefore the equation of motion cannot be solved without the help of a computer. As noted above, our analysis will reveal the important result that the natural frequency

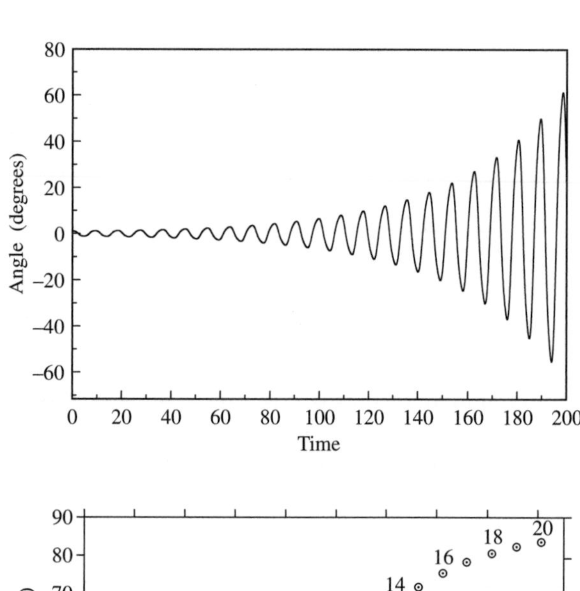

Fig. 3.15
Time series for the angular displacement as calculated by the computer simulation.

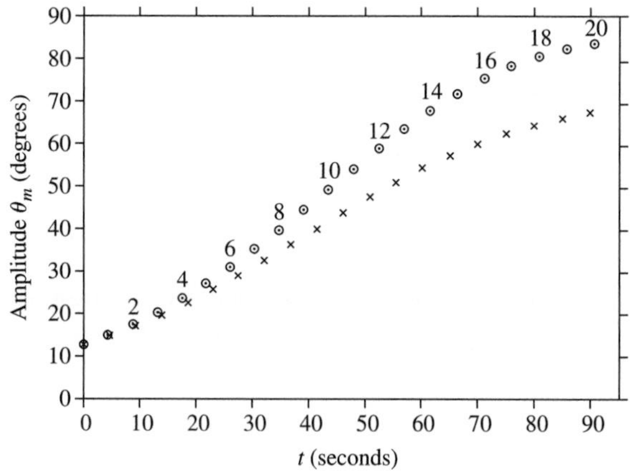

Fig. 3.16
Oscillation amplitude for each pumping cycle. Both sets of points are due to mathematical models, the circles representing a more complex model that yields an accurate representation of actual motion of O Botafumeiro. (Reprinted with permission from Sanmartin (1984, p. 942). ©1984, American Association of Physics Teachers.)

decreases with amplitude, a finding that counters Galileo's original observation of approximate isochronism.

3.3 The nonlinearized pendulum

We have seen that the linearized pendulum is a suitable model for many applications. But once the amplitude of oscillation gets past the point where $\sin\theta \cong \theta$, perhaps 10 degrees, the linearization approximation is no longer valid. Certainly the motion of *O Botafumerio*, with its amplitude of over 80°, requires modeling with the complete sinusoidal restoring force. Therefore it is necessary to return to the nonlinearized equation of motion, Eq. (2.1) for the simple pendulum. The mathematical machinery of linear differential equations and linear systems is no longer available to us. Solutions of nonlinear equations require a variety of less straightforward methods of solution, including power series expansion and, when all else fails, numerical solution by computer. Furthermore, the harmonic oscillator approximation of the pendulum is no longer true, and the various analogs of the harmonic oscillator such as the LRC electrical circuit no longer correspond to the large amplitude, nonlinear pendulum.

3.3.1 Amplitude dependent period

One of the most important differences between the linearized pendulum and the nonlinear pendulum is that while the former has a constant natural frequency, the period of the nonlinear pendulum decreases with increasing amplitude. Galileo's original hypothesis of isochronism is found only in an approximation of this equation ($\sin\theta \simeq \theta$). More generally, the period of the motion *does* vary with the amplitude. Derivation of this result is a nontrivial exercise (MacMillan 1927). We begin again with the undamped, unforced equation of motion,

$$\frac{d^2\theta}{dt^2} + \frac{g}{l}\sin\theta = 0. \tag{3.52}$$

Using

$$d\theta = \frac{d\theta}{dt}dt, \tag{3.53}$$

the equation of motion becomes

$$\frac{d\theta}{dt}\frac{d^2\theta}{dt^2}\,dt = -\frac{g}{l}\sin\,\theta\,d\theta. \tag{3.54}$$

The integral of this equation—the so-called *first integral of motion*—is a conservation of energy equation

$$\left(\frac{d\theta}{dt}\right)^2 = \frac{2g}{l}\cos\theta + \text{Constant}. \tag{3.55}$$

The constant of integration may be determined from the total energy, E, thereby yielding the following result,

$$\left(\frac{d\theta}{dt}\right)^2 = \frac{2g}{l}\cos\theta + \frac{2E}{ml^2} - \frac{2g}{l}, \tag{3.56}$$

which simply expresses the fact that the total energy is the sum of the kinetic and potential energies. At this point, three cases are distinguished by their respective regimes of total energy. In the first case the energy is less than a critical value, *2mgl*, the energy required for the pendulum bob to reach the upper vertical position. The second case corresponds to the total energy being equal to the critical value, and the third case occurs in the regime where the energy is greater than the critical value and is therefore sufficient to make the pendulum execute hindered rotary motion.

Case I: $E < 2mgl$. In this regime, the constant in Eq. (3.55) can be determined by letting the angular velocity be zero at some maximum displacement (or angular amplitude α) Then Eq. (3.55) becomes

$$\left(\frac{d\theta}{dt}\right)^2 = \frac{2g}{l}(\cos\theta - \cos\alpha). \tag{3.57}$$

This equation describes the behavior of the low energy pendulum in phase space $(\theta, \dot{\theta})$ when $\alpha \prec \pi$. For very small amplitudes, the phase orbit is approximately elliptical, as is the case for the linearized pendulum shown in Fig. 2.5. For higher values of α the ellipse becomes horizontally stretched. Let us continue to solve the differential equation. Use of the trigonometric identity $\cos\theta = 1 - 2\sin^2(\theta/2)$ leads to the expression

$$\frac{d\theta}{dt} = \pm 2\sqrt{\frac{g}{l}\left[\sin^2\left(\frac{\alpha}{2}\right) - \sin^2\left(\frac{\theta}{2}\right)\right]}. \tag{3.58}$$

We introduce two further variables, φ and k, with the substitutions

$$\sin\frac{\theta}{2} = \sin\frac{\alpha}{2}\sin\varphi \quad \text{and} \quad k^2 = \sin^2\left(\frac{\alpha}{2}\right). \tag{3.59}$$

An integral equation is now formed in which the variables t and φ are now separated:

$$\int_0^{t_0} dt = \sqrt{\frac{l}{g}}\int_0^{\varphi} \frac{d\varphi}{\sqrt{1 - k^2\sin^2(\varphi)}}. \tag{3.60}$$

The zero of time is set when the pendulum is at the bottom of its arc and t_0 occurs when the pendulum reaches the maximum angular displacement, α. The time t_0 then represents one quarter of the period, T. Also, when the pendulum is at its maximum angular displacement, α, then the new angular variable φ is equal to $\pi/2$. Therefore the period of the motion is

$$T = 4\sqrt{\frac{l}{g}} \int_0^{\pi/2} \frac{d\varphi}{\sqrt{1 - k^2 \sin^2(\varphi)}}. \tag{3.61}$$

The integral in the above expression is called an elliptic integral of the first kind. Its value was first calculated by the French mathematician Adrien-Marie Legendre(1752–1833). Legendre worked extensively on elliptic functions and elliptic integrals for over forty years culminating in his book "Traite des fonctions elliptiques," published in 1830 (Hellemans and Bunch 1991). The integrand may be expanded according to the binomial theorem:

$$\begin{aligned}(1 - k^2 \sin^2(\varphi))^{-1/2} =& 1 + \frac{1}{2}k^2 \sin^2 \varphi + \left(\frac{1 \cdot 3}{2 \cdot 4}\right) k^4 \sin^4 \varphi + \cdots \\ & + \frac{1 \cdot 3 \cdot 5 \cdots (2n-1)}{2 \cdot 4 \cdot 6 \cdots 2n} k^{2n} \sin^{2n} \varphi \cdots,\end{aligned} \tag{3.62}$$

and integrated term by term. For the integration we use the formula

$$\int_0^{\pi/2} \sin^{2n} \varphi \, d\varphi = \frac{1 \cdot 3 \cdot 5 \cdots (2n-1)}{2 \cdot 4 \cdot 6 \cdots 2n} \frac{\pi}{2}, \tag{3.63}$$

and the previously defined value for k to obtain the final value for the period:

$$T = 2\pi\sqrt{\frac{l}{g}}\left[1 + \left(\frac{1}{2}\right)^2 \sin^2\left(\frac{\alpha}{2}\right) + \left(\frac{1 \cdot 3}{2 \cdot 4}\right)^2 \sin^4\left(\frac{\alpha}{2}\right) + \cdots\right]. \tag{3.64}$$

The graph of period versus angular amplitude is shown in Fig. 3.17. Note that when the amplitude is zero that the period is exactly equal to the period determined from the linearized version of the pendulum. As the amplitude approaches π radians, the bob approaches verticality and the factor $\alpha/2$ approaches $\pi/2$. The series now diverges and the period becomes infinitely large, as expected. This behavior is treated as the second case.

For *O Botafumeiro* with its angular amplitude of over 80°, the period of motion is longer than that found from a calculation that assumes a small angle of oscillation. One wonders what Galileo would have discovered if he had been sitting in the cathedral in Santiago de Compostela rather than (reportedly) sitting in the cathedral at Pisa. Would he have observed the difference in the pendulum's period as the team of priests gradually built up the amplitude of the censer's motion? Would his pulse have been a sufficiently accurate timer to quantify the difference? From Fig. 3.17 it is apparent that the period for an amplitude of 80° is about 15% more than that of the small amplitude approximation. With O Botafumeiro the formula for the period is not exactly applicable because of a variety of other effects. Use of the small amplitude formula for the period of oscillation leads to $T = 9.1$ seconds. Measurements taken from a graph found in

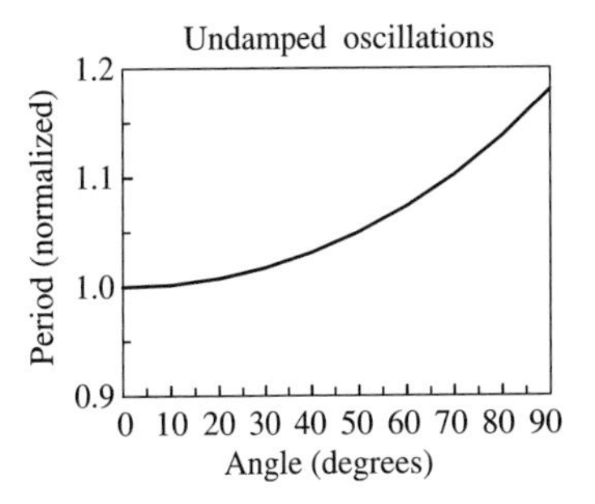

Fig. 3.17
Period of oscillation versus amplitude.

(Sanmartin, 1984) suggest that for maximum swing the period is about 10.2 seconds, a 12% difference. It seems unlikely that Galileo would have attributed much significance to either a 12% or 15% change.[2]

There is another important consequence of the nonlinearity of the pendulum's equation of motion. For the linearized pendulum, the angle and angular velocity of the pendulum behaved as a single sine or cosine wave,

$$\theta = A \sin \omega_0 t. \tag{3.65}$$

where $\omega_0 = \sqrt{g/l} \equiv \omega$. Such is not the case for the nonlinear pendulum. In order to gain some understanding we partially solve the equation of motion, Eq. (3.52), by a method of approximation. We first assume a trial solution, then substitute it into the differential equation, and finally discover a better trial solution. The process is repeated with each increasingly better solution. In this way, the general character of the solution is revealed. We begin with the trial solution Eq. (3.65) and through substitution into Eq. (3.52) obtain

$$\frac{d^2\theta}{dt^2} = -\omega^2 \sin(A \sin \omega t). \tag{3.66}$$

The outer sine function may be expanded such that the angular acceleration includes powers of the trial solution:

$$\frac{d^2\theta}{dt^2} = -\omega^2 \left[A \sin \omega t - \frac{(A \sin \omega t)^3}{3!} + \frac{(A \sin \omega t)^5}{5!} \cdots \right]. \tag{3.67}$$

Each of the powers of the trig functions can be shown to contain a harmonic that corresponds to that particular power. Therefore there are sinusoidal terms in the time series with frequencies that are odd harmonics of the fundamental frequency, ω. If the amplitude of oscillation is relatively small we might only keep the first two terms in the power series expansion and therefore the next approximate trial solution might have the form,

$$\theta = B \sin \omega t + C \sin 3\omega t. \tag{3.68}$$

This process would be repeated with ever better refinements of the amplitudes of each harmonic. The point is that while the motion of the pendulum is oscillatory, it is no longer described by the simple harmonic motion of a single sinusoidal function. A complete description would require a Fourier spectrum of the form (depending on initial conditions)

$$\theta = \sum_{i=0}^{\infty} A_{2i+1} \sin(2i+1)\omega t. \tag{3.69}$$

For the nonlinearized pendulum the natural frequency changes with amplitude and therefore the response curve is not static with continued forcing at a certain frequency. If a system is initially forced at its low amplitude resonant frequency, the increase in amplitude will eventually

[2] There is some evidence to suggest that Galileo would have not been inclined to regard any difference as credible—at least for an ideal pendulum. Isochronism may have been a matter of faith as well as of imperfect observation (Drake 1978).

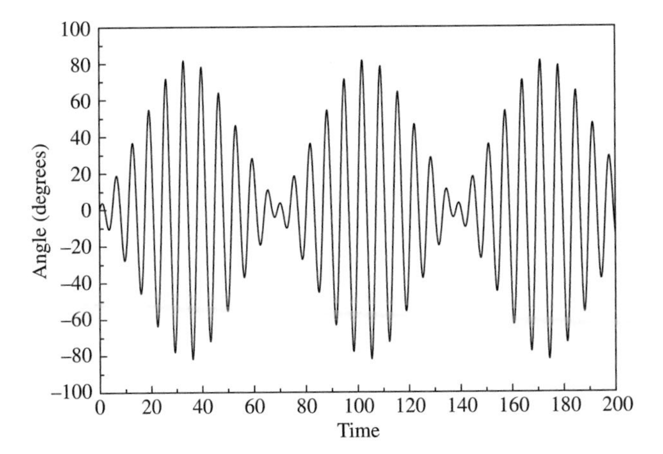

Fig. 3.18
Response of the pendulum to forcing at its small angle resonant frequency. For larger amplitudes, the forcing frequency is increasingly different from the pendulum's natural frequency and the forcing becomes counterproductive.

detune the pendulum, and thereby lower its resonant frequency from that of the originally tuned driving frequency. The forcing goes out of step with the momentary resonant frequency. As the forcing becomes further out of synchronization with the pendulum, the pendulum's amplitude of motion will diminish back to zero. The phenomenon gives rise to a periodic amplitude modulation of the pendulum's motion, as shown in Fig. 3.18. Therefore the rider on a swing must automatically account for the varying response by adjusting her pumping rate to achieve maximum response. In essence, the rider's efforts form part of a feedback loop with the swing, a phenomenon known as *autoresonance* . Similar considerations hold true for the pumping of the large amplitude oscillations of O Botafumeiro. The team of priests form a feedback loop such that their pumping frequency must slow as the amplitude increases.

Let us now consider the case for which the pendulum has just sufficient energy to reach an inverted position.

Case II: $E=2mgl$. For this condition, the expression for energy simplifies considerably. We start with Eq. (3.57) and using the same trigonometric substitution, obtain the result,

$$\left(\frac{d\theta}{dt}\right)^2=\frac{4g}{l}\left(1-\sin^2\left(\frac{\theta}{2}\right)\right), \tag{3.70}$$

which may be simplified to

$$\frac{d\theta}{dt}=2\sqrt{\frac{g}{l}}\cos\left(\frac{\theta}{2}\right). \tag{3.71}$$

Therefore the time interval t between the pendulum being at the bottom of its arc $\theta=0$ and being at some finite angular displacement, θ, is found by integration,

$$t=\frac{1}{2}\sqrt{\frac{l}{g}}\int_0^\theta \sec\frac{\theta}{2}\,d\theta=\sqrt{\frac{l}{g}}\ln\left[\tan\left(\frac{\theta}{4}+\frac{\pi}{4}\right)\right]. \tag{3.72}$$

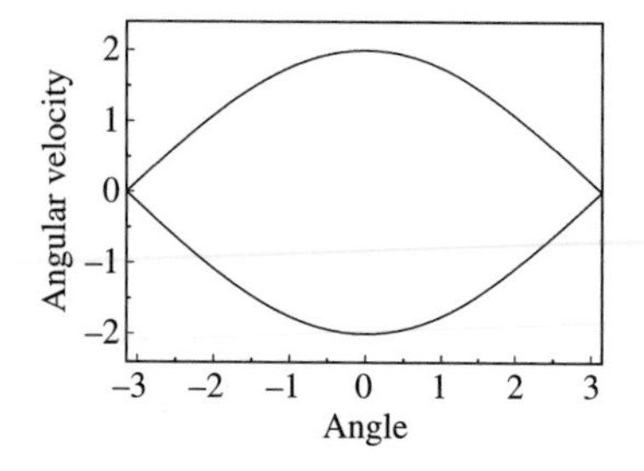

Fig. 3.19
Phase diagram for a pendulum with just enough energy to reach the upright vertical position.

The pendulum approaches verticality as θ approaches π radians and therefore the time interval becomes infinite. We expected this result from the discussion of the previous case. Eq. (3.71) can also be used to plot the trajectory of this special orbit in phase space as shown in Fig. 3.19 when the phase orbit passes through $\theta = \pm\pi$. This is a unique orbit since it characterizes the pendulum's behavior at the critical energy. Note that there is a discontinuity in the derivative of the phase orbit at $\theta = \pm\pi$. This special phase orbit is called the *separatrix*; it marks the boundary between oscillatory and hindered rotary motion of the pendulum.

Case III: $E \rangle 2mgl$. The total energy is

$$E = \frac{1}{2}ml^2\left(\frac{d\theta}{dt}\right)^2 + mgl(1 - \cos\theta). \tag{3.73}$$

The angular velocity term may then be isolated such that

$$\begin{aligned}\left(\frac{d\theta}{dt}\right)^2 &= \frac{2E}{ml^2} - 2\frac{g}{l}\left(1 - 1 + 2\sin^2\frac{\theta}{2}\right) \\ &= \frac{2E}{ml^2} - 4\frac{g}{l}\sin^2\frac{\theta}{2}.\end{aligned} \tag{3.74}$$

Therefore

$$\frac{d\theta}{dt} = \sqrt{\frac{2E}{ml^2}}\sqrt{1 - k^2\sin^2\frac{\theta}{2}}, \tag{3.75}$$

where

$$k = \sqrt{\frac{2mgl}{E}}. \tag{3.76}$$

The differential equation is inverted and put in integral form. The integral over angular displacement, from 0 to π, describes motion for one half an orbit and therefore the integral gives the time for one half rotation. The period of the motion is then double this time. The integral is

$$t = \sqrt{\frac{2ml^2}{E}}\int_0^{\pi}\frac{d(\frac{\theta}{2})}{\sqrt{1 - k^2\sin^2\frac{\theta}{2}}}. \tag{3.77}$$

Finally the period of the motion may be expanded, as before, to give

$$T = 2t = \pi\sqrt{\frac{2ml^2}{E}}\left[1 + \left(\frac{1}{2}\right)^2 k^2 + \left(\frac{1\cdot 3}{2\cdot 4}\right)^2 k^4 \cdots\right]. \tag{3.78}$$

The motion is that of a hindered rotation—faster at the bottom and slower at the top. If E is very large compared to the critical energy of $2mgl$, then the expression for the period reduces considerably. Suppose $E = Nmgl$ where N is a very large number; then k becomes very small and we can ignore all but the first term in the expansion. The period then behaves as

$$T = \pi\sqrt{\frac{2l}{Ng}}, \tag{3.79}$$

which approaches zero for large N. The behavior of the period over the entire range of energies is sketched in Fig. 3.20.

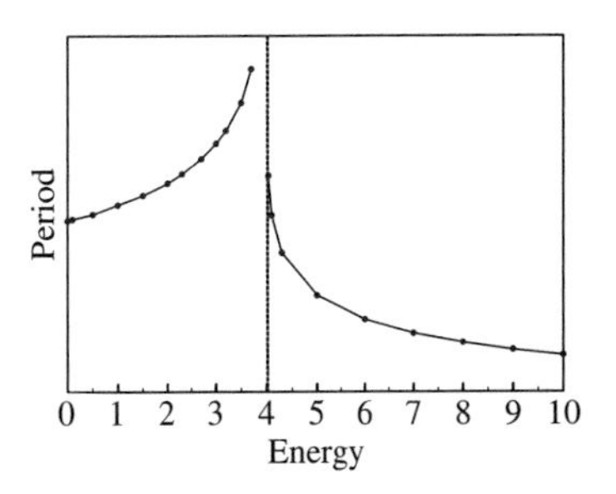

Fig. 3.20
Graph of period versus the energy of the pendulum. The vertical line represents the critical energy.

3.3.2 Phase space revisited

In Chapter 2 we introduced phase space as a means of visualizing the totality of the pendulum's motion. With the nonlinear frictionless pendulum phase space becomes more complex. Phase space naturally divides into two parts that correspond to the two regimes of energy, $E < 2mgl$, and $E > 2mgl$, as indicated by the boundary region (critical energy) shown in Fig. 3.19. For $E \ll 2\,mgl$, (inside the critical orbit or *separatrix*) the phase orbit is somewhat elliptical because the pendulum is reasonably approximated by its linearization. However, for larger energies that are still less than the critical energy, $E = 2mgl$, the ellipses become slightly squashed even as they become bigger. At the critical energy, the phase orbit (*separatrix*) has a discontinuity in slope at $\theta = \pm\pi$, where the period of the pendulum becomes infinite. Now, the pendulum's motion in time is always clockwise around the subcritical orbits. For supercritical orbits, the motion will just continue to the right as the angular displacement grows monotonically. Thus, the flow of the motion may be indicated by arrows, and for the critical orbit, we may think of arrows going into and coming out of the points where $\theta = \pm\pi$ and $\dot{\theta} = 0$. Because some of the arrows are going into these points and others are pointing away, these special points are called *saddle points*, and the arrows indicate whether the pendulum tends to go toward or away from the saddle point along the particular orbit. Trajectories that point toward the saddle point are called *stable* and those that point away are called *unstable*. Once the energy is greater than $E = 2mgl$, then the motion becomes that of a hindered rotor and the pendulum's motion is no longer confined to a finite region of phase space. (A non-hindered rotor would have a constant angular velocity $\dot{\theta}$ and its phase portrait would be a horizontal line.)

The addition of damping creates a new scenario. Now the motion of the pendulum decays and, for small amplitude oscillations the ellipse becomes a spiral toward the center as the motion dies away. For larger amplitudes the motion also decays and therefore the final state of any of the pendular motions is one of rest. The motion is always attracted to the center and therefore the center is called an *attractor*. Each center with coordinates $[\theta = 2n\pi, \dot{\theta} = 0]$ is an attractor, and which attractor the pendulum motion goes to depends upon the region of phase space from which the motion originates. The attractors alternate with the saddle points. The region of phase space that corresponds to a particular attractor is called its *basin of attraction*. Figure 3.21 illustrates a typical phase space for the damped pendulum.

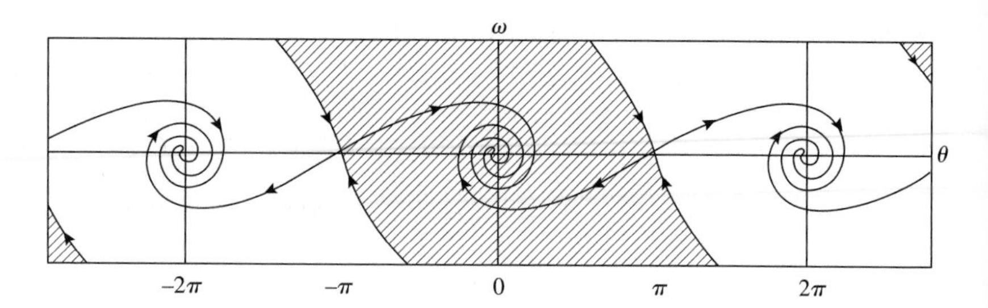

Fig. 3.21
Phase space diagram for the damped pendulum. The shaded and unshaded regions are basins of attraction. All points within a particular region are attracted to the central focal point within the basin. (From (Baker and Gollub 1996). Reprinted with the permission of Cambridge University Press.)

The attractors and the saddle points are located at what are called *fixed points* . These points are coordinates of phase space for which the derivatives of the coordinates vanish. For example, a typical second order differential equation for the damped pendulum

$$\frac{d^2\theta}{dt^2} + 2\gamma\frac{d\theta}{dt} + \frac{g}{l}\sin\theta = 0 \tag{3.80}$$

can be rewritten as two coupled first order differential equations

$$\begin{aligned} \frac{d\dot{\theta}}{dt} &= -2\gamma\dot{\theta} - \frac{g}{l}\sin\theta \\ \frac{d\theta}{dt} &= \dot{\theta} \end{aligned} \tag{3.81}$$

in the variables $\dot{\theta}$ and θ. The fixed points occur when the left side derivatives are zero at coordinates $[\theta = n\pi, \dot{\theta} = 0]$. These are the saddle points ($n = odd$) and the attractors ($n = even$). The arrows in Fig. 3.21 indicate whether the fixed point is stable.

Finally, if a small amount of constant or periodic forcing is added to a damped pendulum, an equilibrium state will be reached in which the added energy from the forcing just compensates for the energy dissipation. Equation (3.80) is then augmented by a forcing term *F(t)*, such that

$$\frac{d^2\theta}{dt^2} + 2\gamma\frac{d\theta}{dt} + \frac{g}{l}\sin\theta = F(t). \tag{3.82}$$

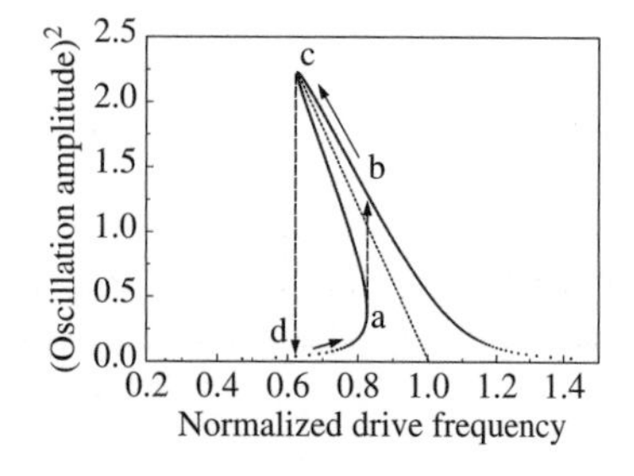

Fig. 3.22
A pendulum's response to change in forcing frequency. The response is different depending on whether the forcing starts well above the resonance or from well below resonance as explained in the text.

The phase space orbit will again be a closed curve, although not necessarily elliptical in shape. The closed orbit is sometimes called a limit cycle. Since the phase orbit tends to this motion—for a particular set of system parameters—no matter what the initial conditions, the orbit is also an *attractor*.

A graph of the intensity of the pendulum's response to mild periodic forcing as a function of forcing frequency is shown in Fig. 3.22 (see, for example, fig. 1 in (Miles 1988*b*)). In creating this diagram the nonlinear potential has been approximated by its linear term θ and the first nonlinear term $-\theta^3/6$ in the expansion of the pendulum's $\sin\theta$ restoring torque. For mild forcing and light damping the nonlinear approximation adequately portrays the pendulum's behavior. The cubic term provides anharmonicity and the negative sign for the nonlinearity indicates that the equivalent linear spring has been effectively "softened" such that the natural frequency is lowered, and the peak response is shifted to the left of the small amplitude natural frequency. (A positive sign indicates that the spring has "hardened" with a commensurate shift to the right in the response peak.) A decrease in forcing frequency from above the small amplitude frequency follows the upper curve, whereas an increase from below follows the lower curve. But when each curve reaches its endpoints as indicated by the vertical dashed line the pendulum's response to forcing will change abruptly and move to the correspondingly opposite curve. Thus there is *hysteresis* in the response curve.[3]

On the other hand, strong nonresonant forcing—at a frequency different from the natural frequency—will precipitate a host of motions whose phase trajectories are complex. Some of these motions are periodic as in Fig. 3.23 and, as we will see in Chapter 6, some of the motions do not repeat and are therefore called *chaotic*.

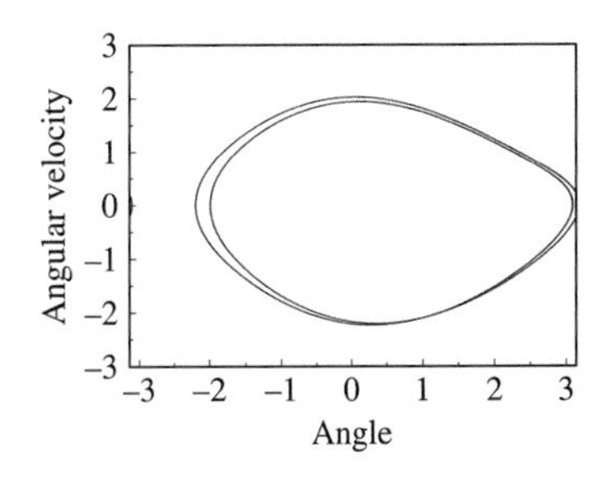

Fig. 3.23
Phase plane diagram of a pendulum whose period is twice that of the forcing period. Note that the boundary conditions on the angle are periodic and therefore there are discontinuities at $\theta = \pm\pi$.

3.3.3 An electronic 'pendulum'

Earlier in this chapter, we showed that the undamped, unforced, *linearized* pendulum was equivalent to a simple *LC* circuit. With the addition of damping of the pendulum, a resistance *R* was added to the *LC* circuit. Finally, the electronic circuit was augmented by a forcing voltage and the corresponding differential equation for the circuit was given by Eq. (3.23) which, in turn, is the electronic analog of Eq. (3.38) for the pendulum. In the same spirit we now present an electronic analog of the *nonlinear*, damped, forced pendulum, Eq. (3.82). It will be evident that the existence of nonlinearity adds considerable complexity.

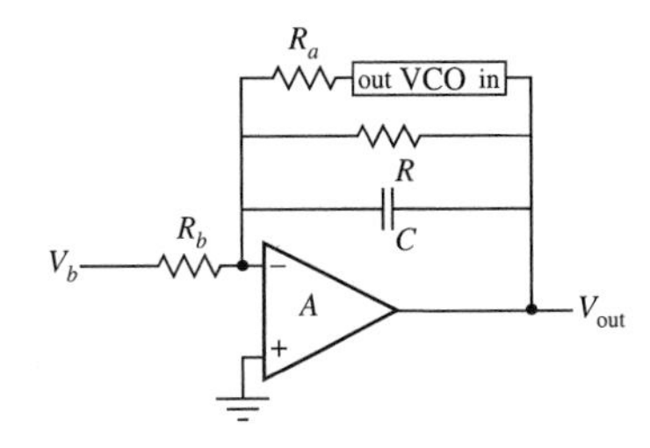

Fig. 3.24
Single op-amp circuit which is an electronic analog of a driven pendulum. The block labelled VCO is a voltage controlled oscillator (note the orientation of the input and output terminals).

Consider the apparently simple circuit depicted in Fig. 3.24. Fundamentally, this is an electronic integrator (see any standard electronics text such as (Sedra and Smith 2004)) to which is added a voltage controlled oscillator (VCO) subcircuit whose complete schematic is given in Fig. 3.25.

To analyze the overall operation of the simulator, we begin with the VCO. Operational amplifier A_1 is configured as an integrator, so its output is governed by the usual expression

$$V_1 = -\frac{1}{(20\ \text{k}\Omega)(0.1\mu\text{F})}\int V_X\, dt, \tag{3.83}$$

where V_X is the output of the multiplier module. Notice that V_X will be the product of the input voltage and either $+8$ V or -8 V, depending on the state of switch *SW*1. As can be seen in the figure, the other voltage-controlled switch, *SW*2, is set by the polarity of the input voltage.

For the purpose of this discussion, let us take a positive input voltage. *SW*1 and *SW*2 are set as shown in the diagram. The input to A_1 will be $+8 \times V_{\text{in}}$; hence according to Eq. (3.83), its output will be a linear ramp

[3] See (Jackson 1989, p. 310).

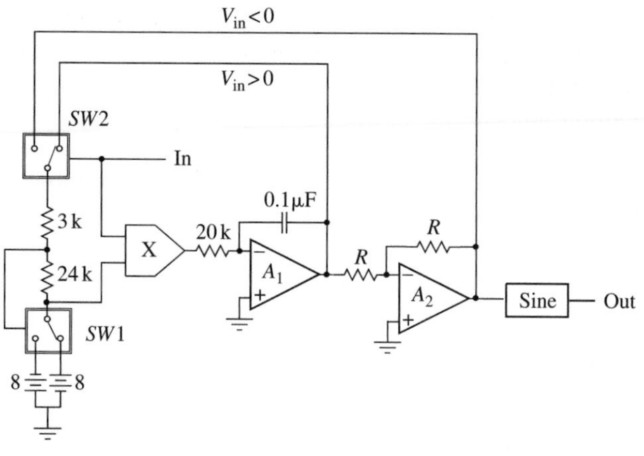

Fig. 3.25
Schematic of the VCO subcircuit. The block marked with an X is a voltage multiplier module. $SW1$ and $SW2$ are voltage-controlled switches.

with slope $-500 \times 8 \times V_{\text{in}}$ (V/s). This is fed to the top node of the 3 kΩ/24 kΩ resistor chain; the bottom of the chain is for the moment at +8 V. Note that the point between the 3 kΩ resistor and the 24 kΩ resistor will reach zero volts when the top of the chain drops to exactly −1 V. Therefore, when the negative-going ramp from amplifier A_1 reaches −1 V, switch $SW1$ will flip, now applying −8 V to the multiplier and causing the slope of the integrator ramp to change sign. This positive going linear output from A_1 will continue until it reaches +1 V, at which point $SW1$ will flip back to its original position as depicted in Fig. 3.25.

To summarize then, the output of A_1 will be a sawtooth voltage waveform with an amplitude ±1 V and a period which is twice the time required for a ramp to cover the range ±1 V (that is, a change of 2 V). From an earlier remark, the reciprocal of the slope of the output ramp is $(4000 \times V_{\text{in}})^{-1}$ (s/V) and so the *period* is $2 \times 2 \times (4000)^{-1} = 1$ ms for an input of 1 V. Altering the input voltage correspondingly changes the period (or frequency) of the oscillation.

Operational amplifier A_2 is a simple inverter which, combined with the action of $SW2$, guarantees proper operation of the VCO whenever a negative input voltage is present. The block labelled sine in Fig. 3.25 denotes a subcircuit whose purpose is to convert the sawtooth waveform to a sinusoidal waveform. Typically, such a function can be achieved by means of an array of resistors and diodes.[4]

Taking all these facts together, the VCO can be characterized by the functional relation

$$V_{\text{out}} = a \sin\left(2\pi k \int V_{\text{in}} dt\right), \tag{3.84}$$

where, for the component values used here[5], $k = 1000$ Hz/V.

[4] See, for example (Sedra and Smith 2004, fig. 13.31, p. 1204).

[5] The value of the amplitude coefficient a depends on the internal details of the sine converter circuit. A typical value might be approximately 0.7. For comparison, the amplitude of the fundamental Fourier component of a symmetric triangle waveform is $8/\pi^2 = 0.81$.

Turning now to the first circuit Fig. 3.24 and noting that the inverting input is a virtual ground, it is apparent that the conditions at the summing point of the op-amp dictate

$$\frac{V_b}{R_b} = -C\frac{dV_{\text{out}}}{dt} - \frac{V_{\text{out}}}{R} - \frac{VCO_{\text{out}}}{R_a}. \tag{3.85}$$

If we define $\theta(t) = 2\pi k \int VCO_{\text{in}}\, dt$, then noting that VCO_{in} is the same as V_{out},

$$\frac{1}{2\pi k}\frac{d\theta}{dt} = V_{\text{out}} \tag{3.86}$$

in which case Eq. (3.85) becomes

$$\frac{C}{2\pi k}\frac{d^2\theta}{dt^2} + \frac{1}{2\pi k R}\frac{d\theta}{dt} + \frac{a\sin\theta}{R_a} = -\frac{V_b}{R_b}, \tag{3.87}$$

which has the same form as the equation for a driven, damped, pendulum.

It is sometimes useful to simplify Eq. (3.87) by choosing a normalizing time

$$t* = \left[\sqrt{\frac{2\pi ka}{CR_a}}\right]t \tag{3.88}$$

in which case

$$\ddot{\theta} + \left[\frac{1}{R}\sqrt{\frac{R_a}{2\pi kaC}}\right]\dot{\theta} + \sin\,\theta = -\frac{V_b}{a}\frac{R_a}{R_b}, \tag{3.89}$$

where the dot indicates differentiation with respect to $t*$. Clearly, the right hand term assumes the role of a normalized torque.

The results of a PSpice[6] simulation of this circuit are shown in Fig. 3.26. Because of the negative sign on the right-hand side of Eq. (3.89), a negative-going ramp in bias voltage V_b corresponds to a positive applied torque. The upper trace is the voltage V_{out} which is proportional to the phase velocity. Thus the figure illustrates the circuit equivalent of a pendulum with slowly increasing torque. As the critical value of bias is reached, the pendulum just goes over the top, after which it rotates with the expected fast-slow modulation—a hindered rotational motion.

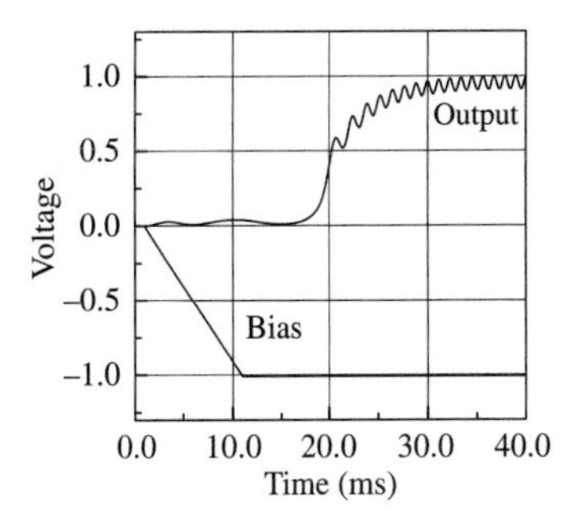

Fig. 3.26
PSpice results for the pendulum simulation circuit. The slowly ramping bias is equivalent to a slowly increasing torque applied to a pendulum.

As this analysis demonstrates, the analog circuit of Fig. 3.24 will replicate the dynamics of a pendulum, but at a much faster time scale—milliseconds in contrast to seconds. The output voltage is proportional to the time derivative of the pendulum phase angle according to Eq. (3.86). The various circuit components can be selected to give any desired equivalent values of pendulum parameters, including the Q (see Appendix A) which for light damping turns out to be

$$Q = R\sqrt{\frac{2\pi kaC}{R_a}}. \tag{3.90}$$

[6] PSpice is software for simulating electronic circuits and is a product of Cadence Design Systems, Inc. Further information about PSpice may be found at www.orcad.com.

While this electronic analog is more complex than the LRC circuit for the linearized pendulum, there is an even more complex electrical analog which we describe in Chapter 9. This second electrical analog to the nonlinear pendulum is a superconducting, quantum mechanical device known as a Josephson Junction.

3.3.4 Parametric forcing revisited

We have already provided a simple model of parametric pumping for the case of the linearized pendulum. Let us now look at some of the complications provided by the nonlinear pendulum for the self-pumped swing and for parametric pendulums such as *O Botafumeiro*.

Pumping a swing

In the case of the playground swing , the rider's pumping action involves movement of his or her body so that the effective length of the swing is periodically altered. One method requires the rider to alternately stand and squat on the swing such that the effective length of the swing shortens and lengthens periodically. The other common method is for the rider to remain seated but to alternately rock his body, again so that the effective length of the swing changes periodically. Variations on either of these methods make use of pulling and pushing on the ropes of the swing to further enhance the motion. A further distinction can be made between methods of (a) initiating the motion, and (b) enhancing and maintaining the motion. In the last three decades a flurry of scholarly analysis has appeared on the physics of pumping a swing. Examples are given in the following references: (Tea and Falk 1968; Siegman 1969; Gore 1970, 1971; McMullan 1972; Walker 1989; Case and Swanson 1990; Wirkus et al. 1998).

In 1968 Tea and Falk of City College of the University of New York analyzed the case of a standing rider pumping the swing by alternately standing and squatting on the swing (Tea and Falk 1968). This activity periodically changes the center of mass of the swing and thereby its effective length in a manner very similar to that shown in Fig. 3.14. The swing may be treated as a simple pendulum with all its mass concentrated at the center of mass. The rider is squatting when the swing descends and then quickly stands as the swing passes through its lowest point. He can choose either to repeat the pattern on the return half of the cycle or simply to remain standing until the beginning of a new cycle.

Energy considerations help us discover the effects of the pumping action. In going from a squatting to a standing position at the point of lowest arc, the force provided by the pumper is radial, and therefore does not provide a torque that changes the angular momentum of the swing and rider. Thus, the angular momentum is conserved, while the center of mass and therefore the effective length of the pendulum are changed. Conservation of angular momentum prescribes that the decrease in length is compensated by an increase in angular velocity. In turn, the increased angular velocity leads to a rapid increase in kinetic energy at the bottom of the arc that converts to an increase in potential energy at the maximum height of the swing. Therefore the objective of increased amplitude of the swing is achieved by the pumping action. Following (Tea and Falk 1968) we develop an equation for the increase of the swing's angular amplitude as a result of the pump action.

Assuming negligible friction, both angular momentum and energy are conserved at various points in the motion. In making the transition from the squatting to the standing position, angular momentum is conserved as follows:

$$ml_0^2\omega_A = ml_1^2\omega_B, \tag{3.91}$$

where ω_A and ω_B are respectively the angular velocities before and after the rider stands up and l_0 and l_1, respectively, are the effective lengths of the pendulum before and after the rider stands. Similarly, total energy is conserved prior to the rider standing, and then an increased energy level (from the work done by the rider) is obtained after the rider stands. That is,

$$mgl_0(1 - \cos\theta_0) = \frac{1}{2}ml_0^2\omega_A^2 \text{ and } mgl_1(1 - \cos\theta_1) = \frac{1}{2}ml_1^2\omega_B^2. \tag{3.92}$$

With some algebraic manipulation it is possible to show that these equations lead to an increase in the maximum angular displacement of the swing:

$$\cos\theta_0 - \cos\theta_1 = \left(\frac{l_0^3 - l_1^3}{l_1^3}\right)(1 - \cos\theta_0). \tag{3.93}$$

Since $l_0 > l_1$ the maximum displacement is indeed increased by the pump action. Eq. (3.93) can be rearranged to generate a map of θ_1 in terms of increasing θ_0 values. Figure 3.27 (amplitude versus no. of cycles from Eq. (3.93)) shows the increase in the swing's angular amplitude as a function of the number of pumping cycles, assuming that the center of mass shortens by 10% during each pump action.It is apparent that the pumping action becomes more effective as the angular displacement increases. (The difficulty of getting the swing started is well known to swing riders.)

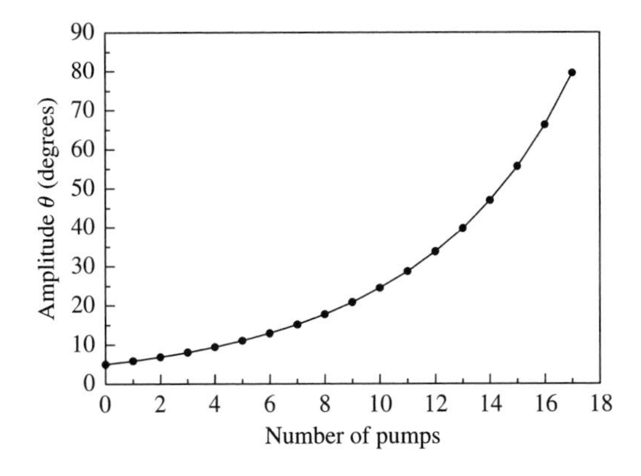

Fig. 3.27
Amplitude versus number of pumps using Eq. (3.93).

The amplitude increases as a result of the rider standing up. Is this amplitude increase lost when the rider squats? While some amplitude is lost, the fact that the squatting occurs at the extremes of the swing motion means that the change in potential energy of the center of mass is less than during the standing motion at the bottom of the arc and therefore the energy change in standing is less by a factor of sin θ.

It is interesting to determine how much energy is required of the rider during the action of standing. The work done is equal to the change in kinetic and potential energy as the center of mass goes from point A to point B. Therefore the expression for the work is

$$W_{AB} = -\int_{l_A}^{l_B} m(g + \omega^2 r)dr. \tag{3.94}$$

Again, we use conservation of angular momentum through the rising motion such that

$$\omega = \frac{r_A^2}{r^2}\omega_A. \tag{3.95}$$

Substitution of this result into the expression for the work and carrying out the integration eventually leads to

$$W_{AB} = -mg(l_A - l_B)\left[1 + l_A\frac{(1-\cos\theta_0)(l_A + l_B)}{l_B^2}\right]. \tag{3.96}$$

The work done increases as the swing goes higher because of the cosine factor, again a common experience. This increased work is due to greater centrifugal force at the bottom of the arc from the higher velocity of the swing. The force and therefore the work done in rising requires an increasingly greater expenditure of energy by the rider. On the other hand, the graph in Fig. 3.27 suggests that this extra work also produces larger increments in the swing's amplitude.

This simple analysis applies to both standing and sitting pumping although in the latter case the change in the center of mass is smaller and therefore the predicted change in swing amplitude will be less for each cycle. In this analysis, we have not included frictional effects that might be found in the suspension point of the swing or in the surrounding air. We expect, however, that these effects are relatively small and that our discussion gives a reasonable picture of the underlying physics. In 1969, (Siegman 1969) Siegman of Stanford University showed that the model of pumping the swing by changing its center of mass is equivalent to the action of the common notion of a parametric pendulum; namely a pendulum that is energized by the periodic variation of the vertical position of its pivot point. The connection of the pumped swing with parametric devices opens new vistas of related phenomena (Strarrett and Tagg 1995; Louisell 1960).

In 1990, Case and Swanson of Grinnel College took a somewhat different approach (Case and Swanson 1990). They modeled the rider and swing as a compound pendulum, with a massive bob, m_1, at the rider's position on the seat and the rest of the body by two other bobs, m_2 that account for the extension of the body due to the rest of the body parts—arms, legs, head and so on. For simplicity, the two other bobs of equal mass are represented as dumbbells that are positioned symmetrically about the seat of the swing, as shown in Fig. 3.28. Using the notation in this figure, the Lagrangian for the system is, to within a constant, as follows;

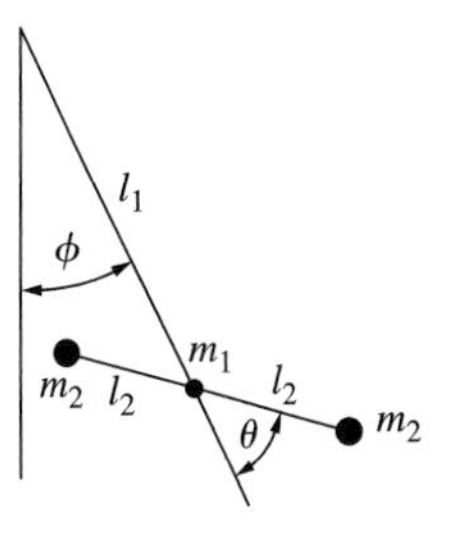

Fig. 3.28
Person on a swing modeled by a pair of point masses.

$$L = T - V = \frac{1}{2}(m_1 + 2m_2)l_1^2\left(\frac{d\phi}{dt}\right)^2 + m_2 l_2^2\left(\frac{d\phi}{dt} + \frac{d\theta}{dt}\right)^2 + (m_1 + 2m_2)l_1 g\cos\phi. \tag{3.97}$$

Our interest is in the motion of the swing itself. The motion of the rider, relative to the position of the swing seat, is that of a periodic dumbbell and satisfies an equation of the form

$$\theta = \theta_1 + \theta_0 \cos\omega t, \tag{3.98}$$

where ω is the angular frequency of the pumping. The "at-rest" angle is θ_1 and the amplitude of the variation in orientation of the "dumbbell" is θ_0. Because the pendulum is nonlinear, the resonant frequency of the swing decreases as the amplitude of the motion increases. Therefore the rider will naturally vary the pumping frequency in order to match the resonant frequency of the swing. Lagrange's method provides the equations of motion for the swing:

$$\frac{d}{dt}\frac{\partial L}{\partial\dot{\phi}} - \frac{\partial L}{\partial\phi} = [(m_1 + 2m_2)l_1^2 + 2m_2 l_2^2]\ddot{\phi} + [m_1 + 2m_2]l_1 g\sin\phi + 2m_2 l_2^2\ddot{\theta} = 0, \tag{3.99}$$

with substitution from Eq. (3.98). The result is the equation of motion for a harmonically driven, undamped, nonlinear, pendulum,

$$[(m_1 + 2m_2)l_1^2 + 2m_2 l_2^2]\ddot{\phi} + [m_1 + 2m_2]l_1 g\sin\phi = 2\omega m_2 l_2^2\theta_0\cos\omega t. \tag{3.100}$$

While this equation looks complex, its essence is expressed by

$$\alpha\ddot{\phi} + \beta\sin\phi = \gamma\cos\omega t,$$

which is simply an undamped pendulum with sinusoidal forcing. The rotational motion of the dumbbell provides the pumping action and therefore increases the amplitude of the swing. We emphasize that the period of the pendular motion will increase with amplitude since the linearization approximation for small angle is not applied in this equation, and therefore the rider will need to decrease the frequency of the pumping action as the swing's amplitude increases.

Since the model uses a symmetric dumbbell, the center of mass of the swing is always at the position of the central mass, and therefore the parametric mechanism described earlier—periodically varying the center of mass relative to the pivot—does not apply. However, if the dumbbell is allowed to be asymmetric, then the center of mass will change and the parametric energizing mechanism does come into play. But interestingly, Case and Swanson found that the parametric mechanism only dominates as the amplitude becomes large. We might have suspected this from a study of Fig. 3.27 where the amplitude of the swing grows slowly at first and then more rapidly at larger angular displacements.

Finally, Wirkus et al. of Cornell university (Wirkus et al. 1998) incorporated the work of both Falk and Tea, and Case and Swanson to determine an optimum mix of pumping techniques. They concluded that the rider should begin with sitting pumping, and progress to standing pumping at larger amplitudes for the most rapid accumulation of swing amplitude. In the full asymmetric dumbbell case the analysis becomes very complicated. The interested reader is referred to their work for the details. This brief survey of 40 years of study of the motion of playground swings illustrates the continuing fascination and inherent complexity of the subject. Yet the self pumping of playground swings is just one application of parametric forcing. Let us now return to parametric pumping through change in the length of the pendulum. Our example is *O Botafumeiro*.

O Botafumeiro: A simple model

Like the self-pumped swing, O Botafumeiro is a parametric pendulum whose length is periodically shortened. This pumping modifies the effect of gravity such that the downward force on the censer is increased as the pendulum is shortened, and decreased as the pendulum regains its full length. We begin with a nondissipative Lagrangian L, then develop equations of motion for the conservative system, and finally, insert the pumping and dissipative effects to complete the equations of motion. The Lagrangian for the appropriate conservative system is

$$L = T - V = \frac{1}{2}m\dot{r}^2 + \frac{1}{2}mr^2\dot{\theta}^2 - mgr(1 - \cos\theta). \tag{3.101}$$

The censer is approximated by a point mass (m) and the radial distance r is measured from the pivot to the center of mass of the censer. The equations of motion follow from standard operations on L,

$$\frac{d}{dt}\left(\frac{\partial L}{\partial \dot{r}}\right) - \frac{\partial L}{\partial r} = 0 \text{ and } \frac{d}{dt}\left(\frac{\partial L}{\partial \dot{\theta}}\right) - \frac{\partial L}{\partial \theta} = 0, \tag{3.102}$$

that lead to radial and tangential components, respectively, of the motion of the pendulum. These computations yield the following equations of motion,

$$\begin{aligned} m\ddot{r} - mr^2\dot{\theta} + mg(1 - \cos\theta) &= 0 \\ mr^2\ddot{\theta} + 2mr\dot{r}\dot{\theta} + mgr\sin\theta &= 0. \end{aligned} \tag{3.103}$$

The first equation is the radial equation and includes a centrifugal term $mr^2\dot{\theta}$, whereas the second equation is the tangential equation and describes the angular acceleration and the torques causing that acceleration. (The term $2mr\dot{r}\dot{\theta}$ is due to Coriolis force, which is discussed in the next chapter.) We focus on the tangential equation and add a dissipative torque that is proportional to the square of the tangential velocity, $v_T = r\dot{\theta}$. For an extended object moving swiftly in air, the frictional force is typically given by

$$F_{\text{Friction}} = \frac{1}{2}\rho_A S C_D v^2. \tag{3.104}$$

The friction parameters are the density of air ρ_A, the equivalent surface area that the censer presents as it moves through the air S, and the drag coefficient of the censer C_D. The latter two parameters are provided by Sanmartin (1984) as well as other necessary parameter values. See the table below. The modified equation of motion then becomes

$$mr^2\ddot{\theta} + 2mr\dot{r}\dot{\theta} + \frac{1}{2}\rho_A S C_D (r\dot{\theta})^2 r + mgr\sin\theta = 0. \tag{3.105}$$

Dividing Eq. (3.105) by mr^2 yields a new equation

$$\ddot{\theta} + 2\dot{r}\dot{\theta}/r + \frac{1}{2m}\rho_A S C_D \dot{\theta}^2 r + g/r\sin\theta = 0. \tag{3.106}$$

The pumping action is now added. We note that g is affected by the radial acceleration of the censer and therefore both r and g are functions of time. Figure 3.12 shows the approximate path of the censer as caused by the pumping action of the *tiraboleiros*. The period of the pumping is twice the period of the pendulum and the time variation of the pendulum length may be crudely modeled by

$$r(t) = \left(r_0 - \frac{\Delta r}{2}\right) + \frac{\Delta r}{2}\cos\omega_F t. \tag{3.107}$$

The variation in $r(t)$ modifies the effective gravitational field as

$$g(t) = g_0 - \ddot{r} = g_0 + \frac{\Delta r}{2}\omega_F^2\cos\omega_F t. \tag{3.108}$$

The parameter values provided by Sanmartin (1984) are given in the following table.

ρ_A	1.29 kg/m^3
S	$\pi(l/3)^2 = \pi(1.2/3)^2 = 0.503$ m^2
C_D	0.59
m	57 kg
r_0	20.9 m
Δr	2.9 m
$\omega_F(0)$	1.37 rad/s (zero amplitude)
g_0	9.8 m/s^2
ω_0	$\sqrt{g/r_0} = \sqrt{9.8/20.9} = 0.685$ rad/s

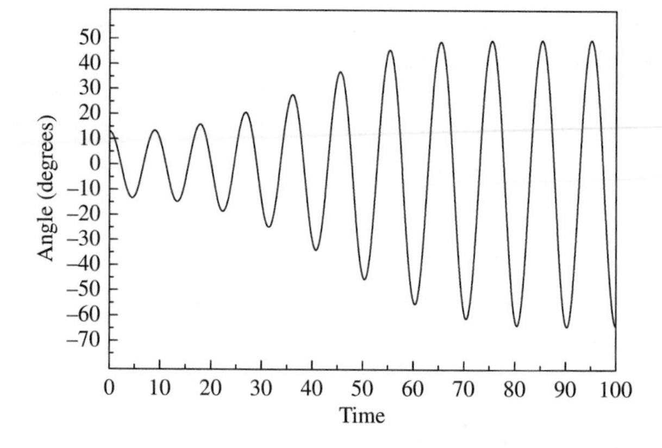

Fig. 3.29
Time series of the angular displacement of O Botafumeiro based upon the model described in the text.

The zero amplitude value of the pumping frequency ω_F is determined by noting that O Botafumeiro requires two pumping cycles for each oscillation. Of course, the pumping team will need to adjust their frequency of forcing as the amplitude increases, and therefore our model incorporates adjustment in the frequency according to a first order approximation of Eq. (3.64):

$$\omega_F = \frac{\omega_F(0)}{[1 + \theta_0^2/16]}.$$

As the amplitude changes the *tiraboleiros* need to modify the amount of pulling on the rope. This factor is more difficult to model. Relatively simple feedback mechanisms tend to lead to instability, and therefore we do not carry the process further. The resulting simulation shown in Fig. 3.29 illustrates the general features of the motion of O Botafumeiro's motion. Like the real pendulum the simulation starts at about 13° where a priest initially pushes the censer. It takes about 80 s to reach full amplitude in about 17 forcing cycles, in agreement with the real pendulum. However, there are two points of discrepancy between the model results and the reality. For the real pendulum the final amplitude is about 82° whereas in the model the amplitude is considerably less. Furthermore the amount of forcing in the model is $\Delta r = 0.9$ which is only about one third of the actual forcing stated in the table. Unfortunately, the model becomes unstable for strong forcing. These difficulties might be somewhat resolved by a feedback mechanism that would tie the forcing strength to the amplitude. As noted, simple feedback mechanisms seem to be unstable and therefore we leave the model with only the forcing frequency adjustment. However, the agreement of many facets of the real pendulum's motion with that exhibited by the simple model is gratifying, and modest deviation in the parameters' values from those given in the table yield results that more closely approximate the buildup in the real pendulum.

As well as the noted discrepancies with the motion of the real pendulum, we also note that the physical complexity of O Botafumeiro is considerably simplified in the model. The mass of the rope is a significant factor at

15.7 kg. The pendulum is not rigid and is perhaps more properly treated as a double pendulum. The thickness of the rope at 4.5 cm is also important and contributes to friction effects beyond those modeled for the censer alone. Sanmartin has included many of these effects in his analysis and consequently achieves a better agreement of theory and experiment. Nevertheless our simple model adequately represents many of the features of O Botafumeiro, one of the world's most unique pendulums.

At this point we defer further technical analysis of the nonlinear pendulum until Chapter 6 in which we discuss the chaotic pendulum. Let us end this chapter with a literary note.

3.4 A pendulum of horror

In 1842, the first American author of tales of horror, Edgar Allen Poe (1809–1849) wrote a short story entitled, *The Pit and the Pendulum* (Poe 1966). Poe's stories often contained a strong element of terror, in part, because he left many of the details quite vague, just as a standard technique of psychological terror is to keep the victim in ignorance as to his ultimate fate. The *Pit and the Pendulum* does exactly that to both the reader and the protagonist. The main, and practically only character, whose name we never know, is brought as a prisoner before the court of the Spanish inquisition in Toledo, Spain. The trial is recalled by the prisoner during a confused dreamlike state. Subsequently, he is carried into the bowels of the earth and flung into a damp and dark dungeon. He attempts to investigate the physical condition of his cell but exhaustion forces sleep upon him. When our hero awakes, he is tied to a low cot with only one hand free with which to feed himself the spiced meat that is mysteriously laid beside the cot. (Fig. 3.30) He now notices a large pendulum high above the cot and observes the start of its oscillations. However, the presence of rats attempting to steal his food distracts his attention from the pendulum. Meanwhile the pendulum continues to swish by overhead and, with each arc, the pendulum bob, now seen to be in the shape of a large sharp metal blade, comes ever closer to his person. The descent of the pendulum is tortuously slow giving our hero a chance to assess his situation. The strap by which he is held to the cot is a long single piece that is wound many times around his body. His first thought is that the pendulum might eventually cut the strap and allow him to free himself before he suffers his apparently inevitable fate. But, to his chagrin, he notes that the only place where the strap does not cover his body lies in the path of the pendulum. Therefore he needs to devise some other method of regaining freedom of movement.

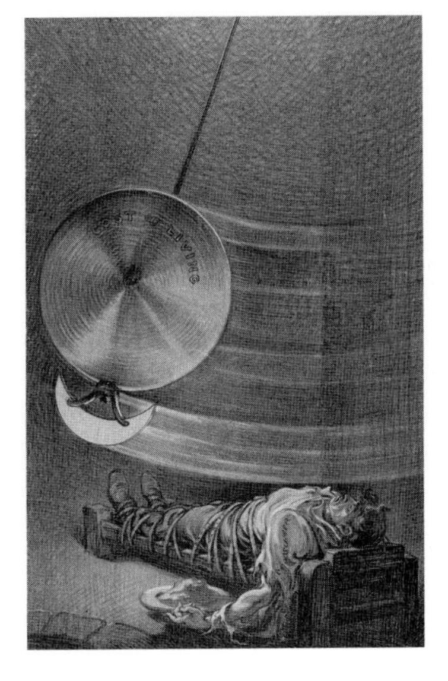

Fig. 3.30
Depiction of the victim in Poe's story "The Pit and the Pendulum." This drawing is actually a political cartoon drawn by Udo Keppler in the 1920s, highlighting the effect of inflation during that decade. Note "Cost of Living" on the blade/bob. (Political Cartoon Collection, Public Policy Papers, Department of Rare Books and Special Collections, Princeton University Library. Reprinted with permission.)

The story continues as the pendulum draws ever closer with increasingly larger amplitudes for its swing. Like O Botafumerio, this pendulum can also be heard to make an ominous swishing sound as it describes its increasing arc. As the pendulum nears the prisoner's body he estimates the total range of the pendulum's motion to be about thirty feet. Given that the room itself is only about forty feet high, this pendular motion is of very large amplitude, again similar to O Botafumerio. At this point we refer readers to the original story in order to learn the fate of the prisoner. However, it is of interest to ask whether Poe's tale is realistic. Does it seem likely that the inquisition would have used something like a motorized pendulum in this context? The answer is probably no. The time frame of the story is not clear but we may impose some limitations. The Spanish inquisition ended in the early nineteenth century and the story itself refers to capture of the city by a certain French General La Salle. Perhaps this allusion is to the Napoleonic wars, but the time could also be much earlier. Scholars (LLorente 1823; Lea 1907) suggest that while torture was a standard and accepted means of learning truth in matters both secular and religious, it was not likely to be used gratuitously by the inquisition once sentence had been passed. And, in the story, there had been a trial and a sentence pronounced. Furthermore, instruments of torture were fairly simple and direct. A driven pendulum that would slowly and noiselessly descend, and slowly increase its amplitude of motion, all inside and above a small dungeon, is relatively complex and hardly worth the effort even if craftsmen could be found to build such a machine. Therefore Poe's story is somewhat unrealistic in this regard. Some believe that the inspiration for Poe's tale was actually a large swinging bell that Poe had observed. Nevertheless, realistic or not, the image of a sharp-edged pendulum makes it frightening. It is the regularity and inexorability of the pendulum's motion that contributes to the climate of terror found in this story. The pendulum's periodicity, that is so important in other contexts, is here made to serve the cause of literary suspense.

With Poe's pendulum we conclude our discussion of the conventional classical physics of the pendulum. This discussion, beginning with the undamped linearized pendulum of the previous chapter, and progressing in complexity to the nonlinear pendulum of this chapter, will help our understanding of the pendulum as we meet more of its facets in the following chapters. In the next two chapters, we discuss important applications of what are, for the most part, linearized pendulums. The nonlinear pendulum is especially important for understanding the chaotic pendulum (Chapters 6 and 7) and the quantum pendulum (Chapter 8).

3.5 Exercises

1. Consider the effect of *constant* forcing of the damped driven pendulum. Start with Eq. (3.18) and let the forcing term be simply Υ.
   (a) Write the new equation of motion.
   (b) One way to think about this system is to make the force (aside from friction) acting on the pendulum to be $F = -mgl\sin\theta + \Upsilon$ and find a "potential" $V = -\int F\,d\theta$. Write the expression for the potential and sketch its graph as a function of θ. Because of the shape of the graph this potential is sometimes called a "washboard" potential.

   If the constant torque is slowly increased from zero the pendulum bob rises (with negligible angular velocity) up to the critical angle of 90°. In this regime the local minima in the washboard potential will keep the pendulum bob from moving out of the local minimum. At $\theta_c = 90°$ the torque is at its critical value of $\Upsilon_c = mgl$.

The washboard potential is now sufficiently steep that the local minima disappear and the pendulum bob can fall out of the local minimum, accelerate and rotate almost freely. Thus there is a sudden jump in the pendulum's angular velocity. (The rotation is slightly hindered by the bumps in the washboard, corresponding to the pendulum slowing as it goes over the top of its motion.) If the constant torque is now gradually decreased, then the pendulum continues the hindered rotational motion because it now has sufficient momentum. Even at relatively low values of the torque where, for increasing torque, the pendulum was previously stationary, it now still rotates. The phenomenon of different behaviors with increasing and decreasing torque is another example of *hysteresis*, and a graph is shown in Fig. 3.31.

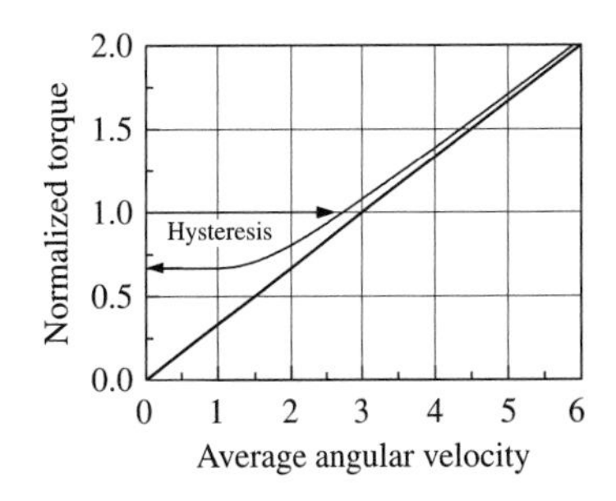

Fig. 3.31
Slowly varying constant torque versus angular velocity. Another example of hysteresis.

2. (a) Substitute the solution given by Eq. (3.6) into Eq. (3.2) and thereby verify Eq. (3.6). (b) Determine the phase angle ϕ in terms of the initial angle $\theta(0)$ and the initial angular velocity, $\dot{\theta}(0)$.
3. With the forced, damped linearized pendulum, the amplitude of the steady state part of Eq. (3.19) $F/\sqrt{4\gamma^2\omega^2 + (\omega_0^2 - \omega^2)^2}$ varies with both frequency ω and damping strength γ. Sketch a few curves of Amplitude versus ω, each being at different values of damping strength. (See the Appendices for samples of such curves.)
4. Physical systems that are linear and dissipative all respond to periodic forcing as indicated in the curves developed in exercise 3. Such curves may be characterized by a "quality" factor Q that is calculated in the following manner. First, we note that the "power" of the pendulum's response is proportional to the *square* of the steady state solution time series, $F^2/((4\gamma^2\omega^2) + (\omega_0^2 - \omega^2)^2)$. Find the value of the power $P(\omega_0)$ when $\omega = \omega_0$. (For light damping, this value of the power should be very close to the maximum value.) Now observe that when $4\gamma^2\omega^2 = (\omega_0^2 - \omega^2)^2$, the power is reduced by one-half. For light damping, where $\omega \approx \omega_0, \omega + \omega_0 \approx 2\omega_0$, we define the half-width of the resonant peak as $\Delta\omega = |\omega - \omega_0|$. The quality factor Q gives a measure of the sharpness of the resonant peak in terms of the relative size of the resonant frequency and the width of the curve at the half power points. That is we define $Q = \omega_0/2\Delta\omega$. Using the information given in this exercise show that $Q = (1/2)(\omega_0/\gamma) = (1/2)\omega_0\tau$. Thus, a sharp resonant curve with light damping has a high Q. Find Q for a pendulum with a period of 1 s and a ring time of 10 s. (Remember the factor of 2π in these calculations.) See Appendix A.
5. The amount of damping is inversely proportional to the "ring time" τ of a pendulum or spring or electric LRC circuit. That is, the less damping the longer the time that the system oscillates. Mathematically, $\tau \approx 1/\gamma$. The calculations made in the previous exercise lead to a very simple relationship between the linewidth $2\Delta\omega$ and the ring time τ. Show that the linewidth of the resonance curve is related to the ring time as $\Delta\omega \cdot \tau \approx 1$. This relationship is similar to the energy version of the Heisenberg uncertainty principle in quantum physics whereby the lifetime of an unstable energy state is inversely proportional to the uncertainty in energy.
6. Eq. (3.13) has the same general form as Eq. (3.2). Make the former exactly match the latter and thereby find the corresponding value for γ. Show that the quality factor for this electrical circuit is given by $Q = \omega_0 L/R$. Find Q for a circuit that is resonant at 50 MHz, has an inductance of $L = 2\,\mu h$, and resistance of 2 Ω.
7. The Laplace transform technique is a powerful one for dealing with pulse systems. However, simpler techniques may be used at least during the time of the first pulse, $0 < t < \tau$. In the equation of motion as given, with initial conditions, by Eq. (3.30), try a solution of the form $\theta(t) = c_1 \sin\omega_0 t + c_2 \cos\omega_0 t + B$. Substitute this trial solution into Eq. (3.30), determine B, and the constants c_1 and c_2. Compare your result to the first term in the comprehensive solution given by Eq. (3.37).

8. Use L'Hopital's rule to justify Eq. (3.27).
9. In the text Eq. (3.96) gives the work done by a rider on a swing during a single pumping action. An identical expression can be calculated with a slightly different approach. Furthermore we take this approach in the context of a pendulum like O Botafumerio that is periodically shortened. We suppose that the pendulum's length is shortened from l_0 to l_1 at the moment the pendulum is at the bottom of its swing. The kinetic energy at the bottom of the swing $K = (1/2)ml_0^2\theta_0^2$ is equal to the total energy E_0 and is equal to the potential energy at the maximum angular displacement, $V = mgl_0(1 - \cos\theta_0)$. At the bottom of the swing the rope is pulled so that the pendulum is shortened to l_1. (a) Use conservation of angular momentum to show that the change in kinetic energy at the bottom of the swing is $\Delta K = (1/2)ml_0^2\theta_0^2\left[(l_0^2 - l_1^2)/l_1^2\right] = E_0\left[(l_0^2 - l_1^2)/l_1^2\right]$. The act of shortening the rope not only increases the kinetic energy, it also increases the gravitational potential energy by ΔV. (b)What is magnitude of this latter increase? (c) Combine both of these energy increases to arrive at a result that has the same magnitude as the work done given in Eq. (3.96). (d) What is the average force exerted by the pullers during this shortening of the pendulum?
10. The derivation of Eq. (3.61) is nontrivial and only an outline is provided in the text. Derive this equation with the following hints. Start with Eq. (3.57) and use the substitution given by Eq. (3.59). Note that $\sin(\theta/2) = k\sin\varphi$ and differentiate this expression with respect to time. Then square both sides of the resulting equation, and multiply the equation by 4. Note that the left side now contains $(d\theta/dt)^2 = \dot{\theta}^2$. Now go back to Eq. (3.57) and multiply both sides of that equation by $cos^2(\theta/2)$. At this point you should have

$$\cos^2\frac{\theta}{2}\dot{\theta}^2 = \frac{2g}{l}\left[\cos^2\frac{\theta}{2}\cos\theta - \cos^2\frac{\theta}{2}\cos\alpha\right] = 4k^2\cos^2\phi\dot{\phi}^2.$$

Using relationships such as $\sin^2\beta + \cos^2\beta = 1$, $\cos\beta = 1 - 2\sin^2(\beta/2)$, and Eq. (3.59), work with the two right-hand members of this equation to obtain

$$\frac{g}{l}\left[1 - k^2\sin^2\phi\right] = \dot{\phi}^2,$$

and the rest of the derivation is obvious.
11. Use the trial solution $\theta = \theta_0\cos\omega t$ in the nonlinear equation of motion, Eq. (3.52), and develop a general expression for the Fourier series solution, similar to that given in the text.
12. Suppose that the effective length of the pendulum (in the Pit) is 10 m and that the amplitude of its swing is 45°. Use Eq. (3.64) to calculate the period of its motion to within 1% of the true value.
13. We wish to prove that the Dirac function pulsing of the swing leads to a steady state motion that is independent of how many oscillations have occurred. Consider the solution given by Eq. (3.43). (a) Write out the solution for the time interval $nT < t < (n+1)T$. (b) Isolate the common factors of $(A/\omega)e^{-\gamma t}\sin\omega t$ and write the remaining quantities as a finite geometric series. (c) Show that for a finite geometric series $1 + x + x^2 + \cdots + x^n = (1 - x^{n+1})/(1 - x)$. (d) Use this result to show that when t is large then $\theta(t) \approx (A/\omega)\left[(e^{-\gamma[t-(n+1)T]}/(e^{\gamma T-1})\right]\sin\omega t$. Let the factor in the numerator be $x(t)$ and show that in the given interval $1 < x(t) < e^{\gamma T}$. (For light damping $\gamma T \ll 1$, so that $x(t)$ does not vary much.) (e) Finally, by redefining the time coordinate as $0 < t' < T$, show that the steady state solution becomes $\theta(t) = (A/\omega)\left[(e^{-\gamma t'})/(e^{\gamma T} - 1)\right]\sin\omega t'$.

4 The Foucault pendulum

> That was when I saw the pendulum.
>
> The sphere, hanging from a long wire set in the ceiling of the choir, swayed back and forth with isochronal majesty . . . I knew the earth was rotating, and I with it, and Saint-Martin-des-Champs and all Paris with me, and that together we were rotating beneath the Pendulum, whose own plane never changed direction, because up there, along the infinite extrapolation of its wire beyond the choir ceiling, up toward the most distant galaxies lay the Only Fixed Point in the universe, eternally unmoving.[1]

With a novelist's hyperbole, Umberto Eco begins his unusual tale entitled *Foucault's Pendulum*. It is a story of innocent diversions that turn into consuming reality, of fictitious plots, hatched by three unlikely dreamers with their word-randomizing computer, that strangely coincide with ancient diabolical schemes. These fictitious schemes would reach fruition in the former church, Saint Martin des Champs—now part of the Conservatoire National des Arts et Métiers (CNAM) in Paris, that has housed a Foucault pendulum since 1855. The plot of Eco's novel has little to do with the physics of the pendulum, except that the pendulum's regularity provides, as with Poe's story, a prominent and ominous symbol. Through symbolism of the pendulum and many classical allusions, Eco's tale builds a crescendo of dread toward the horror of its final conclusion.

4.1 What is a Foucault pendulum?

But let us return to the reality. What is a Foucault pendulum, what is its history, and what physical phenomenon does it demonstrate?[2]

In theory, any earth-based pendulum is a Foucault pendulum. However, a realistic Foucault pendulum is a one that is specially constructed to highlight the *rotation* of its plane of oscillation due to the earth's rotation relative to a frame of reference fixed in the stars. That is, the plane of the pendulum's oscillation tries to stay fixed relative to the stars while the earth rotates underneath it. See Fig. 4.1. All pendulums with point suspensions attempt such rotation, but for most pendulums this behavior is masked by

Fig. 4.1
A Foucault pendulum (similar to the original) hanging in the Pantheon, Paris, where the orginal was first publicly displayed. ©Robert Homes/Corbis/Magma.

[1] See (Eco 1989, pp. 3 and 5).

[2] The remarkable, interesting, and scholarly book by William Tobin about Leon Foucault was very helpful in the writing of this chapter. (See the references.)

Fig. 4.2
Léon Foucault, 1819–1868, inventor of the Foucault pendulum. Foucault made many contributions to science and technology. ©Bettmann/Corbis/Magma.

other more prominent effects. For an ideal Foucault pendulum, the plane of oscillation would be seen as *fixed* by an observer positioned in the stars. (In this discussion we ignore the rotation of the earth around the sun, and the rotation of the sun around the center of the galaxy, and so forth.) Therefore the earthbound observer sees a slow rotation of the plane of oscillation and it is this remarkable feature of the Foucault pendulum which demonstrates, on a large scale, the rotation of the earth.

The inventor, for whom this special pendulum is named, Jean Bernard Léon Foucault (see Fig. 4.2), was born September 18, 1819 (died February 11, 1868), the son of a Parisian publisher. He received his secondary education at the prestigious College Stanislas and later enrolled in the Paris medical school. Yet sometime around 1844 he abandoned medicine to pursue his strong interests in physical science. Many of his early experiments were in optics. These included efforts to improve Daguerre's new photographic processes, studies of the intensity of the sun's light, the phenomenon of the interference of light that is observed by combining coherent beams of light, and studies of the chromatic polarization of light by crystals. Foucault's sometime collaborator in this work was Hippolyte Fizeau, known for his efforts to measure the effects of the Luminiferous "ether" on light transmission. Foucault showed that light travelled faster in air than in water and also measured the absolute (vacuum) speed of light. Perhaps as result of his pendulum work, Foucault invented the precision gyroscope and demonstrated its use as a compass. In 1855, Foucault was appointed physicist at the Paris observatory and there invented techniques that would allow for the manufacture of large mirrors for reflecting telescopes thereby significantly increasing the power and utility of the telescope (Tobin and Pippard 1994). Remarkably, the electric currents that we now refer to as eddy currents were, during the ninteenth century, sometimes known as Foucault currents, since Foucault studied them experimentally (Greenslade 2002). While not a brilliant theoretician,[3] Foucault was an extremely creative experimentalist. This brief description hardly does justice to his contributions.

In early 1851 Foucault had the insight that led to his famous discovery (Foucault 1851). This bit of serendipity occurred during construction of a pendulum clock to regulate the drive of a telescope. In the course of machining this device, Foucault noticed that the *plane* of oscillation of a vibrating rod, that was fixed in a lathe chuck, remained fixed in orientation even as he slowly turned the chuck. This observation inspired him with the notion that the vibrating motion of a freely suspended pendulum might somehow be related to the earth's rotation. In the special case of a pendulum suspended above the north pole, the plane of vibration would appear to complete a full circle in 24 hours, whereas at the equator, the plane of vibration would not rotate; the period of rotation then being infinite. The simple relationship between the period τ of rotation of the plane of oscillation and the latitude of the pendulum β is

$$\tau = 24/\sin(\beta) \text{ hours.}$$

[3] During the oral defense of his doctoral thesis in 1853, Foucault found himself unable to correct some mathematical errors in his thesis. (See p. 131 of (Tobin 2003).)

For the latitude of Paris—Foucault's city—the trigonometric factor is about 0.75 and therefore the plane of vibration rotates about 0.19 degrees/min. (Sher (1969) pointed out that the plane of oscillation of the pendulum must follow the component of the motion of a star along the observer's horizon, leading to the same sin (β) factor.)

In January of 1851, Foucault made his initial observations with a pendulum having a length of 2 m and a bob mass of 5 kg suspended in his mother's basement. While he was able to observe the anticipated effect with the basement pendulum, he quickly realized the benefits of an even longer pendulum. In particular, the change in the maximum displacement of the pendulum after each oscillation would be more noticeable, spurious effects would be minimized, and the pendulum would probably oscillate longer before needing another push. Thus a very long pendulum would provide a more compelling demonstration. His next pendulum was 11 m long. In February, it was hung in the Paris Observatory. Meanwhile the prince-president Louis Napoleon Bonaparte became interested in Foucault's work so that with princely backing, Foucault constructed a mammoth 67 m, 28 kg pendulum under the dome of the Pantheon in Paris in March of 1851 (see Fig. 4.3).

The formula for the period of oscillation T (as opposed to the period of rotation of the plane of oscillation τ) is given by Eq. (2.5)

$$T = 2\pi\sqrt{\frac{L}{g}}, \tag{4.1}$$

where L is the length in meters and g is acceleration due to gravity ($9.8\ \mathrm{m/s^2}$). For Foucault's large pendulum the equation yields an oscillation period of 16.4 s. Foucault had a wooden circle with a 6 m diameter set up with its axis concentric with the vertical wire of the pendulum. The circumference of the circle was divided into fractions of a degree. During one of the slow oscillations, the plane of vibration would turn by about

Fig. 4.3
Public display of the Foucault pendulum. (Engraving from L'Illustration, Paris, 1851).

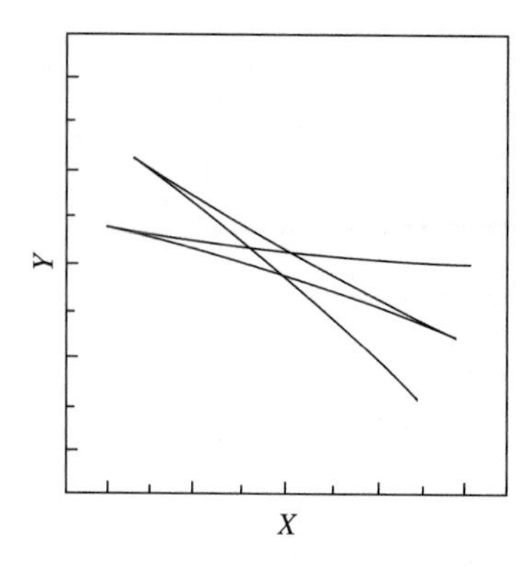

Fig. 4.4
Computer simulation of a few oscillations of the pendulum bob, projected onto the floor. The earth's rotation is artificially speeded up in order to clearly demonstrate the Foucault effect.

0.051°, and the incremental change in position of the the pendulum bob at maximum displacement above the floor was found to be about 2.5 mm for a single oscillation. After a few swings the cumulative change is obvious. In order to register the cumulative effect, fine sand was spread below the bottom of the bob such that the motion of the bob could be recorded in the sand by the spike fixed downward at the bottom of the bob. Figure 4.4 (five consecutive bob paths from a computer simulation) shows a schematic (and much exaggerated) diagram of the pendulum's motion. Foucault's 67 m pendulum, initially hung in the Pantheon in 1851, was fairly soon thereafter taken down when the building again became Saint Genevieve's church when Louis Napoleon staged a coup d'etat and became emperor. At the beginning of the twentieth century, the Astronomical Society of France—presided over by Henri Poincaré—supported having the pendulum again mounted in the Pantheon, once more a secular, public building. This was done and the exhibit opened on June 30, 1902. A year or so later the pendulum was taken down, and it was not until the fall of 1995

Fig. 4.5
Engraving of the impulse mechanism used by Foucault to maintain the pendulum's motion against frictional losses.

that it was brought out of storage and again hung from the dome of the Pantheon (Deligeorges 2000). Not long after the original Pantheon demonstration, a more permanent exhibit was established in Arts and Metiers in 1855. This latter pendulum was driven electromagnetically with an apparatus illustrated in Fig. 4.5. A remnant of this original magnetic drive is now stored at the Paris Observatory (see Fig. 4.6). Two of the 1851 pendulum bobs are currently on display at the Musée des Arts et Métiers in Paris.

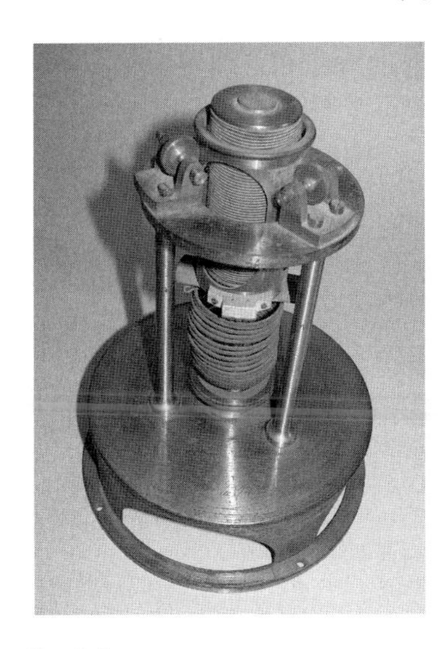

Fig. 4.6
Surviving remnant of Foucault's original electromagnetic pulser.

4.2 Frames of reference

Let us place the phenomenon of the rotating plane of oscillation in the more general context of the observation of motion from different *frames of reference*. In common usage we speak of a frame of reference as a point of view; perhaps the perspective of specific political philosophy or set of religious beliefs. In physics, a frame of reference is the platform from which an observer is located and from which one makes measurements. Observers who are in different frames of reference may see the same event differently.

Consider the mundane example of two people playing a game of ping pong on a transatlantic jetliner traveling at a constant speed relative to the earth. In their frame of reference (the airplane) the motion of the ball looks quite normal to them. But for an earthbound observer the ball's motion would look quite unusual. To the earthbound observer, the ball appears to be traveling at a horizontal speed whose average value is that of the jetliner with additions of small negative and positive velocities added to the high speed average. Events look different from different frames of reference. Often certain frames of reference are chosen as a way of singling out a certain aspect of an object's motion from a complex of motions. In our example of the ping-pong game, the motion looks complex from the earthbound observer's frame of reference. By introducing a frame of reference attached to the airplane, the dynamics of the game appear to be simpler. A more sophisticated example of the use of various frames of reference occurs in studies of nuclear magnetic resonance (NMR). NMR provides the theoretical basis of the medical diagnostic tool of magnetic resonance imaging (MRI).

(For centuries Newton's mechanics satisfactorily accounted for observations made by observers from different frames. But the notion of a frame of reference in physics became more prominent with the introduction of Einstein's special theory of relativity in 1905 (Einstein 1905). Like Newtonian mechanics, Einstein's special theory of relativity explained observations of events that occurred in one frame of reference as seen from another frame that was moving with constant velocity relative to the first frame. The novelty of Einstein's work was the dramatic and unsuspected effects predicted to occur at high speeds due to the existence of a finite and fixed speed of light, the fastest speed with which information can be transmitted. This was a consequence of the proposition that the speed of light must be the same for all observers, no matter what the motion of their frame of reference.)[4]

For the case of the ping pong game on the airplane, the two frames of reference—that of the airplane and the ground—are traveling with a *constant velocity* relative to each other. Such frames are called *inertial frames*. But often frames of reference are *accelerated* relative to each other and a new phenomenon occurs. We note, for example, that passengers in an elevator, that is being accelerated upward, feel an extra pull downward. The effect is identical to a momentary increase in the gravitational field. Earlier, we saw how parametric pumping changes the effective gravitational field on a pendulum bob. These phenomena are examples of what Einstein called the *principle of equivalence*; that is, the effect of an accelerating frame is indistinguishable from that of a new force (Einstein 1908). Yet, like the notion of frames of reference, such equivalence is also implicit in the mechanics of Newton.

A particular example of frames in relative acceleration is a pair of frames in relative *rotation*. A passenger in (the frame of reference of) a car, experiences a new force, *centrifugal force*, as the car rounds a corner. This car-bound observer believes that a force is pulling him out of the car away from the direction of the turn. Yet an earthbound observer only sees the passenger accelerate—centripetal acceleration—toward the center of the turn, in going around the corner.

The observer in the rotating frame experiences another, less obvious, force in a rotating frame that is essential for our discussion of the Foucault pendulum. In 1837 the French mathematician Siméon Denis Poisson (1781–1840) published a paper (Poisson 1837) based upon the calculations of his one-time student Coriolis—but not publicly acknowledged by Poisson (Scientific Biography) p. 480—that described what we now call *Coriolis force*. (Gaspard Gustave Coriolis (1792–1843) was a physicist who specialized in various problems of theoretical mechanics and published his theory of rotating frames in the context of their effects upon machine operation (Coriolis 1832)). For example, an observer on a rotating platform (such as a merry-go-round) who attempts to throw a ball to her friend on the same platform, would observe the ball to follow a curved trajectory rather than a straight line. During the ball's time of flight the platform has rotated and the friend, who is supposed to catch the ball, is no longer in the same place. Thus, the Coriolis acceleration is proportional to the ball's velocity and the angular velocity of the rotating platform.

The earth is also a rotating frame of reference relative to the frame of the stars. In this case, the Coriolis acceleration is proportional to an object's velocity and whatever component of that velocity is perpendicular to the axis of the earth's rotation. That is, motions in the plane of the earth's

[4] On the other hand, Newton's mechanics effectively assumed that the speed of light was infinite and that information was passed instantaneously from the event to any observers. Because our common experience is with events whose speeds are very small compared to the speed of light, including observations of the Foucault pendulum, the Newtonian version of relativity is quite adequate for our purpose. In effect, the speed of light is almost infinitely large compared to any motions that we discuss.

surface may be affected by the daily rotation of the earth and therefore the dynamics as appearing to the earthbound observer may be different from that appearing to a "stellar" observer. Mathematically this new force is expressed, in the earth's frame, as

$$\mathbf{F} = 2m\mathbf{v} \times \boldsymbol{\Omega}, \tag{4.2}$$

where m is the mass of the object, $\mathbf{v}$ is the velocity of the object, and $\boldsymbol{\Omega}$ is the angular velocity of the rotating frame of reference.[5] To the earthbound observer, the effect is most noticeable at the poles for motion parallel to the earth's surface, where the vector cross product takes its maximum value. An object in motion will deviate to the right in northern latitudes and to the left in southern latitudes. At the equator, only that component of surface motion that is parallel to the equator is be subject to Coriolis acceleration and this acceleration is perpendicular to the earth's surface.

One common manifestation of Coriolis "force" is that artillery shells can deviate from their targets—to the right in northern latitudes and to the left in southern latitudes. (In fact, Poisson wrote a long monograph on the effect of Coriolis force on projectile motion in 1838 (Poisson 1838).) This phenomenon was dramatically demonstrated in World War I. During a naval battle near the Falkland Islands, located near the southern tip of South America, the British gunners initially compensated incorrectly for Coriolis force by using adjustments that were valid for the Northern Hemisphere. However, once they changed the sign of the correction, their fire became accurate (Crane 1990). Modern guns are less susceptible to Coriolis effects because, while their muzzle velocities are much higher and thus the Coriolis force is stronger, the projectile's time of flight is much less and therefore the time during which the force acts—or the earth rotates—is relatively small.

In his 1837 paper, Poisson mentioned the possibility of a (Coriolis) effect upon a pendulum's plane of oscillation. Since the rotation of the earth is slow and the velocities of typical pendulums are small, the effect of Coriolis force, on a single oscillation, even at the poles, is barely observable. In general, if there is to be any effect at all at some latitude β, it would be due to that component of the earth's rotation vector which would be vertical to the earth at any given point on the earth's surface, $\Omega \sin \beta$. At the equator this term would vanish, whereas at the poles where $\sin \beta$ equals one, its magnitude would be $2\pi/24$ *radians per hour* or $1/24$ of a revolution per hour. Poisson supposed that the effect on the pendulum's plane of oscillation would be unobservable. Certainly this is more or less true for a single oscillation of the pendulum. Foucault, on the other hand, noted in his 1851 paper (Foucault 1851) that the Coriolis effect would be *cumulative* as each oscillation contributed additively to the effect and therefore the resultant would be quite observable under physically realizable conditions.

[5] In the case of the earth's rotation the vector $\boldsymbol{\Omega}$ points outwardly from the surface of the earth at the North pole.

4.3 Public physics

Thus, in February of 1851, Foucault boldly invited contemporary scientists to a demonstration at the Paris Observatory with the words "Vous êtes invités à venir tourner la terre..." (Deligeorges 2000) (you are invited to come to the turning of the earth...). See Fig. 4.3. The much longer Pantheon demonstration, rapidly constructed for a March opening, produced a flood of visitors. Foucault's pendulum had indeed gone public.

Reaction to Foucault's demonstration was widespread. A few examples will suffice. Two entries from the *Proceedings of the Royal Society* (London) are instructive. The English geologist John Philips (1800–1874), a fellow of the Royal Society (FRS) communicated work that Thomas Cooke, an optician in the city of York, had done with his own version of the pendulum (Philips 1851). Cooke used a 52 foot pendulum with two different bobs, one being an oblate spheroid, weighing eight pounds, and the other being a prolate spheroid weighing 20 pounds. The pendulums performed as expected except that in some experiments, there was a small but noticeable elliptical aspect to the path of the pendulum's motion and a veering of the elliptical motion. Philips also did experiments in his house with much smaller pendulums and found that the spurious effects were more pronounced.[6] The contemporary British Astronomer Royal George Airy FRS (1801–1892) provided these authors with a partial explanation of the so-called clockmaker's formula that quantifies the rate of turning of the ellipse of a pendulum of length L and period T,

$$\text{Rate} = \frac{3(\text{Area of ellipse})}{4L^2T}, \tag{4.3}$$

expressed in radians per second. As we shall learn, this precession is due to the anharmonic effect of the nonzero amplitudes of the pendulum's oscillations in perpendicular directions.

In that same issue of the *Proceedings* (Wheatstone 1851) there was also a contribution from Charles Wheatstone, the English experimentalist and fellow of the Royal society (1802–1875) to whom generations of physics students have mistakenly attributed the invention of the Wheatstone bridge, a device used for precision measurements of electrical resistance.[7] Wheatstone emphasized (as we described earlier) that Foucault had come to the idea of his pendulum's rotation of the plane of oscillation by noting the effect of coupling one motion with another. Wheatstone described several other instances of this type of interaction and then built a mechanical device that coupled together vibration of a spring, rather than a pendulum, with rotation of the earth. Wheatstone's apparatus is

[6] These observations portend the fact that construction of a well behaved Foucault pendulum is difficult. While all earthbound pendulums are potentially Foucault pendulms, most exhibit spurious motions that overwhelm the foucault effect.

[7] Charles Wheatstone was a prominent British physicist in the field of instrumentation. His most famous invention, the electric telegraph with various improvements, eventually earned him a knighthood in 1868.

shown in Fig. 4.7. One end of the spring is attached to a rider on the vertical semicircular arch that is fixed to a circular turntable. The other end of the spring is connected to the center of the turntable. If the spring is vertical then its transverse plane of oscillation will remain fixed even if the turntable is moving. From the viewpoint of an observer on the turntable, the plane of the spring's vibration appears to rotate with a period equal to that of the turntable. As the angle is changed the period changes with the sinβ factor exactly as does the Foucault pendulum's period of veering. The point that Wheatstone made with this device, was that gravity was not necessary to achieve motion analogous to the Foucault effect.

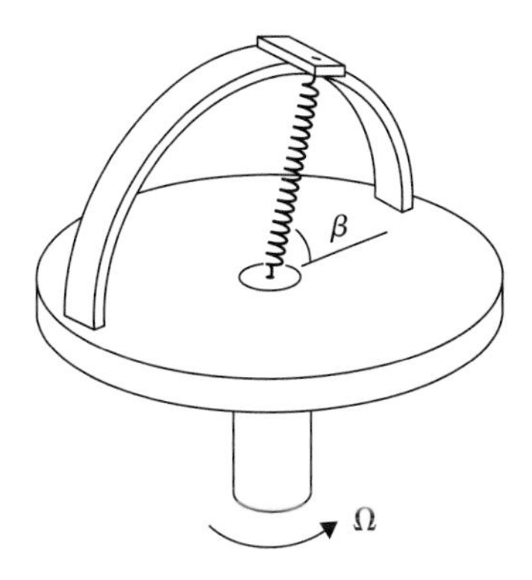

Fig. 4.7
The spring is vibrated sideways, and as the turntable is rotated, the spring's plane of oscillation rotates relative to the turntable with frequency of $\Omega \sin \beta$. This apparatus demonstrates that the veering effect is not intrinsically related to gravity. (From (Tobin 2003, p. 171). Reprinted with the permission of Cambridge University Press.)

On the other side of the Atlantic ocean a veritable "pendulum mania" occurred with the building of over thirty large Foucault pendulums at diverse locations along the eastern coast of the United States. (Tobin 2003, p. 149.) Scholarly discussion in American journals was also forthcoming. For example, the *Journal of the Franklin Institute* in Philadelphia, records several contributions on the subject in 1851. Charles Allen provided a geometric analysis of the pendulum's motion (Allen 1851). In the same volume, J. S. Brown also proposed an analysis (Brown 1851). He prefaces his communication with the provocative statement that "in all the explanations of Foucault's pendulum . . . not one has given . . . a full exposition of its phenomena, and many have made false statements." In a footnote, the journal's editor takes polite but firm issue with Brown and, in reference to Allen's contribution, he states that it "appears to us to be thoroughly correct."

Finally there is the much quoted note on the rotation of the earth to the editor of a contemporary issue of *Punch* magazine. (Quoted on p. 153 in (Tobin 2003).)

> Sir,—Allow me to call your serious and polite attention to the extraordinary phenomenon, demonstrating the rotation of the Earth, which I at this present moment experience, and you yourself or anybody else, I have not the slightest doubt, would be satisfied of, under similar circumstances. Some sceptical and obstinate individuals may doubt that the Earth's motion is visible, but I say from personal observation it's a positive fact.
>
> I don't care about latitude or longitude, or a vibratory pendulum revolving round the sine of a tangent on a spherical surface, nor axes, nor apsides, nor anything of the sort. That is all rubbish. All I know is, I see the ceiling of this coffee-room going round. I perceive this distinctly with the naked eye—only my sight has been sharpened by a slight stimulant. I write after my sixth go of brandy-and-water, whereof witness my hand,
> Swiggins
> Goose and Gridiron, May 5th, 1851.

The Foucault pendulum was certainly science with a high profile.

4.4 A quantitative approach

The idealized motion for the Foucault pendulum consists of (a) its back and forth oscillation with a relatively short period T and (b) the slow rotation of the plane of oscillation, period τ, caused by the earth's rotation. Let us develop a mathematical basis for these motions. From this discussion we will also find other motions whose relative importance depends on a variety of factors. In some cases these other motions may obscure the desired Foucault effect.

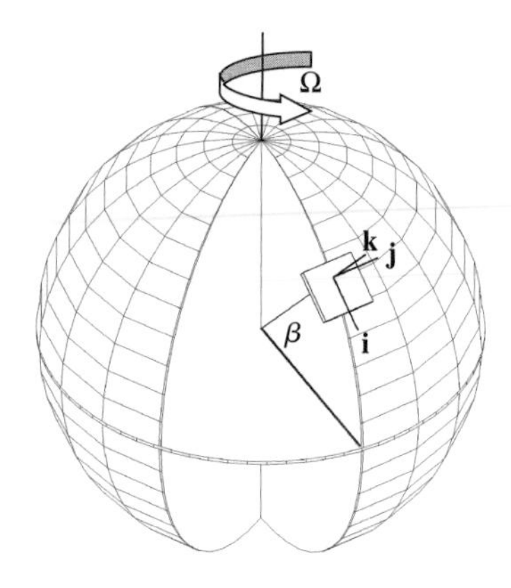

Fig. 4.8
Coordinate system located at latitude β on the rotating earth.

The appropriate coordinate system is shown in Fig. 4.8 with the origin placed at the site of the pendulum bob.

The equation of motion for a particle, the bob of the pendulum, acted upon by gravity $mg\mathbf{k}$, and the tension in the wire $\mathbf{T}$, is given by

$$m\frac{d^2\mathbf{r}}{dt^2} = \mathbf{T} - mg\mathbf{k}, \tag{4.4}$$

where $\mathbf{r}$ is the position vector of the particle and $\mathbf{k}$ is a unit vector directed vertically upward. However, for a particle in a rotating frame of reference (Synge and Griffith 1959) the acceleration vector is transformed as

$$\left(\frac{d^2\mathbf{r}}{dt^2}\right)_{\text{Inertial}} \Longrightarrow \left(\frac{d^2\mathbf{r}}{dt^2} + 2\boldsymbol{\Omega}\times\frac{d\mathbf{r}}{dt} + \boldsymbol{\Omega}\times(\boldsymbol{\Omega}\times\mathbf{r})\right)_{\text{Rotating}} \tag{4.5}$$

(See problem no. 4.) The rightmost term, on the right side, is centrifugal acceleration and the middle term is Coriolis acceleration. The angular velocity vector $\boldsymbol{\Omega}$ points upward, toward the North star, above the North Pole. Thus, the earth's angular velocity, in terms of components at latitude β, is

$$\boldsymbol{\Omega} = -\mathbf{i}\Omega\cos\beta + \mathbf{k}\Omega\sin\beta, \tag{4.6}$$

where β is the latitude of the observer. The equation of motion of the position vector $\mathbf{r} = x\mathbf{i} + y\mathbf{j} + z\mathbf{k}$ in the earth's rotating frame becomes

$$m\frac{d^2\mathbf{r}}{dt^2} = \mathbf{T} - mg\mathbf{k} - 2m\boldsymbol{\Omega}\times\frac{d\mathbf{r}}{dt} - m\boldsymbol{\Omega}\times(\boldsymbol{\Omega}\times\mathbf{r}). \tag{4.7}$$

In the linear approximation, the tension vector is given by

$$\mathbf{T} = -mg\frac{x}{L}\mathbf{i} - mg\frac{y}{L}\mathbf{j} + mg\frac{L-z}{L}\mathbf{k} \tag{4.8}$$

or

$$\mathbf{T} = -m\omega^2 x\mathbf{i} - m\omega^2 y\mathbf{j} + m\omega^2(L-z)\mathbf{k} \tag{4.9}$$

using the notation, $\omega = \sqrt{g/L}$. Finally, the components of the equation of motion for small amplitudes become

$$\begin{aligned}
\frac{d^2x}{dt^2} &= -\omega^2 x + 2s\Omega\frac{dy}{dt} + s^2\Omega^2 x + sc\Omega^2 y \\
\frac{d^2y}{dt^2} &= -\omega^2 y - 2s\Omega\frac{dx}{dt} - 2c\Omega\frac{dz}{dt} + \Omega^2 y \\
\frac{d^2z}{dt^2} &= -\omega^2 z + 2c\Omega\frac{dy}{dt} + cs\Omega^2 x + c^2\Omega^2 z.
\end{aligned} \tag{4.10}$$

$(s = \sin\beta$ and $c = \cos\beta)$

These are complicated coupled differential equations and their complete solution is only found through computer simulation. However, given the relatively slow angular velocity of the earth, Ω, and the much more rapid oscillatory motion of the pendulum, several simplifications may be made. One approximation is to assume that the centrifugal force, which is quadratic in Ω, is negligible. Then terms containing Ω^2 may be ignored. In this approximation the equations of motion simplify to

$$\begin{aligned}\frac{d^2x}{dt^2} &= -\omega^2 x + 2s\Omega\frac{dy}{dt} \\ \frac{d^2y}{dt^2} &= -\omega^2 y - 2s\Omega\frac{dx}{dt} - 2c\Omega\frac{dz}{dt} \\ \frac{d^2z}{dt^2} &= -\omega^2 z + 2c\Omega\frac{dy}{dt}.\end{aligned} \tag{4.11}$$

The equations are still coupled and therefore the oscillatory motion of the pendulum will still be complex. A further approximation resides in the fact that the pendulum is typically quite long compared to the z component of displacement and therefore z, dz/dt, and d^2z/dt^2, may also be safely set equal to zero. This new approximation results in the equations

$$\begin{aligned}\frac{d^2x}{dt^2} &= -\omega^2 x + 2s\Omega\frac{dy}{dt} \\ \frac{d^2y}{dt^2} &= -\omega^2 y - 2s\Omega\frac{dx}{dt},\end{aligned} \tag{4.12}$$

which, although still coupled, are now readily solvable with analytic techniques.

One convenient approach to the solution of these coupled differential equations is to let the x and y components be the real and imaginary parts, respectively, of a complex variable, $Z = x + iy$. The two equations are then combined into the single equation

$$\frac{d^2Z}{dt^2} + 2i\Omega'\frac{dZ}{dt} + \omega^2 Z = 0, \tag{4.13}$$

where $\Omega' = \Omega \sin\beta$. This linear homogeneous equation may be solved with a standard trial solution of the form $Z = Z_0 e^{\lambda t}$. Substitution of the trial solution into Eq. (4.13) yields a quadratic equation in the unknown λ

$$\lambda^2 + 2i\lambda\Omega' + \omega^2 = 0. \tag{4.14}$$

Again we emphasize that the earth's rotation rate is much slower than the pendulum's oscillation frequency, $\Omega \ll \omega$, and therefore the solution to the quadratic equation in λ may be approximated by

$$\lambda = -i(\Omega' \pm \omega), \tag{4.15}$$

which, in turn leads to the solution,

$$Z = e^{-i\Omega' t}\left(Ae^{i\omega t} + Be^{-i\omega t}\right). \tag{4.16}$$

A and B are complex numbers that depend on the initial position and velocity of the pendulum. Clearly there are two periodic motions in this result; one has a period corresponding to a single oscillation of the pendulum

$$T = \frac{2\pi}{\omega} = 2\pi\sqrt{\frac{L}{g}}, \tag{4.17}$$

whereas the other motion corresponds to complete rotation of the plane of oscillation and has a very long period,

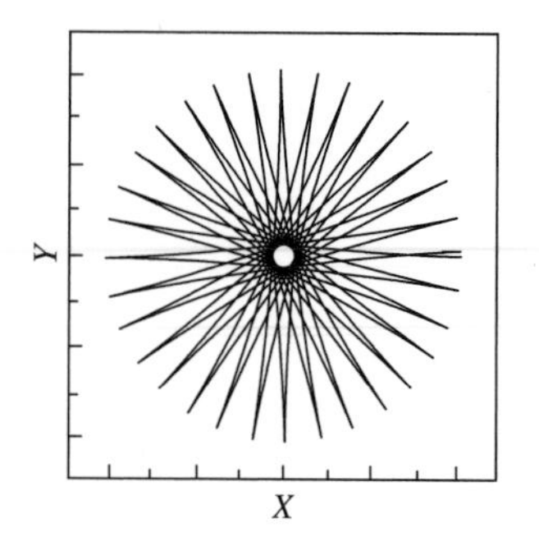

Fig. 4.9
Computer simulation showing the veering of the pendulum for one half of the period τ. The rotation of the earth is artificially increased in order to clearly show the effect.

$$\tau = \frac{2\pi}{\Omega \sin l} = \frac{24 \text{ hours}}{\sin l} \tag{4.18}$$

as predicted above. Therefore we see a mathematical basis for the two desired motions.[8] Figure 4.9 shows an exaggerated view of the Foucault veering.

4.4.1 Starting the pendulum

Within the above approximations we have established the existence of the two primary motions. But other effects may occur unless care is taken in starting the motion. Consider the motion of the pendulum without the rotation of the earth. Beginning with Eq. (4.16), we delete the Coriolis term and see that in a Newtonian frame

$$\begin{aligned} Z &= Ae^{i\omega t} + Be^{-i\omega t} = x + iy \\ &= (A + B)\cos\omega t + i(A - B)\sin\omega t. \end{aligned} \tag{4.19}$$

This equation provides the parametric equations for the following trajectory in the xy plane:

$$\frac{x^2}{(A+B)^2} + \frac{y^2}{(A-B)^2} = 1. \tag{4.20}$$

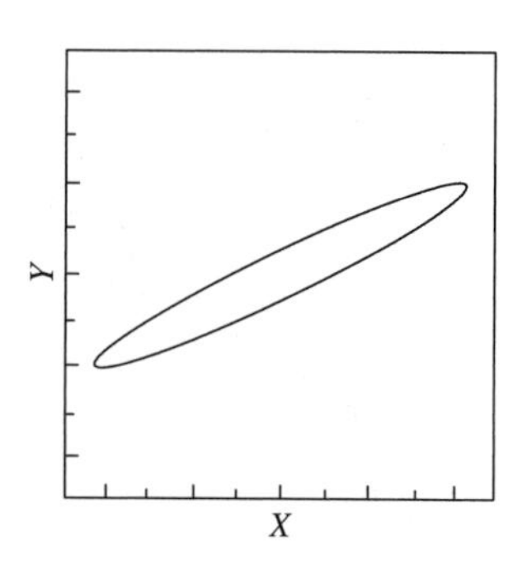

Fig. 4.10
Path of the pendulum bob in an inertial frame if the pendulum is started in a nonspecial way.

The path is clearly elliptical except when $(A - B) = 0$, in which case, the motion simplifies to a straight line. Thus, for all but a *special* set of initial conditions, the pendulum's motion is elliptical as shown in Fig. 4.10.

Let us explore this elliptical motion further. By differentiation the velocity vector becomes

$$\dot{Z} = -\omega(A + B)\sin\omega t + i\omega(A - B)\cos\omega t. \tag{4.21}$$

Because there is no preferred direction, general initial conditions impose directionality according to

$$\begin{aligned} x(0) &= (A + B), \quad y(0) = 0, \\ \dot{x}(0) &= 0, \quad \dot{y}(0) = \omega(A - B), \end{aligned} \tag{4.22}$$

[8] One can also get some sense of the $\Omega \sin\beta$ factor by noting that it is the component of the stars' motion along the horizon. As previously noted, the swing of the pendulum follows the fixed stars rather than the earth's motion (Tobin and Pippard 1994).

and therefore the direction in which the pendulum is first displaced becomes the x-direction. Similarly, the elliptical orbit occurs because there is an initial component of velocity in the y-direction equal to $\omega(A - B)$. Since, a general elliptical motion will interfere with the Foucault effect, it is important to start the pendulum without imparting any velocity to the pendulum that is perpendicular to the initial displacement. This smooth start is equivalent to setting A equal to B. This constraint lead to Foucault's original method. He secured the initial displacement of the pendulum by a string or thread attached to a fixed position. The zero velocity initial condition ($A = B$) is achieved by burning through the string so that the pendulum smoothly starts its motion in the direction of its displacement. The pendulum will then, to a large extent, oscillate in a plane.

Yet even with this precaution, there may still be a slight elliptical character to the pendulum's motion because, of course, the pendulum is really in the earth's rotating frame, as opposed to the inertial frame. The earth's rotation actually imparts a small perpendicular component of velocity. But this component can usually be ignored and the effect is typically less than 0.1% of the Foucault precession (Pippard 1991).

Asymmetry

Up to this point we have assumed that the pendulum behaved equally no matter what its plane of oscillation. That is, the x- and y-directions of pendular motion are identical and the x-direction is defined only by the initial displacement. However, unless great care is taken, there is some small measure of asymmetry. For example, the pendulum may be suspended from a beam that has more "give" in one direction or the other. Air currents around the pendulum such as those found in public museums may also cause asymmetry. Foucault pendulums are often located in the stairwells of museums and, interestingly, currents from one-way traffic patterns of spectators going down the stairs, more than up the stairs, is a source of asymmetry. Open doors and windows can also cause problems (Pippard 1991). The net result is that $\omega_x \neq \omega_y$.

We can think about the situation as follows. If the pendulum is "perfect"—totally symmetric—then there are two possible identical fundamental motions of the pendulum in the two mutually perpendicular directions. In this case the system is said to be (in mathematical terms) *degenerate*, and all motions of the pendulum are just linear combinations of these two fundamental motions. Neglecting for the moment the earth's rotation, if the pendulum is started with some particular combination of the two fundamental motions then it will maintain that particular state. On the other hand, if the pendulum is not "perfect" then the two fundamental states are not the same and the system is said to be *nondegenerate*. Any initial motion is still made up of a linear combination but now the motion will not remain in the initial state. The difference in frequencies leads to a periodic transfer of energy back and forth between the two fundamental

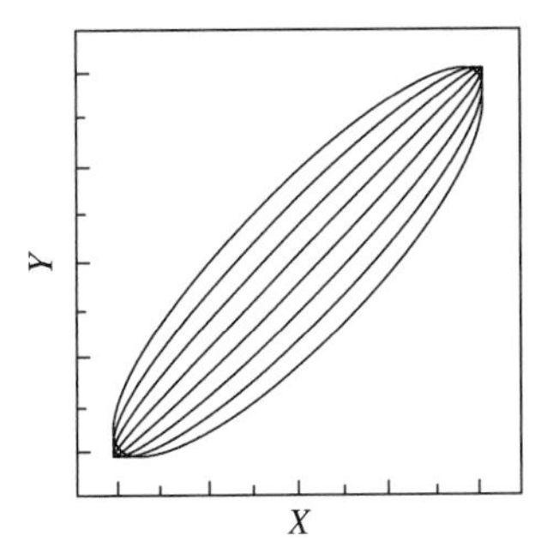

Fig. 4.11
Effect of asymmetry on the pendulum in an inertial frame. The inequality of oscillation frequencies causes cyclical flattening and thinning of the orbit. A few oscillations are shown.

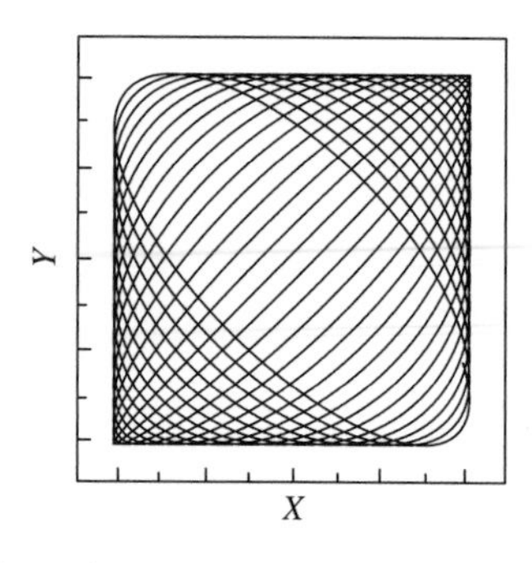

Fig. 4.12
The elliptical oscillation orbit caused by assymetry gradually fattens to become circular. Then the orbit becomes thinner again, but in the direction perpendicular to the initial oscillation. The pattern is repeated.

Fig. 4.13
Photograph of H. Kamerlingh Onnes.

states at a rate that depends upon the frequency difference, or beat frequency, $\omega_x - \omega_y$. For example, if the pendulum initially travels equally in the x- and y-directions it will soon display elliptical motion, then circular, then elliptical, and so forth in the same pattern. This is the motion illustrated by the Lissajou figures of two mutually perpendicular oscillations at the slightly differing frequencies, ω_x and ω_y. Planar oscillations are only found in two preferred orientations (the so-called principal axes of the structure) and even these motions are susceptible to changing air currents. The plane of oscillation will have rotated by 90° (Figs. 4.11 and 4.12 illustrate the effect).

The time for transition between the two states is about $t = \pi/(|\omega_x - \omega_y|)$. For long pendulums with very low frequencies this effect is less pronounced as the frequency differences can be made relatively small. However, it is a serious problem for short pendulums and, typically, the asymmetry effect overshadows the Foucault effect in a relatively short time. The first mathematical analysis of this asymmetry effect was done by Heike Kamerlingh Onnes (1853–1926)—the future discoverer of superconductivity—as part of his 1879 Ph.D. dissertation, a quarter century after Foucault's original demonstration of the earth's rotation (Fig. 4.13). The problem was suggested to Kamerlingh Onnes by a former teacher, Gustav Kirchoff, whose laws of electrical circuitry are well known to undergraduate physics students (Schulz-DuBois 1970). Kamerlingh Onnes not only did the analysis, but he also used his theoretical work as the basis for construction of a Foucault pendulum with a mass distribution that corrected the asymmetry. In essence, he equalized the moments of inertia of the pendulum in the two mutually perpendicular directions, thereby equalizing the resonant frequencies. Figure 4.14 shows an exploded view of the pivoting arrangement for the pendulum. Kamerlingh Onnes also placed the whole apparatus in a gas-tight container to negate the effects of air drafts, and developed a special optical system to observe the pendulum as shown in Fig. 4.15.

The photograph shown in Fig. 4.16 was taken after the equipment was partially reassembled. The enclosure piece is at the side in order that the pendulum may be seen. Unlike the typical very long Foucault pendulum, this one is quite short in order that Kamerlingh Onnes could effectively study the oscillations changes provided by his adjustments to the pivoting arrangement. Kamerlingh Onnes' careful experimental work in this context was a precursor to his later and more famous experimental contribution to low temperature physics.[9] (More will be said about this latter contribution in Chapter 9.)

Kamerlingh Onnes' pendulum was actually a rigid rod and therefore a full analysis would consider the differences in the moments of inertia in the two perpendicular directions. However, following Schulz-DuBois (1970)

[9] At this writing Kamerlingh Onnes apparatus is in storage at the Boerhaave Museum in Leiden, near the university where Kamerlingh Onnes made his famous discoveries in low temperature physics.

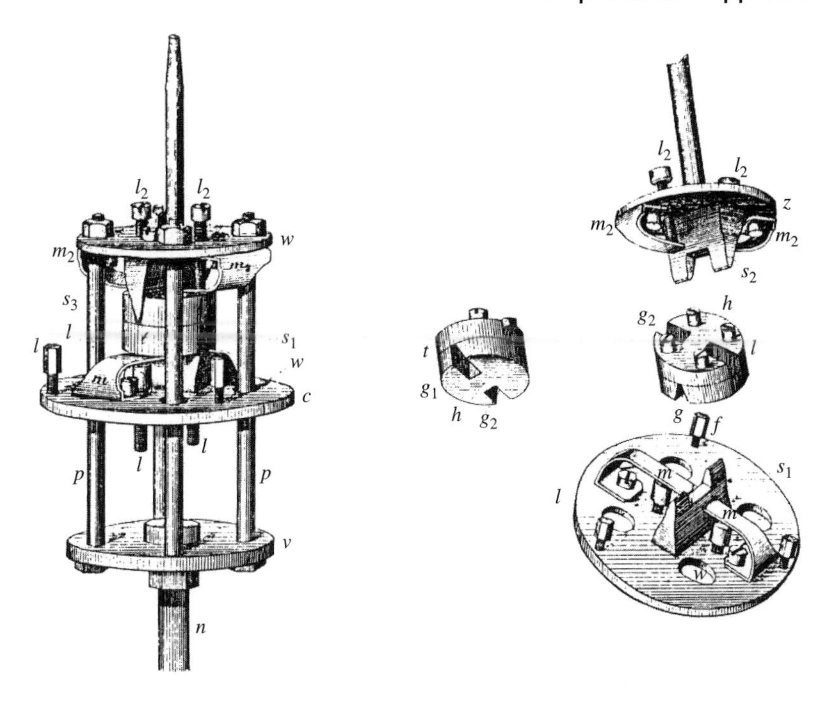

Fig. 4.14
Several views of the support for the pendulum. Adjustments could be made in the pivoting to exaggerate or minimize asymmetry.

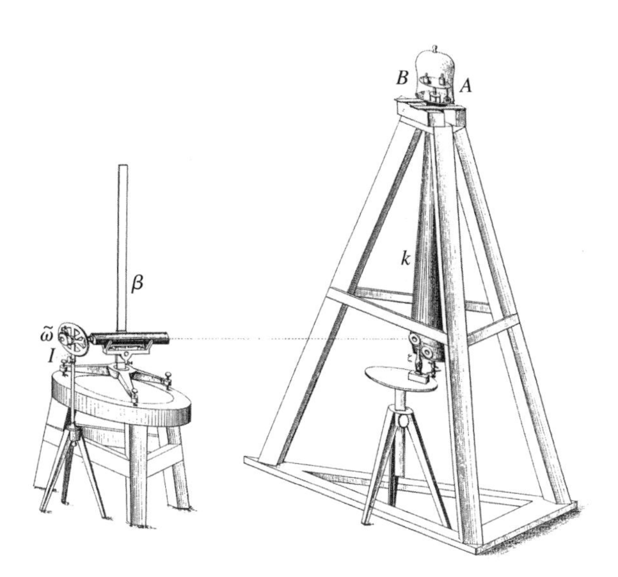

Fig. 4.15
A drawing from Kamerlingh Onnes' thesis depicting the arrangement of the complete apparatus. The apparatus was less than 2 m in height.

we use, for our discussion, the simpler model of a pendulum with a point mass bob, and introduce the asymmetry by having different effective lengths, L_x and L_y for the two directions. In this way the x and y components of the forces in the harmonic approximation are modified to become

$$F_x \simeq -mgx/L_x \text{ and } F_y \simeq -mgy/L_y. \tag{4.23}$$

Fig. 4.16
A recent photograph of the Kamerlingh Onnes pendulum, provided by the Boerhaave Museum, Leiden. (Reproduced by permission of the Boerhaave Museum.)

This approach leads to modification of the frequencies in Eq. (4.12);

$$\begin{aligned}\frac{d^2x}{dt^2} - 2\Omega'\frac{dy}{dt} + \omega_x^2 x = 0\\ \frac{d^2y}{dt^2} + 2\Omega'\frac{dx}{dt} + \omega_y^2 y = 0,\end{aligned} \tag{4.24}$$

where $\Omega' = \Omega\sin\beta$ as before. The differences in the frequencies prevent us from using the complex notation. Therefore we modify our approach and substitute the following trial solutions into the equations of motion:

$$\begin{aligned}x = a\cos[(\omega + \Delta)t + \varphi]\\ y = b\sin[(\omega + \Delta)t + \varphi],\end{aligned} \tag{4.25}$$

where $\omega = \omega_x - \delta = \omega_y + \delta$ is an average of ω_x and ω_y. (Note the assumption that the components of the resultant motion will have identical frequencies, $\omega + \Delta$.) We can make some simplification by noting that the frequency difference between the two directions is small and that the pendulum's frequency is much greater than the earth's rotation frequency; that is, Ω', Δ, $\delta \ll \omega$. Simplifications involving these parameters and substitution of the trial solutions in the new equations of motion leads to the following set of homogeneous equations in the variables a and b:

$$\begin{aligned}a(\Delta - \delta) + b\Omega' = 0\\ a\Omega' + b(\Delta + \delta) = 0.\end{aligned} \tag{4.26}$$

Nontrivial solutions exist only when the determinant of the coefficients is zero. This constraint leads to two possibilities (eigenvalues) for Δ, that is $\Delta = \pm\sqrt{\Omega'^2 + \delta^2}$. We then substitute each of these eigenvalues back into the homogeneous algebraic equations and solve for a and b. Because the equations are homogeneous only the ratio a/b is specified and there is some freedom in choosing the magnitude of a and b. We normalize so that $a^2 + b^2 = 1$. Each eigenvalue of Δ yields its own pair of x and y functions. These form two-dimensional vectors (eigenvectors). With much algebraic manipulation, it can shown that the two eigenvectors or eigenfunctions are

$$\Psi_1 = \begin{pmatrix} x \\ y \end{pmatrix} = \begin{pmatrix} \frac{-\Omega'}{\sqrt{2\Delta(\Delta-\delta)}}\cos\left[\left(\omega + \sqrt{\delta^2 + \Omega'^2}\right)t + \varphi_1\right] \\ \frac{\Omega'}{\sqrt{2\Delta(\Delta+\delta)}}\sin\left[\left(\omega + \sqrt{\delta^2 + \Omega'^2}\right)t + \varphi_1\right] \end{pmatrix} \tag{4.27}$$

and

$$\Psi_2 = \begin{pmatrix} x \\ y \end{pmatrix} = \begin{pmatrix} \frac{\Omega'}{\sqrt{2\Delta(\Delta+\delta)}}\cos\left[\left(\omega - \sqrt{\delta^2 + \Omega'^2}\right)t + \varphi_2\right] \\ \frac{\Omega'}{\sqrt{2\Delta(\Delta-\delta)}}\sin\left[\left(\omega - \sqrt{\delta^2 + \Omega'^2}\right)t + \varphi_2\right] \end{pmatrix}. \tag{4.28}$$

These eigenfunctions describe elliptical motion for the pendulum bob. The rotation of the ellipse for Ψ_1 is opposite to the rotation of the ellipse for Ψ_2.

As a check on these results, let us examine the eigenfunctions for the simpler cases when the lengths and therefore the frequencies are equal; that is, $\omega_x = \omega_y = \omega$, then $\delta = 0$. Then Eqs. (4.27) and (4.28) take the simple form

$$\Psi_1 = \begin{pmatrix} x \\ y \end{pmatrix} = \begin{pmatrix} \frac{-1}{\sqrt{2}}\cos[(\omega + \Omega')t + \varphi_1] \\ \frac{1}{\sqrt{2}}\sin[(\omega + \Omega')t + \varphi_1] \end{pmatrix}$$
$$\Psi_2 = \begin{pmatrix} x \\ y \end{pmatrix} = \begin{pmatrix} \frac{1}{\sqrt{2}}\cos[(\omega - \Omega')t + \varphi_2] \\ \frac{1}{\sqrt{2}}\sin[(\omega - \Omega')t + \varphi_2] \end{pmatrix} \quad (4.29)$$

and the eigenfunctions are circles that rotate in the opposite sense with slightly different frequencies due to the earth's rotation. Simplifying even further we might let the pendulum be symmetric *and* let the earth's rotation effect $\Omega' = \Omega \sin\beta$ be zero (as it is at the equator). Then both eigenfrequencies are equal and the system consists of two oppositely circulating orbits. An equal linear combination of the two modes would then have the pendulum oscillate in a straight line.

Nevertheless, for parameter values that approximate those used by Foucault and even a one percent different in the frequencies, the effect of the asymmetry is still quite marked. Care must be taken to minimize this source of asymmetry in real pendulums.

Finally, an examination of the comprehensive equations of motion, Eq. (4.10), shows that there is a slight asymmetry between the equations for the accelerations in the x- and y-directions in the terms involving Ω^2, which are contributions to the centrifugal effect. For the x-direction, there is the usual harmonic term that gives rise to the main oscillatory motion, but there is also another term $x\Omega^2 \sin^2\beta$, whereas for the y-direction, the extra term has the form $y\Omega^2 \cos^2\beta$. Only when the latitude is $\pi/4$ would there be complete symmetry. However, the effect of these centrifugal terms is small and may be ignored.

Anharmonicity

The usual harmonic approximation is only true if the amplitude of the pendulum's oscillation is quite small. Equation (4.10) is the linearization approximation. Yet for any finite amplitude the period is amplitude dependent. To a first approximation beyond the linear one, the period is

$$T \approx T_0 \left[1 + \frac{\alpha^2}{16}\right],$$

where α is the angular amplitude. The Foucault veering causes the formation of a very narrow ellipse whose period of oscillation is longer in the direction of the major axis (length $2a$) than it is in the direction of the minor axis (length $2b$). The difference in periods of the perpendicular motions—caused by the anharmonic effect—leads to a precession of the ellipse, in the direction that the pendulum is traveling. The rate of this precession is the rate of rotation of the tip (apse) of the ellipse and is given by the so-called clockmaker's formula (Synge and Griffith 1959) alluded to by Airy in 1851.

$$\text{Rate} = \frac{3(\text{area of the ellipse})}{4L^2 T}. \tag{4.30}$$

How much of a problem is this extra precession caused by anharmonicity? The effect can be readily approximated. Let the initial displacement be a. In one oscillation of time T, the veering of the Foucault pendulum creates a semi-minor axis of

$$b \approx a\frac{T}{\tau}, \tag{4.31}$$

where $\tau = 2\pi/\Omega\sin\beta$ is the Foucault period. Since the area of an ellipse is πab, then the rate of veering due to anharmonic effects is

$$\text{Rate} = \frac{3\pi a^2 T}{4L^2 T\tau} = \frac{3\Omega\sin\beta}{8}\left[\frac{a}{L}\right]^2. \tag{4.32}$$

The interesting thing about this formula is the ratio a^2/L^2. If the pendulum is short, the ellipse rotates rapidly, whereas if the pendulum is quite long, the effect is quite small. Thus there are two motions that involve rotation of the plane of oscillation of the pendulum in earth's frame of reference; the Foucault effect and this new "area" effect. As well as reducing this effect with a long pendulum, the area effect may sometimes be distinguished from the Foucault effect. The area effect is such that the rotation of the apse occurs in the direction of the motion and therefore depends upon the conditions under which the pendulum motion is initiated. On the other hand, the Foucault effect is always clockwise in the northern hemisphere, and counterclockwise in the southern hemisphere.

Now that we are aware of some of the distinctions between the "spurious" and Foucault effects, let us ignore the spurious effects and return to the original complex expression of the motion, Eq. (4.16), as seen from the earth's frame of reference, and develop a solution for the x and y components. This is the motion in the earth's frame without the complication of asymmetry or anharmonicity. In the earth's frame of reference, the initial conditions obtained with the burning thread method are

$$\begin{aligned} Z(0) &= a = A + B \\ Z'(0) &= 0 = (\omega - \Omega')A - (\omega + \Omega')B. \end{aligned} \tag{4.33}$$

In terms of the initial displacement X, the unknown constants become

$$A = \frac{(\omega + \Omega')a}{2\omega} \quad \text{and} \quad B = \frac{(\omega - \Omega')a}{2\omega}. \tag{4.34}$$

These values can be substituted into the solution given by Eq. (4.16) and, after some trigonometric manipulations, yield the x and y components

$$\begin{aligned} y &= a\left[-\sin\Omega' t\cos\omega t + \frac{\Omega'}{\omega}\cos\Omega' t\sin\omega t\right] \\ x &= a\left[\cos\Omega' t\cos\omega t + \frac{\Omega'}{\omega}\sin\Omega' t\sin\omega t\right]. \end{aligned} \tag{4.35}$$

These expressions seem complicated, but we again note that the ratio Ω'/ω is quite small, and therefore the components can be decomposed into primary and secondary terms. The primary, dominant earthbound motion is given by

$$\begin{aligned} x &= a\cos\Omega' t\cos\omega t \\ y &= -a\sin\Omega' t\cos\omega t, \end{aligned} \tag{4.36}$$

which is planar oscillatory motion at frequency ω modulated by the Foucault rotation of the plane of oscillation as shown in Fig. 4.9. Of course, the terms in Ω'/ω that we ignored do provide a slight complication, but the burning string method is still a good way to start the pendulum.

For very long pendulums the spurious effects are small, and the main concern is the dissipation of energy as the pendulum gradually losses amplitude. However, for short pendulums the spurious effects are not negligible. After the following literary *divertissement*, we note some ways that builders of Foucault pendulums have overcome the complicating effects of these limitations and thereby produced workable pendulums that are much smaller than Foucault's original giant creation. For the moment, let us consider another literary dimension of the Foucault pendulum.

4.5 A darker side

In 1981, novelist Catherine Aird, wrote a mystery entitled *His Burial too* (Aird 1981) that uses the Foucault pendulum as a murder weapon.[10] In Aird's story a certain Richard Tindall, owner of an engineering company and father of the requisite—for detective stories of a certain vintage—attractive daughter, has been murdered. Various characters are developed who have greater or lesser motives for promoting Tindall's demise and thus a range of suspects is created. Yet their alibis seem airtight and certain aspects of the execution of the crime remain a mystery for much of the story. We quote.[11]

At 11:30 PM in the belltower of the old Saxon church at Randall's bridge, a huge statue toppled and smashed. Heavy blocks of broken marble now lay up against the door barring entry or exit. The solitary window was too narrow for a man to pass through; the belfry high above led only to the steep roof which rose beyond the reach of any ladder. When Detective Inspector C. D. Sloan put his shoulder to the door (blocked by the broken statue) it opened only a crack. Through it he could clearly see the room was empty—except for the bells, the debris of shattered marble... and the protruding arm of a dead man. How could a murderer have escaped from this sealed tower? Sloan's only clues: a spent match, a black thread (burned at the end), a pendulum hanging in the center of the tower. And an emerald earring.

[10] This novel was discussed by Crane (1995) in one of his many papers on the Foucault pendulum. Prof. Crane of the University of Michigan is a longtime enthusiast and innovator of the Foucault pendulum.

[11] Excerpts from the back cover and pages 185 and 186 of Aird (1981).

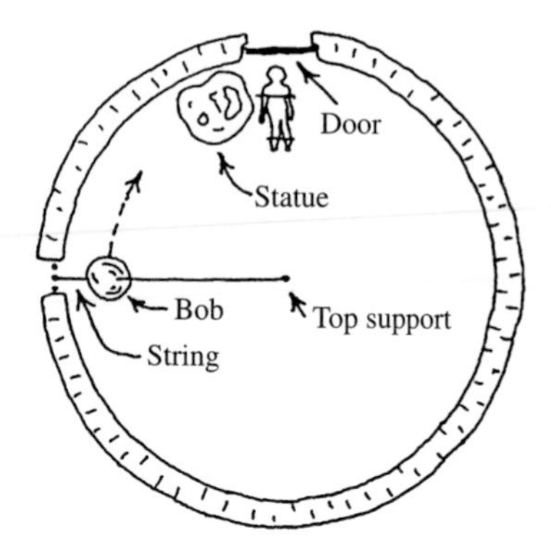

Fig. 4.17
Schematic of the crime scene in Aird's novel. (Reprinted with permission from Crane (1990, p. 265). ©1990, American Association of Physics Teachers.)

Later in the course of the investigation the fog of ignorance begins to dissipate.

> From Sloan's superior: "What's all this business about pendulums, then?"
>
> Sloan shook his head. "I don't think the pendulum would do the trick, Sir. The sculpture was off-line from the window."
>
> Cold trickled down Sloan's spine. Then his head came up.... There was such a thing as a pendulum that turned. He'd seen it. In a museum... It was coming back. What the pendulum needed was a smooth start, and half an hour later it was off course...
>
> The smooth start had been important. He remembered watching while a man from the museum started it by burning the anchoring string in two with a match.

And so the investigation proceeds to its now inevitable conclusion. The pendulum oscillation was initiated some 5 hours before the actual murder. And therefore the criminal had a chance to develop an alibi for his whereabouts at the time of the killing. During the intervening 5 hours the plane of oscillation rotated about 60° until it hit a massive statue that was delicately balanced to fall on the previously tied up Mr. Tindall. When the statue fell, it killed the victim and sealed the door to the chamber. The other clues are explainable once the basic physics of the Foucault pendulum is understood. The general positioning of the tools of the murder is illustrated in Fig. 4.17 (Crane 1990). Two features of this story seem slightly implausible. First, it is very unlikely that the pendulum used in this story could have maintained its oscillations for 5 hours. The second difficulty lies with the fact that the trigger for the destabilization of the statue was the very small change in the plane of oscillation. The murderous statue must have been balanced very precariously indeed.

4.6 Toward a better Foucault pendulum

We saw that the effects that mask the Foucault effect can be minimized by using a very long pendulum. Foucault's dramatic 1851 presentation in the Pantheon worked because his pendulum was long—67 m—and because he used a massive bob of 28 kg to give the pendulum plenty of inertia so that it would oscillate long enough to illustrate the desired effect. Furthermore, asymmetry in long pendulums in not especially problematic. But long pendulums can have a safety problem. A month or two after Foucault swung his pendulum in the Pantheon, the suspending wire broke at the point of suspension. In this circumstance the tension in the wire causes the wire to act like a whip on any nearby spectators. Therefore Foucault designed a "parachute" that would support the wire just below the suspension in the event of breakage. The parachute is a kind of cylinder that is clamped to wire and, in the event of breakage, falls a short distance to a supporting ring. Foucault's original parachute, together with the suspension structure, is stored in the Paris Observatory.

For short pendulums, all of the previously described effects are significant.

Modern solutions to the anharmonic and asymmetry effects—both of which contribute to spurious precession—exhibit pragmatic creativity. In 1931, the asymmetry problem was partially solved by placing a ring, now known as the Charron ring, just below the pivot point such that it would surround the suspending wire (Charron 1931). The diameter of the ring was made slightly less than the full swing of the wire, and therefore the wire would contact the ring at the extremes of its swing. This delicate impact would arrest the spurious precessional motion of the pendulum. However, as is well known, the ring only succeeds partially in that it does introduce a faster dissipation of the pendulum's energy. Thus a source of energy is required to keep the pendulum in motion.

There have been various solutions to the energy problem. We noted that Foucault himself built an electromagnetic drive for the 1855 pendulum.

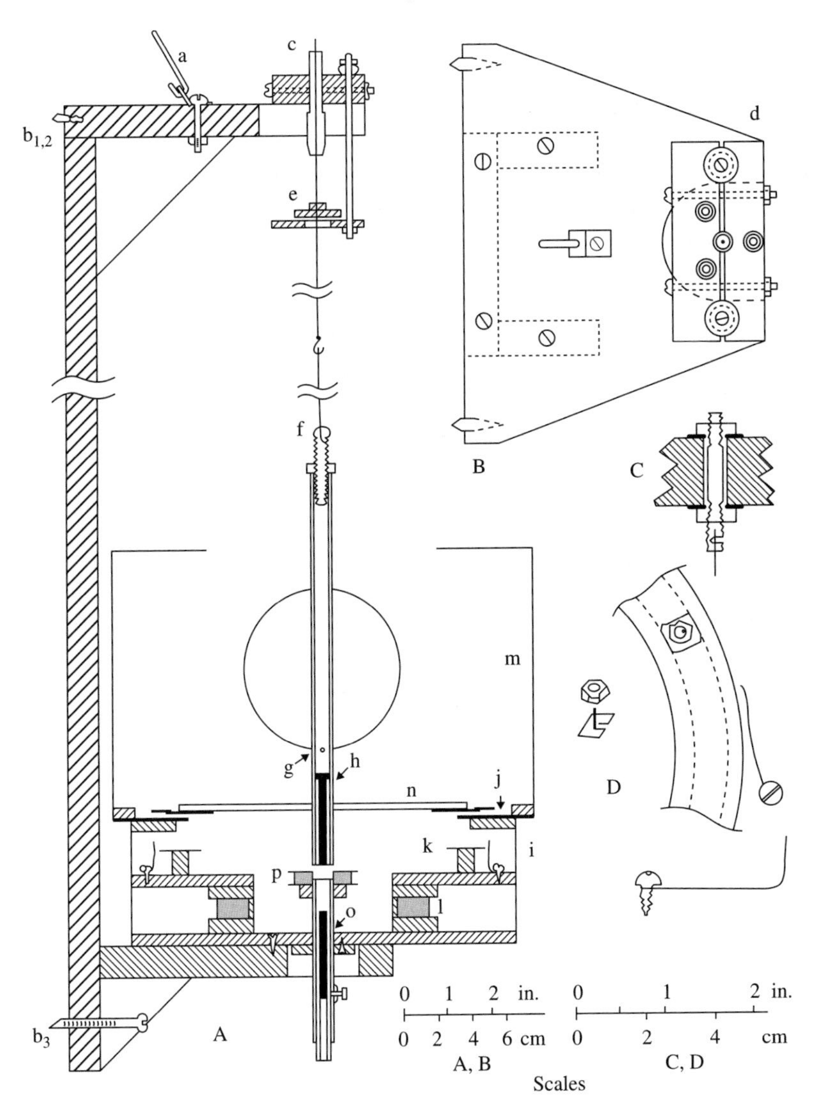

Fig. 4.18
Schematic diagram of Crane's short Foucault pendulum used in building a clock. (Reprinted from (Crane 1995, p. 35). ©1995, American Association of Physics Teachers.)

Fig. 4.19
Photograph by one of the authors (JAB) of the bob of the short Foucault pendulum at the University of Guelph.

A more sophisticated and modern design was used by Crane (1981). He placed a small permanent magnet in the bob of the pendulum that interacted with a combination of a permanent and time varying electromagnet fixed in the rigid structure below the pendulum. The electromagnet gave a pulse that changed orientation every quarter of an oscillation. In this way the magnet in the bob received a push outward when the bob is in the middle and an inward pull when the bob is at its extreme displacement. With this setup Crane was able to eliminate spurious precession and, at the same time, give a periodic push to the pendulum bob, thereby ensuring continued oscillations. Crane built a viable short pendulum with a length of only 70 cm. Several years later he described the use of this pendulum as the timing mechanism for a clock. Figure 4.18, taken from his 1995 publication (Crane 1995), is a detailed schematic of his pendulum. A Foucault pendulum made to this design is on view at the University of Guelph.[12] The lower portion of the pendulum is shown in Fig. 4.19.

Pippard has shown that parametric pumping of the pendulum, such as is done with O Botafumeiro, helps to eliminate the tendency of the pendulum toward elliptical motion and, of course, keeps the pendulum in motion through the periodic infusion of energy. The pendulum installed in the British Science museum in 1988 was built with such a mechanism according to a design by Pippard (1988). In the course of developing his Foucault pendulum, Pippard carried out a detailed study of the improvements that are provided by parametric pumping. Part of the reason for such an analysis was to discover if this pendulum, or one like it, could be used to confirm a prediction of General Relativity theory, called the Lense–Thirring effect. In general relativity, the gravitational field is specified by a four by four matrix of terms. Some of these terms have not been adequately measured near the earth's surface; in particular, terms corresponding to that part of the field known as the gravitomagnetic field. Crudely put, the gravitomagnetic field arises because the earth's rotation "drags" the otherwise inertial frame of reference that exists near the earth's surface. This dragging is only 0.22" per year at the poles. Theoretically, it might be possible to measure this effect by careful observation of an extremely well constructed Foucault pendulum. But it is necessary that the pendulum be operated at one of the earth's poles as the effect is diminished when the measurement is taken at nonpolar latitudes. In 1984 Braginsky et al. (1984) proposed that construction of such a pendulum at the South Pole might do the trick. However, Pippard's analysis (Pippard 1988) suggested that present Foucault pendulum technology was not yet capable of such a test. Nevertheless, the possibility that a pendulum could find yet another use in testing General Relativity is intriguing.

A crude Foucault pendulum was constructed and tested at the South Pole by three adventurous experimenters, Michael Town, John Bird, and Alan Baker (Baker et al. 2001) but the experiment was of modest precision and only seemed to confirm the conventional predictions for a Foucault pendulum (Johnson 2002). The pendulum was erected in a six-story staircase

[12] www.physics.uoguelph.ca/foucaulti2.html (1999).

of a new station that was under construction. Conditions were challenging; the altitude was about 11,000 ft (atmospheric pressure of 660 m bar) and the temperature in the unheated staircase was −90 degrees Fahrenheit. The pendulum had a length of 33 m and a 25 kg bob. Interestingly, it is against the Antarctica Treaty to have an open flame at the South Pole and therefore the authors could not use the burnt thread technique. Instead, they became adept at dropping the bob so that it would oscillate in a plane. Installing even a crude pendulum in these conditions was challenging. But once they overcame the obstacles, these researchers did confirm a period of 24 h ± 50 min as the rotation period of the plane (see Fig. 4.20).

Fig. 4.20
Photograph of two of the South Pole group with their Foucault pendulum bob. (Courtesy of John Bird.)

4.7 A final note

Because the Foucault pendulum provides a dramatic and obvious demonstration of the earth's rotation, it is found in many museums, schools, and other public buildings. A representative list is given in Table. 4.1

Some of the pendulums are very long and have massive bobs, a combination that allows them to partially sustain their motion against the inevitable loss of energy due to friction in the bearing, air resistance, and so forth. Others pendulums, often with smaller dimensions, use some sort of drive mechanism to supply energy that compensates for the friction losses. However, the indefinitely sustained oscillations can sometimes reinforce misconceptions.

Donald Ivey, a popular lecturer at the University of Toronto, once told a story regarding such a misconception and the Foucault pendulum. During a visit to New York city, Ivey joined a tour of the UN building. At the site of its Foucault pendulum there was a fairly large crowd observing the slow back and forth motion of the pendulum and hoping to discern the veering motion of the plane of oscillation. At some point, a woman within earshot of Ivey stated emphatically that the phenomenon of the ongoing oscillations was an example of perpetual motion. Ivey began to squirm. Of course, the assumption of perpetual motion for a real pendulum

Table 4.1 Random selection from the many extant Foucault pendulums

Location	Length (m)	Mass (kg)
Pantheon, Paris	67	47 (1995 bob)
UN building, NYC	23	90
Franklin Institute, Philadelphia	26	410
Univ. of Guelph, Canada	0.83	4.5
Univ. of Maryland	14	28
Yale University	11	12
Griffith Observatory, Los Angeles	12	100
Ryerson Library, Michigan	22	14
Ottawa High School, Michigan	9	14
Univ. of Wisconsin, Madison	14.5	30
Morrison Planeterium	9	87
Science Museum, London	20	9

Fig. 4.21
Engraving from L'Illustration, Paris, 1851.

contravenes the physical principles of both conservation of energy and the second law of thermodynamics. Eventually Ivey could not contain himself. He turned to the woman and quietly explained that there was no such thing as perpetual motion. She flatly and noisily disagreed with him. Ivey then, and perhaps foolishly, revealed himself to be an expert, namely a professor of physics. At this point, the woman loudly shared with the crowd her low opinion of experts in general and university professors in particular. In the face of this verbal assault and possible public ridicule, Professor Ivey decided to retreat from the field of "intellectual" combat.

The degree of "disconnect" between Ivey, the scientist, and the unknown tourist is not unique. Figure 4.21 illustrates corresponding attitudes in the nineteenth century. Quoting from the original caption:

> This tableau, which appeared soon after . . . (the original 1851 demonstration of the Foucault pendulum) . . . reveals some of the typically French attitudes toward science and nature. The "Philosopher" (or scientist) pays no attention to the beauty of the scene, seeing neither the vegetation nor the pretty girl (on) the swing. What he observes is abstracted to a geometrical problem, free from passion and life.
>
> The girl's response to his mathematical analysis is "True, but who cares?" To the philosopher, science *is* measurement; to the young girl, it is irrelevant because it omits life.[13]

[13] See (Williams 1978, p. 6).

Yet whatever the perception, the harvest of scientific knowledge, even from the seemingly unremarkable pendulum, is rich with unexpected and exciting insight, as we will continue to see in our next chapter.[14]

4.8 Exercises

1. Compare the Foucault effect for a long pendulum of length L_L with that for a short pendulum of length L_S. Assume that both pendulums oscillate with the same small angular amplitude, θ_0. Show that the ratio of the displacements $\Delta s_L/\Delta s_S = (L_L/L_S)^{3/2}$ where Δs is the amount by which the bob of the pendulum is displaced relative to the earth after one oscillation. (Assume that $\theta(0) = \theta_0$.) Hint: Note that there are two effects to include; (a) the difference in the linear displacements of each of the pendulums, and (b) the difference in the periods of the two pendulums.
2. If Foucault's long pendulum ($L = 68$ m) were located at the north pole, and had an angular amplitude of 3°, what would be its displacement relative to the earth after one oscillation.
3. The situation of the ping-pong game in the airplane may be analyzed by the so-called Galilean transformations. These transformations are equations that describe events or vectors such as position, velocity, and acceleration, as seen from two frames of reference between which there is a relative constant velocity, **u**. If the frames are labelled 1 and 2 then the equations that describe quantities given in frame 2 relative to frame 1 are

$$\mathbf{r}_1 = \mathbf{r}_2 + \mathbf{u}t$$
$$\mathbf{v}_1 = \mathbf{v}_2 + \mathbf{u}$$
$$\mathbf{a}_1 = \mathbf{a}_2.$$

Now let us generalize the example of the accelerating elevator in the same way. Assume that frame 2 is accelerating with a constant acceleration relative to frame 1. Using the above set of equations as a model, write the transformation equations for the position, velocity, and acceleration vectors.
4. In the analysis of the Foucault pendulum we introduced Eq. (4.5) that gives the relationship between accelerations as observed in inertial frames and rotating frames. In this exercise we try to make this relation plausible by providing the main steps of the derivation.[15] Suppose that a coordinate system R rotates relative to coordinate system I with angular velocity Ω about a common z-axis. A point P with coordinates (x_I, y_I, z_I) is located in the inertial frame and the same point, with coordinates (x_R, y_R, z_R), is located in the rotating frame.

(a) Using trigonometry show that

$$x_I = x_R \cos\Omega t - y_R \sin\Omega t$$
$$y_I = x_R \sin\Omega t + y_R \cos\Omega t$$
$$z_I = z_R.$$

where Ωt is the angle of rotation about the z-axis.

(b) Differentiate these transformations twice with respect to time to obtain the following acceleration transformations. (Note that $\mathbf{\Omega}$ is a constant vector pointing along the common z-direction.)

[14] The Foucault pendulum seems to be a favorite subject for analysis from various mathematical points of view. Examples include work by (Hecht 1983 and Opat 1991).

[15] See (Kittel et al, 1962) pages 84 and following for further discussion.

$$\ddot{x}_I = (\ddot{x}_R \cos \Omega t - \ddot{y}_R \sin \Omega t) - 2\Omega(\dot{x}_R \sin \Omega t + \dot{y}_R \cos \Omega t)$$
$$-\Omega^2(x_R \cos \Omega t - y_R \sin \Omega t)$$
$$\ddot{y}_I = (\ddot{x}_R \sin \Omega t + \ddot{y}_R \cos \Omega t) + 2\Omega(\dot{x}_R \cos \Omega t - \dot{y}_R \sin \Omega t)$$
$$-\Omega^2(x_R \sin \Omega t + y_R \cos \Omega t).$$

(c) Now work backwards from Eq. (4.5) to arrive at the result in (b). Note that any vector in frame R can be projected onto either x_I or y_I. For example, the arbitrary vector g_R in the rotating frame with component (g_{Rx}, g_{Ry}) has an x_I component $(g_{Rx} \cos \Omega t - g_{Ry} \sin \Omega t)$ and a y_I component $(g_{Rx} \sin \Omega t + g_{Ry} \cos \Omega t)$. Find the various parts of the equation

$$\ddot{x}_I = (\ddot{\mathbf{r}}_I)_{x_I} + 2(\mathbf{\Omega} \times \dot{\mathbf{v}}_R)_{x_I} + [\mathbf{\Omega} \times (\mathbf{\Omega} \times \mathbf{r}_R)]_{x_I}$$

and similarly for $\ddot{y}_I$. If the cross products and differentiations are correct, the result should match those given in part (b).

5. (a) Suppose that you are hunting for bears at the North Pole with a rifle that fires a bullet whose average velocity is 600 m/s. By how much will the bullet appear to deviate from the target whose distance is 300 m? (b) Suppose now that you are in naval battle, not too far from the North Pole, firing a canon whose average velocity is 300 m/s. What is the deviation for a target that is 15 km away? If, during the Falkland Island battle, the gunners used the correction for the Northern Hemisphere instead of that for the southern hemisphere, by how much would they have deviated from their target? (c) Develop an expression for the deviation D as a function of the angular velocity ω, the distance to the target x, and the average velocity of the projectile, ν. Comment on the effect of higher projectile velocities.
6. What is the percentage correction to the acceleration due to gravity at the equator that may be simply attributed to centrifugal acceleration? (Assume that the earth is spherical. Its radius is 6.37×10^6 m.)
7. Use the Internet or the table in the text or some other source to find the location (latitude), and dimensions of a Foucault pendulum near you. Calculate the period of its oscillation, the period of its rotation, and the size of the angular deviation for each oscillation.
8. The first law of thermodynamics states that the energy of a closed system is a constant. The second law states that no process is possible for which the sole result is the absorption of heat from a reservoir and its complete conversion into work. Because of these two laws we often distinguish between two kinds of perpetual motion. Perpetual motion of the first kind is that which violates energy conservation, and perpetual motion of the second kind violates the second law of thermodynamics. If the Foucault pendulum at the UN building in New York was actually a perpetual motion machine, in what specific ways would it be violating the laws of thermodynamics. (In 1866 a certain W. Leaton proposed a perpetual motion machine that was an oscillating pendulum and used a clock-like mechanism to provide power. (http://www.chem.unsw.edu.au/staff/hibbert/perpetual/1stviol.html)
9. Show that for pendulums with a small angular displacement α, the period may be approximated by $T \approx 2\pi\sqrt{L/g}\left[1 + \alpha^2/16\right]$.

The torsion pendulum

5

The restoring forces responsible for harmonic motion in a conventional pendulum are provided by the local gravitational field. A grandfather clock, standing on the surface of the Moon, will run slow. A grandfather clock, floating in interplanetary space, will not "tick" at all. Gravity on the Moon is reduced; in free space it is zero.

But if, by the term pendulum, we mean to suggest the rather general situation of a hanging mass executing periodic motion, then in principle different types of pendula could be created by devising configurations for which other forms of restoring force arise. For example, in a *torsion pendulum*, a mass-structure is suspended from a thin fiber that is anchored securely at its other end as shown in Fig. 5.1.

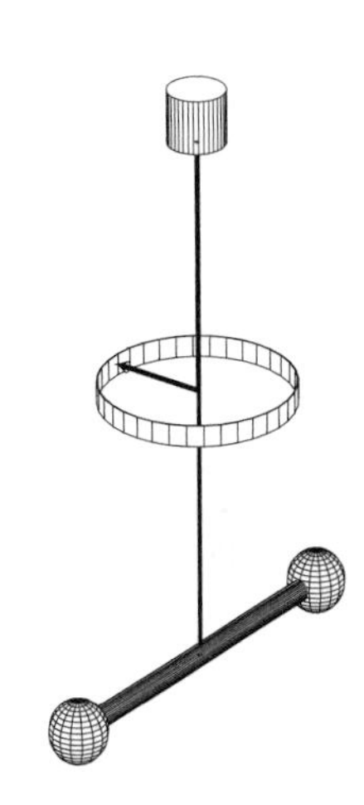

Fig. 5.1
The elements of a torsion pendulum. As the apparatus turns, the fiber twists and the pointer rotates to indicate the angular shift.

Any twist in the fiber results in a counter torque that leads to the desired back-and-forth oscillations (clockwise, counterclockwise, clockwise, etc.). Note that the plane in which this motion takes place is orthogonal to the direction of gravity, which therefore plays no role other than to give the pendulum weight and thus place the fiber under tension.

The torsion pendulum has proved to be an exceptionally sensitive and indispensable instrument in the determination of a surprisingly varied range of natural phenomena, and has provided (and continues to provide) the means to precisely measure some of the fundamental physical constants. The long and distinguished history of the torsion pendulum must begin with an accounting of its key component—the fiber.

5.1 Elasticity of the fiber

The key element in this apparatus is the suspension fiber which can be regarded as a long, thin, solid rod. The hanging pendulum payload will cause the fiber to stretch (longitudinal strain). If a torque is applied at the payload, then the fiber also will twist (angular strain). Both of these types of deformation will now be discussed.

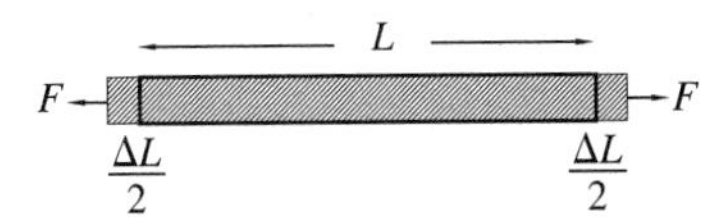

Fig. 5.2
Stretching of a thin rod by an applied force-pair at the ends. The outlines of the original sample are in bold.

The process of stretching is illustrated in Fig. 5.2. With the application of forces F at the ends, the sample elongates by a net amount ΔL above its original length L. The strain ϵ_ℓ is defined as the relative elongation

$$\epsilon_\ell = \frac{\Delta L}{L}.$$

If the cross-sectional area of the sample is A, then the applied stress is just the force per unit area F/A.

Perfectly elastic materials (an ideal closely approximated by many real solids) obey Hooke's Law which says that deformation is proportional to force—in other words stress and strain are linearly related. This also implies that when the stress is removed, the sample will return to its original shape and size. In real materials, a limit exists (the *elastic limit*) beyond which a sample will not fully recover when the load is removed.

The proportionality constant relating stress and strain is called *Young's modulus* (or elastic modulus) Y; it is defined by

$$Y = \frac{\text{applied stress}}{\text{resulting strain}}.$$

Alternately,

$$F = \left[Y\frac{A}{L}\right]\Delta L. \tag{5.1}$$

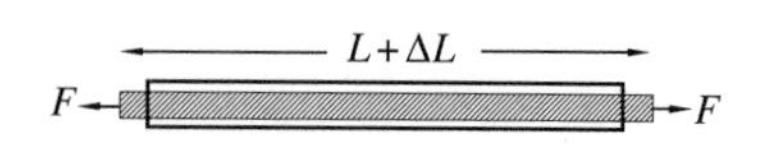

Fig. 5.3
Stretching of a thin rod, showing the accompanying decrease in cross sectional dimension. The outlines of the original sample are shown in bold.

Of course, as a wire stretches, it also tends to become thinned down. That is, its cross-section decreases as suggested in Fig. 5.3. The transverse strain is defined by the expression

$$\epsilon_t = -\frac{\Delta w}{w}.$$

The ratio of thinning to stretching is an important parameter known as *Poisson's Ratio* σ

$$\sigma = -\frac{\epsilon_t}{\epsilon_\ell}.$$

The twisting of a suspension fiber can be modeled by considering the response of a solid cylinder to an applied torque, as in Fig. 5.4.

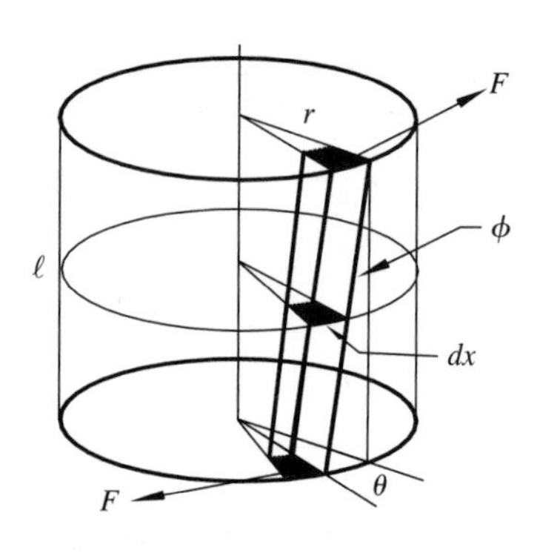

Fig. 5.4
A solid cylinder of length ℓ twisted through an angle θ by the application of a torque.

The shaded sector is located between radii r and $r + dr$, and is bounded by two arc segments of length dx. In the drawing, the sectors happen to be shown at their outermost position, with $r = R$, where R denotes the cylinder radius. F is the force couple[1] acting on the shaded patches at the ends of the twisted column in the diagram.

The *modulus of rigidity* η is defined by the ratio of shearing stress to shearing angle, or

$$\eta = \frac{F/(dr \cdot dx)}{\phi}.$$

In the limit of small θ and small ϕ, $\ell\phi = r\theta$, since each measures the length of the same arc segment. Thus,

$$\eta = \frac{F\ell}{dr\,dx\,r\,\theta}.$$

[1] A couple is a particular form of torque consisting of a pair of force vectors that are oppositely directed and symmetrically placed with repect to the axis of rotation. A couple thus induces no translation—only twisting.

Table 5.1 Elastic constants of some materials

Material	Y (dynes/cm^2)	η (dynes/cm^2)	σ
Steel	20.8×10^{11}	8.12×10^{11}	0.287
Nickel	20.2×10^{11}	7.70×10^{11}	0.309
Platinum	16.8×10^{11}	6.10×10^{11}	0.387
Gold	8.0×10^{11}	2.77×10^{11}	0.422
Aluminum	7.1×10^{11}	2.67×10^{11}	0.339
Quartz	7.5×10^{11}	3.0×10^{11}	

The total applied torque Γ is calculated by integrating rF over the extent of an end face.

$$\Gamma = \frac{\eta\theta}{\ell}\iint r^2\, dr\, dx.$$

For a solid cylinder the limits on the integrals are $x : 0 \to 2\pi r$ and $r : 0 \to R$. Hence

$$\Gamma = \left[\eta\frac{\pi R^4}{2\ell}\right]\theta. \tag{5.2}$$

The important point of the discussion is that this expression links the twist angle θ to the applied torque Γ through the dimensions of the cylinder and its modulus of rigidity, a constant of the material. The combined terms within brackets are a proportionality constant, called the torsional rigidity, which is generally defined by $\tau = \Gamma/\theta$. Note the similarity of the role played by η in the torque–twist relationship (Eq. (5.2)) to that of Young's modulus Y in the stress–strain relationship (Eq. (5.1)).

A significant consequence of Eq. (5.2) is that the amount of twist induced by any given torque varies as the inverse fourth power of the fiber radius. Therefore, suspension fibers become rapidly more sensitive as they are made thinner, a gain that is offset by their increasing fragility.

Table 5.1 (Newman 1961) gives elastic constants for some important and representative materials.

5.2 Statics and dynamics

In the most common applications of the torsion pendulum (as will be discussed in detail later), the objects at the ends of the horizontal beam are acted upon either by gravitational forces produced by attraction to some other object(s), or by electrostatic attraction (or repulsion) produced by charges on the pendulum and on neighboring objects. As depicted in Fig. 5.5, the resulting force-couple twists the pendulum beam.

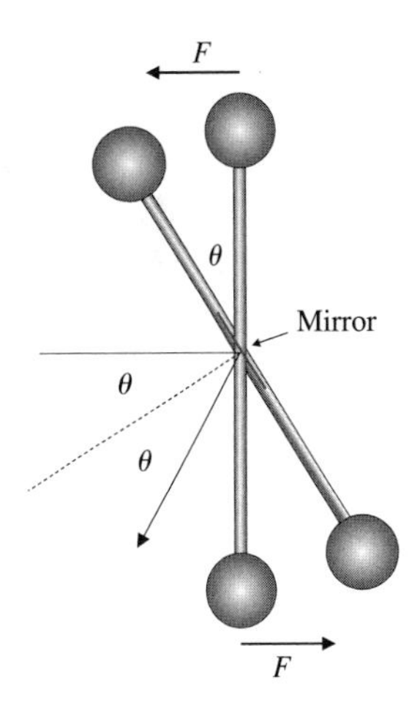

Fig. 5.5
Top view of a torsion pendulum twisted through an angle θ by an external force couple F. An incoming light beam is deflected by 2θ from its original direction.

The instantaneous deflection angle can be sensed by a so-called optical lever, as shown in the diagram. A small mirror attached to the apparatus will cause an incoming light beam to be deflected at twice the twist angle. A suitable photodetector then can determine the orientation of this reflected beam. If the distance from the mirror to detector is D, then the arc

Table 5.2 Specifications for a torsion pendulum

η	5×10^{11} dyne/cm^2
R	12 μm
ℓ	1 m
L	10 cm

length of the beam-shift at the detector will be approximately $2D\theta$. A very small angular deflection can thus be "multiplied" into a considerably larger scale reading.

If the pendulum is allowed to come to rest, then the condition of equilibrium dictates that there be no net torque so

$$FL = \Gamma$$

with L denoting the overall length of the horizontal pendulum beam. Since the reaction torque from the twisted fiber is given by Eq. (5.2), then the reaction force is

$$F = \left[\eta \frac{\pi R^4}{2\ell L}\right]\theta. \tag{5.3}$$

Therefore, a determination of the equilibrium deflection angle becomes a measurement of the external force.

The sensitivity of this method of measuring force is given by

$$S = \frac{dF}{d\theta}$$

and so

$$S = \frac{\eta\pi R^4}{2\ell L}. \tag{5.4}$$

Consider, as an example, the specifications[2] in Table 5.2 for a possible torsion pendulum:

The first three numbers are determined by the choice of suspension fiber, while the final parameter is the length of the pendulum beam. Substituting these values into Eq. (5.4), one obtains for the force sensitivity $S \approx 1.6 \times 10^{-3}$ dynes per radian, or about 2.8×10^{-5} dynes per degree. Very small forces induce measurable rotations in the apparatus. To appreciate the smallness of such forces, the mass of a single grain of rice is of the order of a few milligrams, so its weight (force) would be around 10^{-5} N or 1 dyne.

As noted earlier, the sensitivity varies as R^4, so the best performance is achieved with ultrafine suspension wires (see, for example, Gillies and Ritter 1993, p. 293).

5.2.1 Free oscillations without external forces

If one imagines a small mass attached to an axle via a massless rod of length L, and if a force F acts on that mass in a direction orthogonal to both connecting rod and axle, then Newton's law of motion $F = m(dv/dt)$ can be

[2] Rather than adopting standard MKS units for all parameters, each parameter is quoted in its most commonly used dimensional form.

seen to imply $FL = mL(d/dt)[L(d\theta/dt)]$. In this, $d\theta/dt$ is the angular velocity of the mass as it spins around the axle. The left hand side of this expression is just the torque Γ exerted by the applied force, and on the right, the term ML^2 is the moment of inertia I of the mass with respect to the axis of rotation. Hence

$$\Gamma = I\frac{d^2\theta}{dt^2}.$$

This equation is actually quite general and may be applied to extended objects such as wheels, in which case I is the total moment of inertia of all rotating components and Γ is the net torque acting on the object.

A torsion pendulum consists of a light connecting rod of length L with equal masses m at its ends (Fig. 5.1), suspended from a fiber. The moment of inertia is $I = mL^2/2$. The torque of Eq. 5.2 is now a restoring torque, and therefore is negative. Using Newton's second law we obtain the following equation of motion,

$$I\frac{d^2\theta}{dt^2} + \left[\eta\frac{\pi R^4}{2\ell}\right]\theta = 0. \tag{5.5}$$

This is the differential equation for a simple harmonic oscillator. In this case all of the previously developed results for the linearized pendulum now apply. If the initial conditions are $\theta = \theta_0$ and $d\theta/dt = 0$, then the solution is

$$\theta = \theta_0 \cos(\omega_0 t)$$

with the natural frequency given by

$$\omega_0 = \sqrt{\frac{\eta\pi R^4}{2\ell I}}. \tag{5.6}$$

The free oscillation period is therefore

$$T_0 = \frac{2}{R^2}\sqrt{\frac{\pi 2\ell I}{\eta}}. \tag{5.7}$$

Using the parameters given in the previous example, and assuming the pendulum consists of 10 gm masses mounted at the two ends of the connecting rod of length $L = 10$ cm, Eq. (5.7) gives an oscillation period of about 1000 s (a frequency of 1 mHz). This is very slow motion indeed. Note from Eq. (5.7) that thicker suspension fibers lead to shorter periods.

An application of this last expression is immediately apparent, especially if it is rearranged in the form

$$\eta = \left[\frac{4\pi 2\ell I}{R^4}\right]\frac{1}{T_0^2}. \tag{5.8}$$

All quantities on the right-hand side are dimensional properties of the apparatus, except for the oscillation period T_0, which can be measured experimentally. This method for determining the modulus of rigidity of the fiber material (see Table 5.1) is the first of many practical uses of the torsion pendulum.

5.2.2 Free oscillations with external forces

Now suppose that the masses are acted upon by forces as depicted in Fig. 5.5. The equation of motion changes from Eq. (5.5) to

$$I\frac{d^2\theta}{dt^2} + \left[\eta\frac{\pi R^4}{2\ell}\right]\theta = \pm FL, \tag{5.9}$$

where + in ± would be used when the constant external force F induces a torque in the same sense (clockwise or counterclockwise) as the convention adopted for a positive angular coordinate. The negative sign would be used in the opposite case. The solution of this modified equation of motion is

$$\theta = \theta_0 \cos(\omega_0 t) \pm \frac{FL}{\omega_0^2 I}.$$

Using the previous expression for ω_0,

$$\theta = \theta_0 \cos(\omega_0 t) \pm F\frac{2\ell L}{\eta\pi R^4}.$$

The oscillations now have an offset or "dc" component

$$\theta_{\text{dc}} = F\frac{2\ell L}{\eta\pi R^4},$$

which is just the same as the earlier static result Eq. (5.3). Therefore, a measurement of this dc or time-averaged deflection becomes, not surprisingly, a measurement of the external force.

5.2.3 Damping

There are no lossless dynamics—in the macroscopic world motion is always accompanied by dissipation (the only exceptions being superconductivity and superfluidity). Therefore, in the absence of external forces, a more realistic description of the torsion pendulum might be

$$I\frac{d^2\theta}{dt^2} + \beta\frac{d\theta}{dt} + \left[\eta\frac{\pi R^4}{2\ell}\right]\theta = 0, \tag{5.10}$$

which can be compared to Eq. (5.5). Friction generates a resistive counter-torque that is proportional to the instantaneous rate of rotation, the proportionality specified by the constant β. (Like the original restoring torque, the resistive torque counters the motion and is therefore negative as it would originally appear on the right side of Newton's equation of motion.) This is the most common model for friction, but it should be said

that in reality this is a complex phenomenon, and that bearing friction, air friction, etc. can depend to an extent on higher powers of the angular velocity. The dominant behavior is, however, reasonably described by the linear relationship included in Eq. (5.10).
Equation (5.10) is, essentially, the same as Eq. (3.2) and so the earlier discussion of the three possible damping regimes (under, critical, over) applies here, noting the equivalence $\beta/(2I) = \gamma$.

5.3 Two historical achievements

The torsion pendulum began its illustrious history with two profoundly important scientific applications that occurred late in the eighteenth century. These nearly concurrent developments were the creative products of the emerging technological cultures of France and England. Although, as is nearly always the case, many persons contributed to the final achievements, the most celebrated individuals were Charles Coulomb (Gillmor 1971) and Henry Cavendish.

5.3.1 Coulomb and the electrostatic force

Charles Augustin Coulomb (1736–1806) was a French military engineer. He was a student at the engineering school at Mézières (now called Charleville-Mézières, a town north-east of Paris near the border with Belgium)[3], an establishment mostly reserved for minor nobility, but occasionally, as in Coulomb's case, open to other students with clear promise (Fig. 5.6). After graduating in 1761 as an officer in the Corps of Engineers, he spent less than three years within France before being posted to the island of Martinique in February of 1764. There he was in charge of the design and construction of a new defensive fort at Port Royal. In spite of bouts of ill health, his stay lasted eight years. He returned to France, finally, in June 1772.

Fig. 5.6
Portrait of Charles Augustin Coulomb, probably around 1804. ©Bettman/Corbis/Magma.

Remarkably, Coulomb was able to pursue his interests in science and engineering during his protracted stay in the West Indies, so that by the spring of 1773 he was in a position to read two memoirs to the Academy in Paris. The subject of this first work was the application of statics to problems in architecture. For the next nine years following his return to France, Coulomb was posted to various military sites, ending in 1781 with a permanent transfer to Paris. His rank then allowed him a degree of freedom to pursue his interests in science, most particularly in physics, but also including public health and education, canals and navigation, structural engineering, and military topics.

Coulomb's accomplishments in electricity and magnetism ranged from his early (1777) memoir[4] entitled "Investigations of the Best Method of

[3] Charleville was also the birthplace of the poet-prodigy Arthur Rimbaud (1854–1891) whose writings in the brief period 1870–1874 established him as a leader in the Symbolist movement and as one of the true originals in French poetry.
[4] See (Gillmor 1971, p. 299) for a full reference to the 1777 magnetism memoir.

Making Magnetic Needles," through a famous series of seven memoirs read to the Academy from 1785 through 1791. In the First (1785) and Second (1787) of these, the torsion balance was introduced and the force law for electrostatics was established. The Third Memoir treated charge leakage, while the remaining memoirs dealt with charge distributions on conductors of various shapes, and magnetism.

Coulomb resigned from military service on April 1, 1791. He lived in Paris during the period 1781 to July 1793 when he moved to his country house north of the city. Thus he escaped the nightmare of the Reign of Terror led by Robespierre in 1794. Soon after, for safety, he moved to a property near Blois where he remained until at last he returned to Paris around 1797. He died in his home on August 23, 1806.

The professional life of Charles Augustin Coulomb spanned some of France's most significant and turbulent years. The Bastille was stormed in July of 1789 and shortly thereafter Louis XVI and the royal family were imprisoned in the palace of the Tuileries. Louis was guillotined in the Place de la Concode on January 21, 1793 (about six months before Coulomb left the city). Napoleon rose to power in the interval 1795–1799 when he became First Consul, and then in 1804, Emperor of France. War with Britain was an ongoing affair, with Toulon besieged in 1793 and blockaded again by Nelson at the beginning of new hostilities in 1803, the Battle of the Nile in 1798, the Battle of Trafalgar in 1805. War with Russia and Austria culminated at the Battle of Austerlitz in December, 1805. Coulomb did not live to see Napoleon's forced abdication in 1814, nor his second and final defeat at Waterloo in 1815. It is remarkable that such scientific achievements were possible in a career woven through such upheavals and dangers (the great chemist Lavoisier was guillotined in May of 1794).

The theory of a torsion pendulum was presented earlier in this Chapter. Coulomb's version of a practical apparatus is shown in Fig. 5.7.

This instrument was used to determine the dependence of the period of the pendulum on the suspension wire material (iron and brass were tested), wire length and diameter, and tension. He concluded that the tension was not significant as an influence, but that the period varied as the square root of the length of the wire and as the square root of the inverse fourth power of its diameter. Thus he correctly deduced the functional form of the earlier expression Eq. (5.7).

The most famous achievement of Charles Augustin Coulomb was his application of the torsion pendulum to the problem of determining the nature of the law governing the attraction or repulsion of electrical charge. A proper perspective on this accomplishment must take account of what was known about charge in the late eighteenth century, or, more to the point, what was not known. Still ahead and unforeseen were the discoveries of Michael Faraday during the 1820s and 1830s regarding electromagnetism and the law of electrolysis (Serway et al. 1989). The identification of negative charge with electrons and the measurement of the charge-to-mass ratio of the electron was the work of J. J. Thomson in 1897. The precise determination of the charge of the electron by means of the celebrated oil

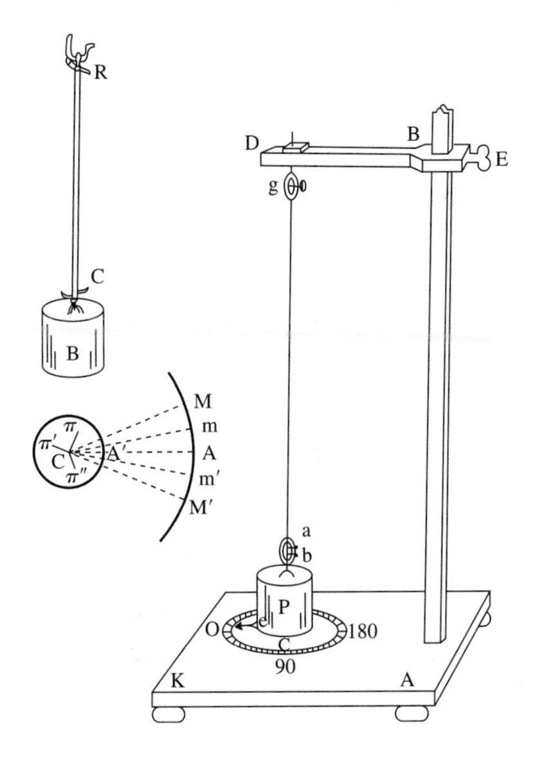

Fig. 5.7
Coulomb's torsion pendulum from 1784. Reprinted by permission of Princeton University Press from (Gillmor 1971, p. 153), ©1971 Princeton University Press, 1999 renewed PUP.

drop experiment was achieved of Robert Millikan in 1909. Thus it was to be more than one hundred years before the true nature of negative charge was determined. The fundamental particle carrying the basic unit of positive charge we now know to be the proton, but this understanding only emerged with the clear picture of the atom as a small massive and positively charged nucleus surrounded by orbiting electrons. This nuclear model was established by Ernest Rutherford in 1910. All of this lay ahead in the unknown future.

There was, nevertheless, wide interest in the phenomenon of static electricity. Benjamin Franklin's celebrated kite experiments in thunderstorms took place in 1752. Franklin conceived a two-fluid picture of electricity in which "positive" and "negative" charges either attracted or repelled one another. Static charge was usually produced by the rubbing of dissimilar materials.[5] The fortuitous discovery in about 1745 by Pieter van Musschenbroek and by Georg von Kleist of what became known as the Leyden jar made it possible to accumulate and store charge. The Leyden jar was the forerunner of the modern capacitor.

One hundred years earlier, in the Principia of 1687, Isaac Newton had established the foundations of mechanics and had discovered the universal

[5] *Triboelectricity*: "a charge of electricity generated by friction, e.g. by rubbing glass with silk," from Greek tribein-to rub. *The New Penguin English Dictionary*, Penguin Books (London 2000).

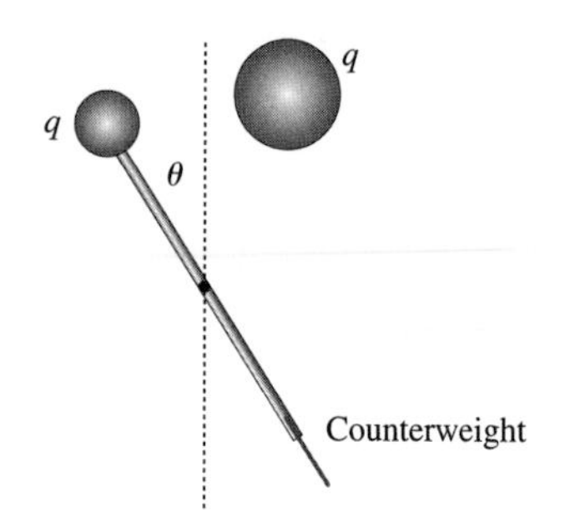

Fig. 5.8
Basic principle of Coulomb's torsion balance (seen looking down the suspension axis). When the two spheres have equal charges q, the spheres repel, as shown.

law of gravitation

$$F = G\frac{m_1 m_2}{r^2}, \tag{5.11}$$

which states that the force between masses m_1 and m_2 is proportional to the product of the masses and is inversely proportional to the square of their separation. Thus it would have been quite natural to wonder if the law governing electrostatic attraction or repulsion might be in some way similar. Of course gravity can only produce mutual attraction—there are no "positive" or "negative" masses.

Coulomb adapted his torsion pendulum to the task. The basic idea is illustrated in Fig. 5.8.

An equal charge is placed on two spheres, one fixed and the other located at the end of an arm that is suspended from a thin fiber.[6] The electrostatic repulsion causes a twist, which is then measured. According to the earlier Eq. (5.3), the force is then determined:

$$F = \left[\eta\frac{\pi R^4}{\ell L}\right]\theta, \tag{5.12}$$

where a factor of 2 has been introduced to account for the fact that the pair of repelling spheres is only at one end of the suspended beam.

The apparatus employed by Coulomb is depicted in Fig. 5.9.

Overall dimensions of the apparatus were about 12″ in diameter by about 36″ in height.

In the drawing, the suspension fiber (silver, copper, or silk) is seen to be attached to the horizontal beam at (P). The ends of the beam were occupied by a paper-disc counterweight (g) which also served as an air-damper, and a pith-ball (a).[7] A second pith-ball was located at the lower end of an insulated rod that extended down through the top plate at (m). Two angular scales were engraved on the apparatus, one around the wall of the large glass cylinder, and the other as part of the fiber attachment at the top.

To begin, the suspension head was then rotated, turning the fiber and beam until the two pith-balls were just in contact. By contacting the (temporarily joined) pair of pith-balls to a Leyden jar, equal charge became distributed on each, causing a force of repulsion and leading to a twisting away of the beam. The amount of charge was of course, unknown.

In equilibrium, the angular deflection of the beam (ϕ) could be read from the scale attached to the side of the glass cylinder. Now if the suspension head had been manually rotated by some amount (ξ), as depicted in Fig. 5.10, then the net twist of the fiber θ would be the sum ($\phi + \xi$).

[6] If the two spheres are identical in size, this is easily done: a static charge is placed on one sphere after which the pair is brought into contact. The initial charge is equally divided between the pair, which can then be separated again.

[7] Pith ball (pith, as in a pith helmet, originally was dried material from the spongewood tree of Bengal or other similar plants). Coulomb employed gilded elderwood spheres (about 1/6″ diameter) for this purpose.

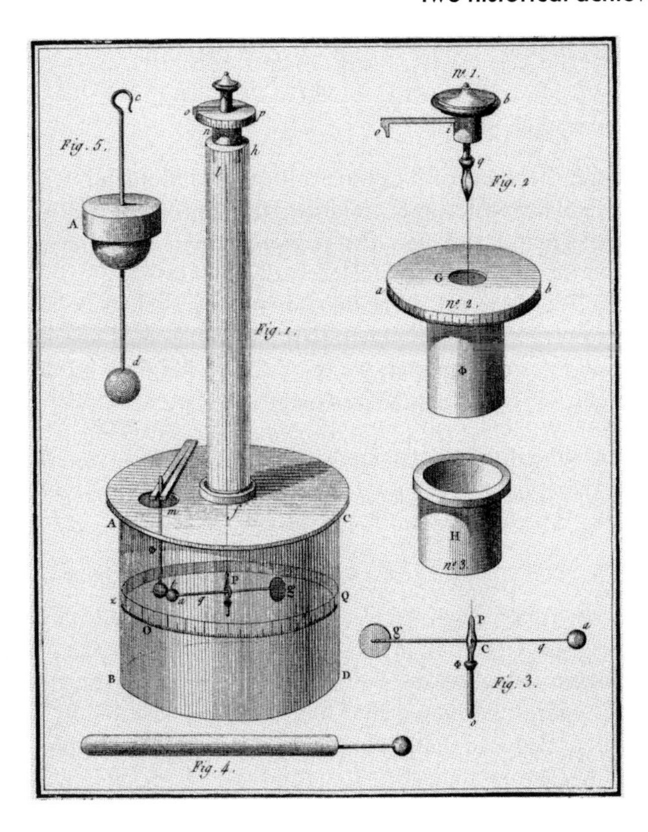

Fig. 5.9
Coulomb's torsion balance (1785) used for determining the law of electrostatic repulsion. ©Bettmann/Corbis/Magma.

Some reflection shows that these readings are related according to

$$\text{separation proportional to} : (\phi + \varphi)$$
$$\text{force proportional to} : (\phi + \xi),$$

where φ is the angular diameter of a pith-ball. This must be included to account for the fact that in the uncharged, undeflected reference position, the balls are in contact with their centres still a diameter apart. For Coulomb's apparatus, this amounted to about 1.8 degrees. In (Gillmor 1971), the following data values[8] are quoted (p. 185):

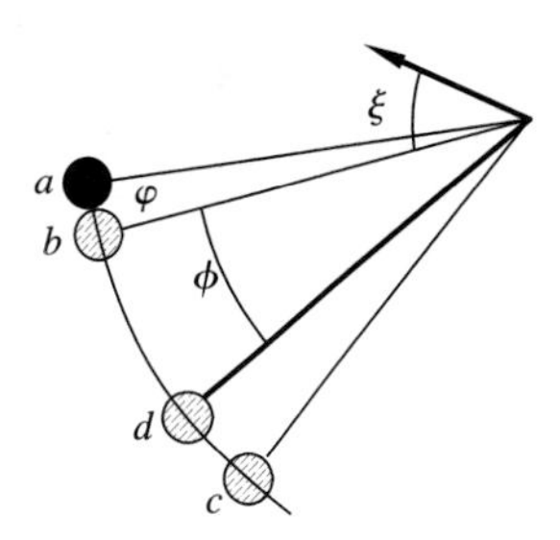

Fig. 5.10
Geometry of the Coulomb apparatus. The object shown with solid shading (a) is the fixed pith ball; the hatched sphere is the moveable pith ball shown at its uncharged (b) and charged (c) positions. When the clamp holding the torsion fiber is rotated by an angle ξ, the charged pith ball will shift to position (d).

Experiment	No. 1	No. 2	No. 3
ϕ	36°	18°	8.5°
ξ	0°	126°	567°

In Fig. 5.11, the effective force which is proportional to $\theta = (\phi + \xi)$ is plotted versus angular separation $(\phi + \varphi)$. An inverse square law for

[8] The surprisingly large value of ξ in the third trial is correct; a soft suspension fiber can require twists of more than a full turn.

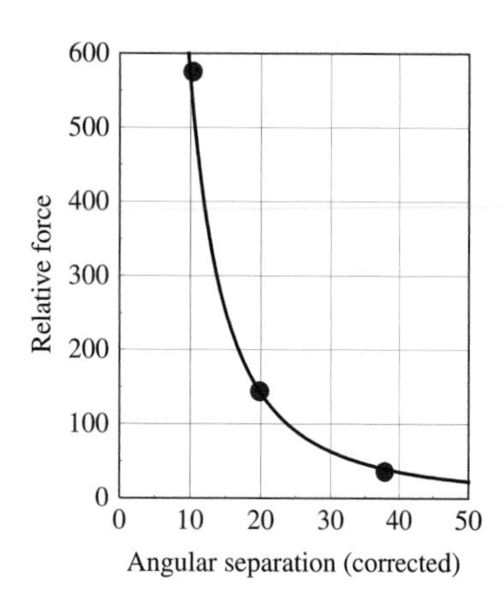

Fig. 5.11
Three data points from Coulomb's original experiment, and a fitted inverse square law relationship.

electrostatic repulsion would be expressed in the form

$$y = \frac{A}{(\phi + \varphi)^2}.$$

For the figure, the value of the single adjustable parameter was chosen to give agreement with Coulomb's middle data point (A = 56624). The continuous curve passes through the remaining two experimental points, showing that the inverse law was clearly demonstrated.

An especially elegant feature of this classic experiment is that the form of the law was precisely demonstrated without the need of specific quantitative information on the apparatus (e.g. suspension fiber parameters), or on the amount of charge actually involved.

How many Coulombs did Coulomb have?

In modern terminology, the electrostatic force law is

$$F = \frac{1}{4\pi\epsilon_0}\frac{q_1 q_2}{r^2}, \tag{5.13}$$

where q_1, q_2 are two charges whose separation is r; and $\epsilon_0 = 8.854187817 \times 10^{-12}$ F/m is the permittivity of free space. Taking the first experiment above, with $\phi = 36$ degrees, and using the sensitivity of Coulomb's apparatus (quoted in (Gillmor 1971)) $\approx 4.3 \times 10^{-4}$ dyne per degree, it can be deduced that the experimental charge on the pith-balls was about 10^{-10} C—a total indicating that approximately one billion electrons resided on each sphere.

Note that Coulomb's experiment confirmed only the inverse-square separation dependence in Eq. (5.13), but did not address the proportionality constant that we now write as $(4\pi\epsilon_0)^{-1}$. Moreover, the factor in the numerator—the dependence on the *product* of the charge magnitudes—was not addressed.

Modern methods and experiments have verified the inverse square dependence in Coulomb's law to very high precision.[9] If the separation dependence was expressed in the form r^{-n}, then one now can say that n is known to be almost exactly 2, within an uncertainty of less than a few times 10^{-16}.

5.3.2 Cavendish and the gravitational force

Fig. 5.12
Portrait of Henry Cavendish. ©Bettman/Corbis/Magma.

About Henry Cavendish, (Fig. 5.12), the Funk and Wagnalls New Encyclopedia entry reads

British physicist and chemist, born of British parents in Nice, France, and educated at Peterhouse College, University of Cambridge. His earliest experiments involved the specific heats of substances. In 1766 he discovered the properties of the element hydrogen and determined its specific gravity. His most celebrated work was the

[9] These methods are not based on the torsion pendulum. Instead they test Gauss's law, from which Coulomb's law follows, and its implication that static charge must reside on the outside of conductors.

discovery of the composition of water; he stated that "water consists of dephlogisticated air (oxygen) with phlogiston (hydrogen)." By what is now known as the Cavendish experiment, he determined that the density of the earth was 5.45 times as great as the density of water, a calculation very close to the 5.5268 established by modern techniques. Cavendish also determined the density of the atmosphere and made important investigations of electrical currents. Cavendish, Henry (1731–1810)

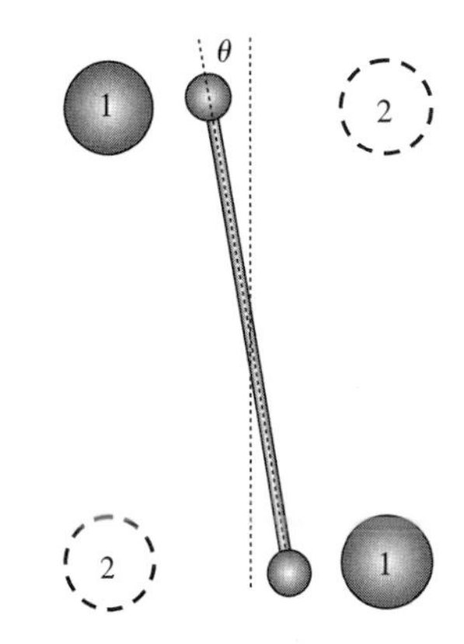

Fig. 5.13
Arrangement of a Cavendish torsion balance for determining the gravitational constant G by the static method. The view is looking down the suspension fiber.

Cavendish and Coulomb were almost exact contemporaries. Just as what we now regard as Coulomb's law was in historical truth not entirely what Coulomb established, so too with the work of Cavendish. It is a widely held notion that Cavendish experimentally verified Newton's gravitational law Eq. (5.11), but as the biographical item above hints, such was not the case. In the first place, Newton himself did not write his "law" in the now familiar form, nor did he introduce the "universal" constant G. One hundred years had to pass after Cavendish before the modern idea of "G" was to appear in a paper entitled "On the Newtonian Constant of Gravitation" by C. V. Boys (Boys 1894). It is also worth keeping in mind that our familiar system of units was not established at the time of Cavendish. For example, the dyne was not introduced until 1873.

A preoccupation of eighteenth century scientists was a determination of the relative *density of the earth*,[10] and it was to this task that Cavendish turned his attention.

The basic idea of the Cavendish experiment is illustrated in Fig. 5.13

In this view down the axis of the suspension fiber, it can be seen that two small masses are fixed to the ends of a horizontal beam of length L. Two large masses are positioned at either pair of sites numbered 1,1 or 2,2. Gravitational attraction will rotate the beam and twist the support fiber through a counterclockwise angle θ when the large masses are in the first position, and through a clockwise angle θ when they are in the second position. The experiment consists of determining the total change in angle 2θ that occurs as the mass positions are switched.

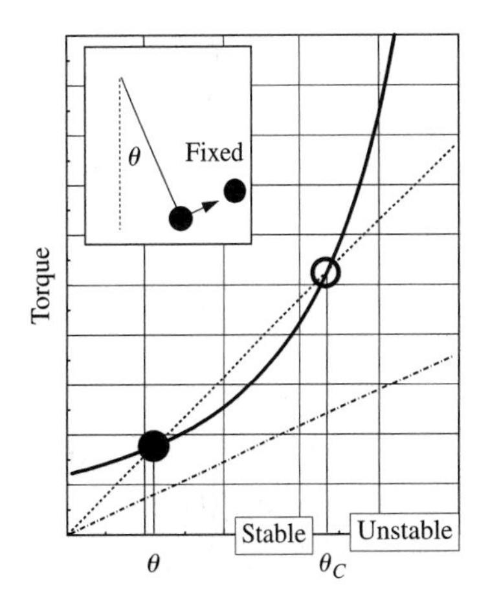

Fig. 5.14
Behavior of the attractive torque (solid curve) for two masses as a function of their relative position, and the counter torque from a twisting suspension fiber. Two possible vaues of fiber stiffness are illustrated (straight lines). For the stiffer suspension, if the movable ball is released from an angle less than θ_C, the system will reach equilibrium at the final angle θ (solid circle). If the release is from an angle greater than θ_C, then the attractive force dominates and causes the balls to come into contact. For the less-stiff fiber, no stable separation exists.

The equilibrium position specified by the angle θ is determined by the balance of the torque produced by the gravitational force of attraction on the small masses toward their fixed large-mass partners, and the opposing torque from the twisted suspension fiber. This is illustrated in Fig. 5.14.

A key point to note is that the fiber torque is a linear function of the twist θ, whereas the gravitationally induced torques vary as the square of the inverse distance between small and large masses, that is, approximately as $(\text{constant} - \theta)^{-2}$. Also, even for zero twist when the fiber is unstrained, the gravitational forces, and thus torques, are nonzero as illustrated in the figure. For a fiber of sufficient stiffness, the linear function can be seen to intersect the nonlinear gravitational curve in two places. As pointed out in the caption, the lower crossing point is the stable equilibrium for the system. For a soft suspension, there may be no crossing point, in which case the gravitational torque will bring the spheres into contact and hold them there. However, as pointed out in the next section, this gravitational "glue"

[10] See Chapter 2.

is exceedingly weak and tiny perturbations such as air currents may be enough to break the bond.

Modern versions of the experiment use a deflected light beam for readout (see Fig. 5.5). At equilibrium, using Eq. (5.2)

$$G\frac{mM}{r^2}L = \left[\eta\frac{\pi R^4}{2\ell}\right]\theta. \tag{5.14}$$

Equation (5.6) shows that the fiber stiffness can be determined from the frequency of free oscillations of the system (in the absence of the large masses), and hence

$$G = [I\omega_0^2]\left[\frac{r^2}{mML}\right]\theta. \tag{5.15}$$

The moment of inertia of a massless beam with m at each end is given by $I = mL^2/2$ and so

$$G = \frac{2\pi^2 Lr^2}{MT_0^2}\theta. \tag{5.16}$$

Therefore an observation of the angular deflection together with a knowledge of the large mass value, the separation between large and small masses, and the free oscillation frequency will yield the universal gravitational constant G.

From G one can deduce the density of the earth (or vice versa). Let the radius and mass of the earth be R_E and M_E, respectively. The acceleration of gravity at the surface of the earth, $g = GM_E/R_E^2$, is known from free-fall experiments and therefore the density of the earth

$$\rho_E = \frac{M_E}{(4/3)\pi R_E^3}$$

is thus

$$\rho_E = \frac{g}{(4/3)\pi G R_E}. \tag{5.17}$$

Weak forces

In the light of what we now know about the gravitational force law and the constant $G = 6.67259 \times 10^{-11}\,\mathrm{m^3/kg\,s}$, it is remarkable that Cavendish was able to obtain meaningful results. Consider the difficulty of the task.

Suppose two 1 kg spheres, each of radius 2 cm, are placed in contact. Then according to Newton's gravitational law Eq. (5.11), the mutual attractive force between them would be about 4×10^{-8} N. Interestingly, the gravitational force turns out to be almost exactly the same magnitude as that found in the Coulomb experiment where the two similar spheres were charged to 10^{-10} C. (In that case, the Coulomb force was about 5×10^{-8} N.) As noted earlier, the weight of a single grain of rice is around 10^{-5} N, a magnitude that is nearly one thousand times larger than the force to be measured in the Cavendish experiment. Fortunately, the torsion balance provides sufficient sensitivity for the task.

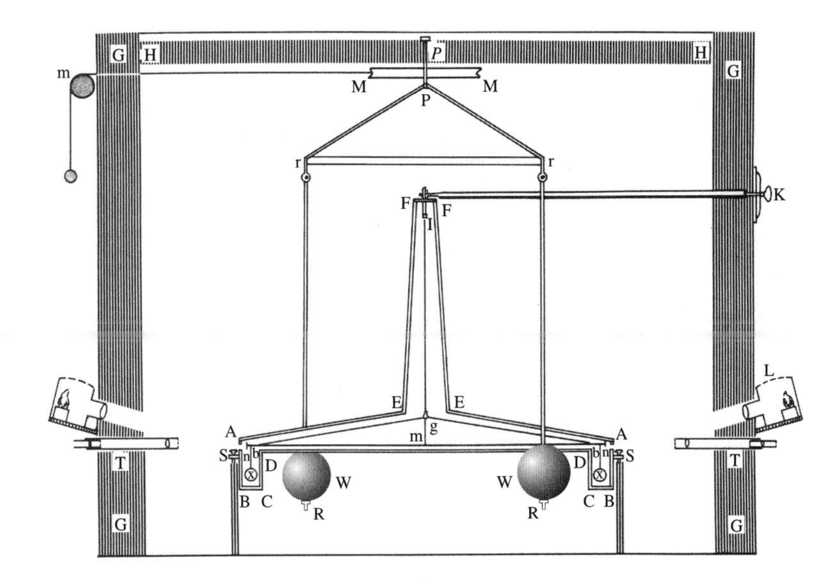

Fig. 5.15
Original apparatus used by Henry Cavendish in 1798 to measure the density of the Earth. Reprinted with permission from (Clotfelter 1987, p. 512). ©1987, American Association of Physics Teachers.

The Cavendish apparatus

It is generally acknowledged that Cavendish improved upon an original concept by his friend and colleague the Reverend John Michell (1724–1793).[11] Figure 5.15 shows the torsion beam built by Cavendish (Clotfelter 1987). It was quite large, with a beam 73.3" in length and large masses that were lead spheres one foot in diameter weighing 348 pounds. The small spheres were 2" in diameter and the center separation distance r was 8.85". The copper suspension wire was 39.25" in length. As suggested in the drawing, the entire apparatus was housed in a room provided with illuminating lanterns and sighting telescopes so that the experimenter remained outside and air currents were reduced. From the historical perspective of Cavendish, for whom there was no "G," Eqs. (5.16) and (5.17) could more meaningfully be written

$$\rho_E = \frac{3g}{8\pi^3 R_E}\left(\frac{MT_0^2}{Lr^2\theta}\right). \tag{5.18}$$

An additional useful conversion can be obtained by recalling that a simple pendulum whose length is chosen to be L/2 would have a small-amplitude period

$$T_p = 2\pi\sqrt{L/2g}.$$

Therefore Eq. (5.18) can be written

$$\rho_E = \frac{3}{4\pi R_E}\left(\frac{MT_0^2}{r^2 T_p^2\theta}\right). \tag{5.19}$$

[11] John Michell was an astronomer who, in 1798, proposed the existence of objects that we today call Black Holes. He made this suggestion without knowledge of a similar suggestion by LaPlace in 1795.

Now all required numerical values are at hand. The radius of the earth was known to eighteenth century scientists; the remaining terms are parameters of the experimental apparatus. Three measured quantities are required: T_p, T_0, θ. Cavendish determined the time for a single swing (what we would now identify as half a period) and obtained 0.97 s for the simple pendulum and 424 s for the torsion apparatus. The gravitational deflection was $\theta \approx 0.0039$ rad (0.22 degrees). From these the mean density is calculated to be 5.72 gm/cm^3. The presently accepted value for the mean density of the earth is 5.5268 gm/cm^3.

5.3.3 Scaling the apparatus

The Coulomb apparatus was significantly smaller than that of Cavendish. Obviously the logic in Cavendish's thinking was to make the gravitational torque as large as practically possible by choosing large masses and a long (and heavy) beam. This, however, necessitated a strong, and therefore thick, suspension wire. As Eq. (5.4) showed, the sensitivity of a suspension fiber varies as the fourth power of its radius—very thin fibers are best. But the load-carrying capacity of a wire goes up as the square of the radius—very thick fibers are best. So Cavendish traded sensitivity for increased force.

This raises the obvious question about optimum designs. It turns out that Coulomb was on the better track. The Cavendish experiment was subsequently improved by making it smaller and using other suspension fibers such as quartz to increase torsional sensitivity. The list of developments includes those of Cornu and Baille in 1878, Boys in 1894, Eötvös in 1896, Burgess in 1901 and Heyl in 1930. A typical modern design would have large masses of a few kg, sphere separations of a few cm, beam lengths of 10 or 20 cm, and suspension fibers with torsion constants of around 10^{-8} Nm/rad.

5.4 Modern applications

5.4.1 Ballistic galvanometer

In its conventional form, a galvanometer measures steady current. The current is made to flow through a multi-turn coil, usually rectangular in shape, which is suspended in the field of a permanent magnet. The interaction of the magnetic field with the moving charges results in Lorentz forces ($\vec{F} = q\vec{v} \times \vec{B}$) that collectively then act on the each side of the coil, producing a twisting torque. In equilibrium the resting orientation of the coil is proportional to the current circulating through it. A pointer then provides the conventional readout against an annotated scale.

A ballistic galvanometer is an adaptation so that a quantity of *charge*, rather than current, is sensed. The basic design remains the same, except for one important factor: the time during which the charge to be measured passes through the instrument must be short compared to the basic time constant of the device.

For a galvanometer coil of area A consisting of N turns carrying current i, positioned in a magnetic field B that is normal to the plane of the coil, the resulting torque will be $\Gamma = NiAB$. Newton's law: $\Gamma = I\ddot{\theta}$ can be expressed in equivalent form: $\Gamma dt = I\, d\omega$. This says that impulsive torque equals change in angular momentum, much as the usual expression of Newton's law, $F = ma$ can be stated: impulse equals change in linear momentum: $Fdt = m\, dv$. Therefore, $NiAB\, dt = I\, d\omega$, and so

$$NAB \int_0^{t_0} i\, dt = I \int_0^{\omega_f} d\omega = I\omega_f.$$

The integral on the left simply yields the total charge passing through the coil in the short interval during which the angular velocity of the coil changes from rest to a final value ω_f. Hence $NABQ = I\omega_f$. If at the end of this initial phase there is no additional charge flowing, then the impulsive Lorentz torque will return to zero. With the clock restarted, and new initial conditions $\theta = 0, \dot{\theta} = \omega_f$, the coil now oscillates clockwise and counter-clockwise on the end of the suspension fiber according to the usual equation Eq. (5.5)

$$I\frac{d^2\theta}{dt^2} + \left[\eta\frac{\pi R^4}{2\ell}\right]\theta = 0$$

when there is no damping. The solution is

$$\theta = \frac{\omega_f}{\omega_0}\sin \omega_0 t.$$

The amplitude of the oscillations, which can be observed experimentally, then gives with the aid of the expression for ω_f

$$\theta_{\max} = \frac{NABQ}{I}\left[\eta\frac{\pi R^4}{2\ell I}\right]^{-1/2}$$

or

$$Q = \frac{\theta_{\max}}{NAB}\left[\eta\frac{\pi R^4 I}{2\ell}\right]^{1/2}. \tag{5.20}$$

Thus the observed maximum angular deflection is a measure of the total charge which passes through the galvanometer.

When small damping is present, the above treatment must be modified along the lines developed in the earlier section Damping. The complication now is that the undamped maximum $\theta_{\max}$ is required for the determination of the charge Q, but the amplitude of the observed back and forth galvanometer oscillations is decaying exponentially. Thus one can determine a sequence of numbers: maximum left angular throw, subsequent smaller maximum right angular throw, subsequent smaller-still maximum left angular throw, etc. From this sequence, it is possible to deduce the required $\theta_{\max}$ and so obtain the value of Q.

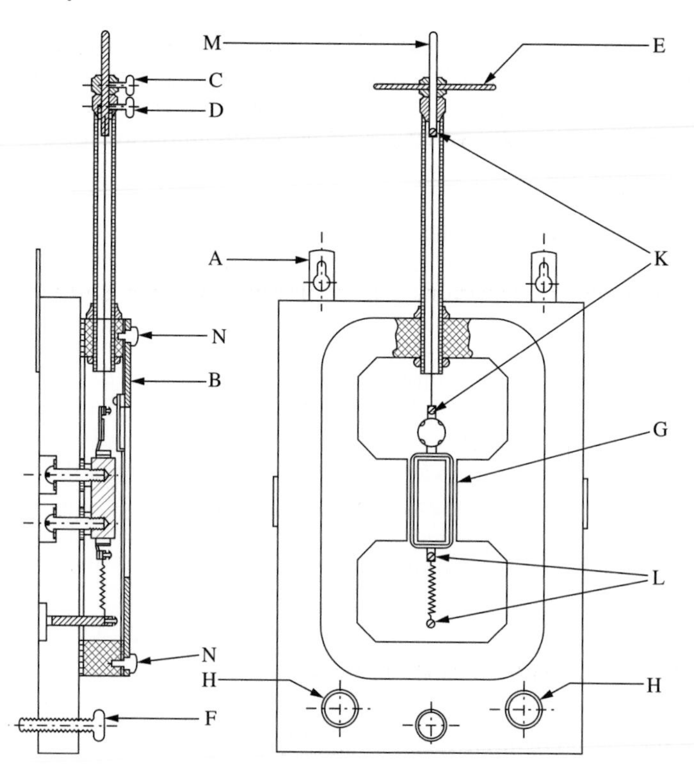

Fig. 5.16
Side and front views of a wall-mounted ballistic galvanometer. Shown are the current coil (G), mirror, and suspension fiber extending between points (K).

Figure 5.16 is an illustration of a commercial[12] wall-mounted ballistic galvanometer.

The rectangular coil (G) hangs freely between pole faces of the permanent magnet in the shape of a large "8". A very fine suspension wire is clamped at the top (K) and just above a small mirror that is in turn fixed to the coil. Not shown is a telescope assembly and large curved scale that allow the rotation of the coil to be determined by sighting at the corotating mirror. The terminals for external connections are marked H. Typical specifications for this type of instrument are: period 2–4 s, ballistic constant 0.03–0.04 μC/cm.

5.4.2 Universal gravitational constant

It was noted earlier that the particular notion of a "Universal Gravitational Constant" was not included in either seventeenth century (Newton) or eighteenth century (Cavendish) formalisms. However by the nineteenth century, modern mathematical language and the idea of "fields"—electromagnetic or gravitational—had taken root. Fundamental constants of nature such as Avogadro's number, assumed a certain preeminence. After more than two hundred years, the torsion pendulum remains the definitive instrument for determining G.

[12] Model 2239, Leeds and Northrup, Philadelphia, PA.

As Eq. (5.16) shows, the Cavendish experiment can give a value for the universal gravitational constant from a *static* measurement, that is from the observed angle of twist θ. Another scheme is now favoured for this task—the *dynamic* time of swing method.

Time-of swing method

Consider first an arrangement as shown in Fig. 5.17. Clearly there is a restoring component F_x of the net attractive force between the m and M pairs. For small oscillations and neglecting the fiber stiffness in this instance,

$$\frac{F_x}{F} = \frac{x}{\sqrt{x^2 + d^2}} \approx \frac{x}{d}.$$

But

$$x = \frac{L}{2}\tan\theta \approx \frac{L\theta}{2}.$$

The equation of motion is then

$$-F_x L = -F\frac{x}{d}L = -F\frac{L^2}{2d}\theta = I\ddot{\theta},$$

which describes harmonic oscillation at a frequency

$$\omega = \sqrt{\frac{FL^2}{2dI}}$$

or period

$$T = 2\pi\sqrt{\frac{2dI}{FL^2}}.$$

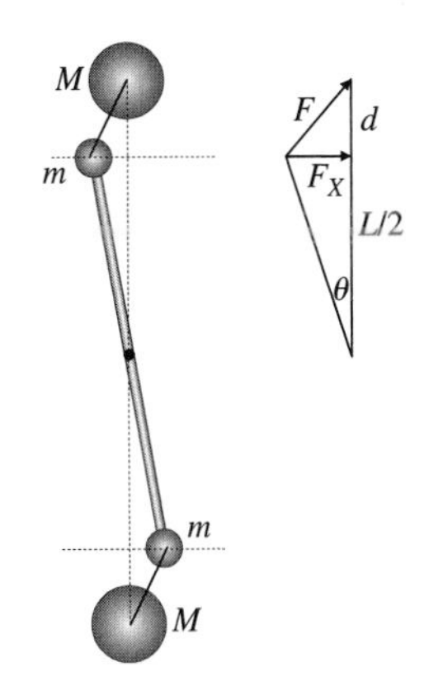

Fig. 5.17
Alternate arrangement of the torsion pendulum for determining the gravitational constant G by the time-of-swing method.

Assuming an inverse square law force, this would predict the following relationship between separation and period; $T \propto d^{3/2}$, a relationship that could be used as the basis of an experimental test.

In the event that the fiber stiffness is also significant (almost always true), then it can be seen that the restoring component of the gravitational torque effectively adds extra stiffness to the suspension, increasing the natural frequency. Suppose the beam was allowed to oscillate first with and then without the attracting masses M present. In the first case the oscillation frequency would be

$$\omega_1 = \sqrt{\left(\frac{\eta\pi R^4}{2\ell I} + G\frac{mM}{Id^2}\frac{L^2}{2d}\right)}$$

while in the second, the frequency would be

$$\omega_0 = \sqrt{\left(\frac{\eta\pi R^4}{2\ell I}\right)}$$

or

$$\omega_1^2 = \omega_0^2 + G\frac{mML^2}{2d^3 I}. \tag{5.21}$$

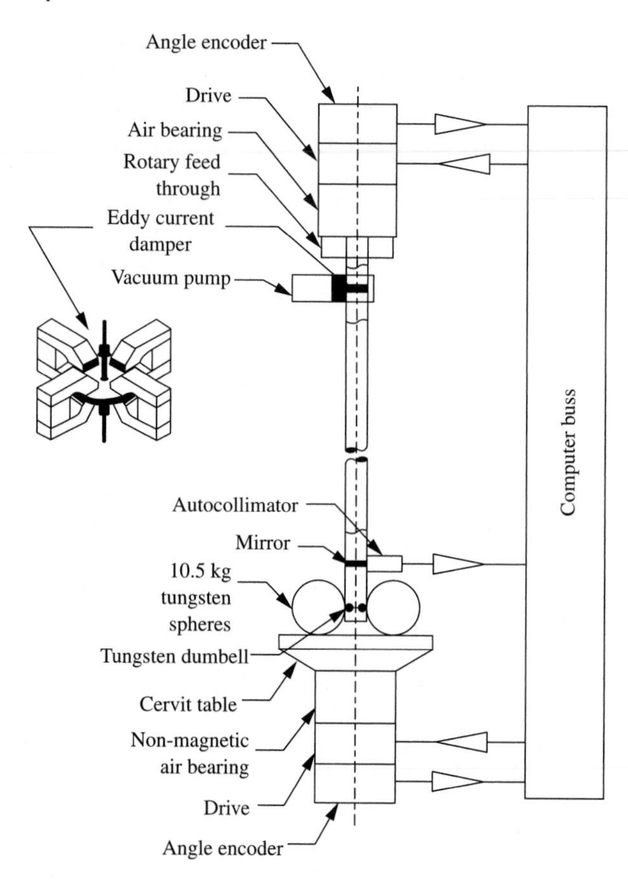

Fig. 5.18
Apparatus of Bagley and Luther used to determine the Newtonian Gravitational Constant. Reprinted with permission from Bagley and Luther, (1997 p. 3048). ©1997 by the American Physical Society.

Therefore a pair of frequency measurements (or, equivalently, periods) can give the gravitational constant G if the masses and apparatus dimensions are known.

The basic principle of the time-of-swing method has been applied to recent very precise determinations of G. The apparatus shown in Fig. 5.18 was used by Bagley and Luther (1997) to measure the universal gravitational constant.

The two large masses were 10.489980 and 10.490250 kg, respectively. The small-mass "dumbbell" was composed of a pair of tungsten discs situated at the ends of a 1.0347 mm diameter rod 28 mm in length. The period of free oscillations of the torsion pendulum with a 12 μm suspension fiber was 210 s. The pendulum was enclosed within a vacuum system and the entire apparatus was housed in an isolated building located on a mesa in New Mexico. In this version of the experiment, oscillation frequencies were measured not with and without the large masses, but instead with the large tungsten masses in either of two positions relative to the dumbbell: in-line and perpendicular. Switching of positions was performed at half-hour intervals with the experimental runs lasting for as long as 40 hours. The reported value of G was $(6.6740 \pm 0.0007) \times 10^{-11}$ m^3/kg/s^2.

Another, similar determination of G was reported by Luo et al. (1998). For this experiment, the two frequencies were measured with and without large attracting masses, which were 6.2513 and 6.2505 kg, respectively, and made from nonmagnetic stainless steel. The geometry was somewhat different from the classic Cavendish arrangement—these large masses were positioned on either side of one of the small masses, with the line of their centers orthogonal to the dumbbell axis. The small mass at the other end of the dumbbell served simply as a counterweight. The two small copper masses were 32.2560 and 32.2858 gm separated by 200 mm. The torsion fiber was 513 mm long 25 μm tungsten wire. The entire apparatus was enclosed in a high vacuum container and was located in room mounted on a shock-proof platform weighing 24 tons which was in turn supported on 16 very large springs. The laboratory itself was in the center of Yu-Jia Mountain in Wuhan, People's Republic of China. The measured periods were 4441 s and 3484 s, with and without the attracting masses, respectively, yielding a final value for the gravitational constant of $(6.6699 \pm 0.0007) \times 10^{-11}\,m^3/kg/s^2$.

It is rather interesting to note that despite these very well engineered designs that make use of the latest techniques in instrumentation, and the almost extreme measures taken to assure accuracy, still the overall precision is only about a few parts in ten thousand. Thus G remains the least accurately known of the universal natural constants.[13]

Quite recently, a new determination was made using a very elegant torsion pendulum as shown in Fig. 5.19.

This design by Jens Gundlach and Stephen Merkowitz at the University of Washington (Gundlach and Merkowitz 2000) employs a simple Pyrex slab in place of the usual dumbbell. External to the fully enclosed pendulum was an array of 8 kg steel balls. The entire pendulum was placed on a rotating turntable. As the pendulum then turned past the steel balls, the suspension fiber would exhibit periodic twists. A further design refinement added feedback control to the turntable so as to accelerate and decelerate the rotation to minimize fiber twist. In the spring of 2000 they reported the best current value for G: $(6.674215 \pm 0.000092) \times 10^{-11}\,m^3/kg/s^2$.

5.4.3 Universality of free fall: Equivalence of gravitational and inertial mass

Newton's law of gravitation describes the mutual attraction of a pair of masses m_1 and m_2. No specification of material substance is provided and in fact (so far as we know) it does not matter if m_1 and m_2 are composed of copper, lead, hydrogen, or anything else. Thus, for example, a pair of one kilogram glass spheres will experience the same attracting force as a pair of one kilogram steel spheres.

[13] See table II : The 1998 recommended values of the fundamental physical constants, from The 1998 CODATA Recommended Values of the Fundamental Physical Constants, *Journal of Research of the National Bureau of Standards*.

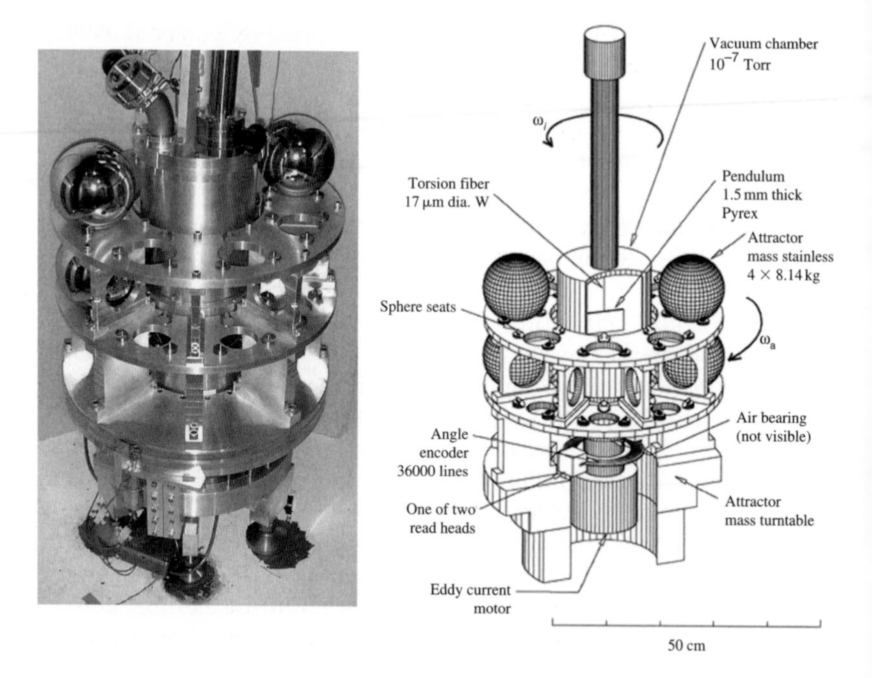

Fig. 5.19
Photograph and cut-away view of the apparatus used by Gundlach and Merkowitz to determine Newton's gravitational constant, G. (Photograph and drawing courtesy of Jens Gundlach.)

This fact has a direct consequence in experiments dealing with falling bodies. At the surface of the earth, the force of attraction of the earth (mass M, radius R) acting on a test mass m is of course

$$F = G\frac{mM}{R^2}$$

neglecting the height of release above the surface in comparison to the comparatively huge value of R. The resulting acceleration downward is thus

$$a = \frac{F}{m} = \frac{GMm}{mR^2} = \frac{GM}{R^2},$$

which is usually assigned the special symbol g.[14] The seemingly trivial final step in which m is cancelled depends on the "equivalence" of the gravitational mass in the numerator and the inertial mass in the denominator; this equivalence is not a matter of small importance. Notice that this acceleration g does not depend either on the test object's own mass, or on its material composition. As Galileo surmised four hundred years ago, all objects fall at the same rate. This principle can appear to be violated in common experience, but when the experiment is performed in a vacuum, the prediction is confirmed—as was vividly demonstrated when an astronaut released an eagle feather and hammer on the surface of the moon. The fact that material composition is absent from such phenomena suggested to Einstein that gravity must be a function of space itself rather than of the particular matter experiencing the force.

[14] The well known value for the acceleration of free fall at sea level at latitude 45 degrees is 980.616 cm/s^2, or 32.17 ft/s^2.

There has always been speculation on the issue of absolute correctness of the equivalence principle. From time to time ingenious experiments have been devised to test the hypothesis of material independence. One of the most celebrated experimenters was Baron Roland von Eötvös of Hungary. He used a torsion balance consisting of a 40 cm long beam suspended from a platinum–iridium wire. Weights made from different materials were placed at the ends of the beam and the apparatus, enclosed so as to reduce air currents, was aligned in an east–west direction. Twisting of the fiber was detected by means of the usual optical lever, that is a light beam reflected from a mirror attached to the fiber/beam.

Since the apparatus was fixed to the surface of the spinning earth, the two beam masses were subject to gravitational forces from the earth and centrifugal forces generated by the rotating reference frame of the laboratory. Any small difference between these two forces as they were experienced by the beam masses would have resulted in a net torque on the torsion pendulum. In experiments carried out between 1889 and 1908, no torque was observed to within the sensitivity limits of the apparatus (Eötvos 1922). This famous null result confirmed the material independence of inertial and gravitational forces, thereby verifying the equivalence principle.

Since this early work, fresh attempts have been made to examine the question. In the early 1960s, R.H. Dicke (1961) and others at Princeton University improved on the original Eötvös torsion pendulum design, using a triangular balance beam with two copper weights and a third mass consisting of a glass container with lead chloride. Copper was chosen because of its almost equal numbers of protons and neutrons (29 and 34), while lead offered a contrast with 82 and 125, the idea being to see if neutron rich matter behaves differently. Great care was taken to minimize convection currents by using a high vacuum enclosure, and by eliminating magnetic impurities in the sample masses because of the possibility of torques caused by interaction with the earth's magnetic field. The need for such care is illustrated by Dicke's remark: "If one of the suspended weights in Eötvös' apparatus contained a single strongly magnetized iron fragment weighing only a few millionths of a gram, it would produce a deflection in the balance 1,000 times greater than the probable error given by Eötvös." This apparatus was placed in a twelve foot deep pit near the Princeton football stadium. In the end, this experiment improved the precision over Eötvös by a factor of 50, and gave once more a null result.

The newest test of universal free fall (Su et al. 1994) involved the torsion pendulum shown in Fig. 5.20.

This pendulum was located in a laboratory situated in a hillside on the campus of the University of Washington. The 20 μm suspension fiber was a gold coated tungsten wire 0.8 m long. The free oscillation period of the pendulum was 695 s. As the diagram illustrates, the entire pendulum was placed on a turntable that continuously and slowly rotated about the vertical axis.

The business end of the pendulum is shown more clearly in Fig. 5.21.

Note that, in contrast to Eötvös who employed two masses, and Dicke who used three, this system has four. Aluminum, beryllium, and copper

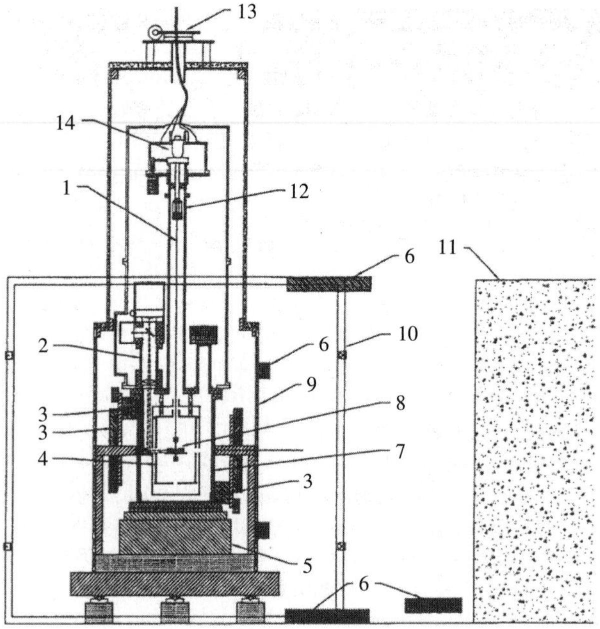

Fig. 5.20
The Eöt–Wash torsion pendulum. Principal components include the suspension fiber (1), the turntable (5), and the pendulum itself (8). The concrete block (11) is 1.23 m high. Reprinted with permission from (Su et al. 1994, p. 3048). ©1994 by the American Physical Society.

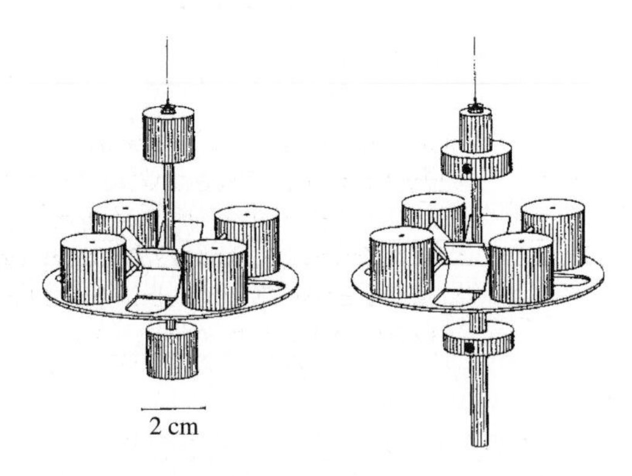

Fig. 5.21
Configuration of the pendulum masses for two versions of the experiment. Reprinted with permission from (Su et al. 1994, p. 3616). ©1994, by the American Physical Society.

objects were used. The details of this experiment are quite complex, but the measurements again confirmed the equivalence principle.

The push to improve torsion pendulum performance is also illustrated by the apparatus (Bantel and Newman 2000) shown in Fig. 5.22.

The figure shows only the payload end of the system. This was located at the bottom of a large dewar which could store liquid nitrogen and liquid helium; its capacity was 95 liters. With liquid N_2 the system temperature can be reduced to 77 K; liquid He can go much lower—to just a few degrees kelvin. For the most precise work, even thermal fluctuations become troublesome because of their effect on the mechanical properties of mirrors used for optical lever readouts (Heidmann et al. 1999). Therefore the low temperature environment was used to reduce both thermal noise effects and thermal sensitivity.

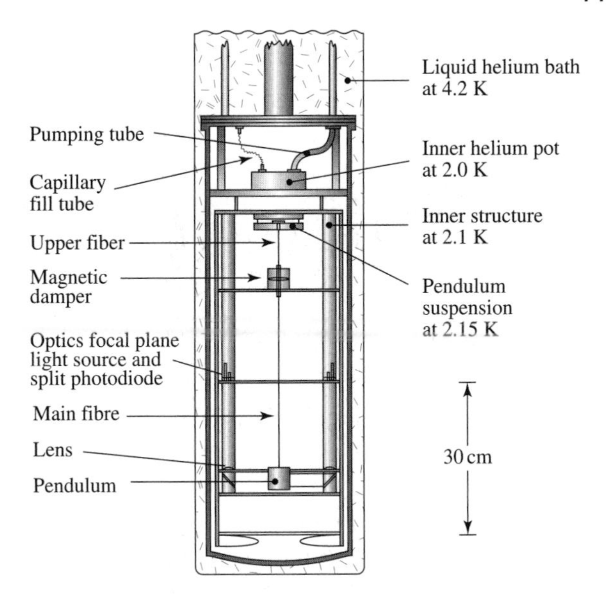

Fig. 5.22
Cryogenic torsion pendulum for gravitational experiments. This drawing shows only the lower part of the complete apparatus which consists of a dewar several meters high. Reprinted with permission from (Bantel and Newman, 2000, p. 2315). Courtesy of the Institute of Physics.

The torsion fiber was 25 μm, 260 mm long hardened aluminum. The pendulum itself was a 17 gm, 16 mm tall aluminum octagon with mirror finished faces. The period of the torsion oscillations was 80 s at room temperature, and 76 s at low temperature (77 K). Large source masses were placed outside the dewar.

As in most of the gravity pendulums already discussed, the location of this apparatus was slightly exotic. The system was placed in an abandoned Nike missile bunker in a federal land preserve near Richland, in eastern Washington state. This isolation offered much reduced seismic noise.

5.4.4 Viscosity measurements and granular media

It is possible to adapt the torsion pendulum to quite a different kind of practical measurement. In fact, Coulomb appreciated the potential of the torsion pendulum for determining the viscosity of fluids. As illustrated in Fig. 5.23, an oscillating disc is placed in the fluid of interest.

In the absence of viscous drag, the disc will simply oscillate at the end of the torsion fiber. The effect of viscosity is to damp these oscillations with resulting motion as described earlier (Eq. (5.10)). For the usual case of underdamping, the solution will be of the form Eq. (3.6).

By observing experimentally the decrease in the magnitude of successive maxima in the left and right deflection angles of the twisting disc, the so-called logarithmic decrement λ can be measured. From two such measurements, one in air and the other in the fluid, the viscosity η can be deduced from theoretical expressions such as (Newman 1961, p.232):

$$\eta = \frac{16I^2}{\pi\rho t_0(R^4 + 2R^3\delta)^2}\left[\frac{\lambda_1 - \lambda_0}{\pi} + \left(\frac{\lambda_1 - \lambda_0}{\pi}\right)^2\right],$$

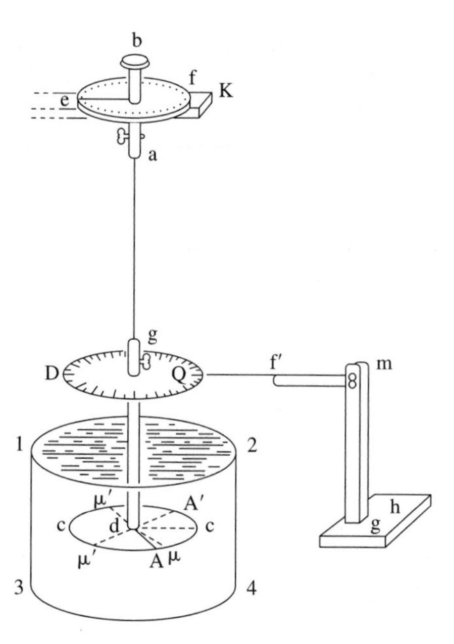

Fig. 5.23
Coulomb's torsion pendulum used for viscosity measurements. Reprinted by permission of Princeton University Press from (Gillmor 1971, p. 171), ©1971 Princeton University Press, 1999 renewed PUP.

where I is the total moment of inertia of the system, ρ is the density of the fluid, t_0 is the period of oscillation in air, R and δ are the radius and thickness of the disc, and λ_0 and λ_1 are the log decrements in air and the fluid, respectively.

A slightly different version of this idea was used in 1946 by Elevter Andronikashvili to determine the fraction of superfluid in liquid helium as a function of temperature. In the experiment a stack of closely spaced disks was attached to the end of the suspension fiber. When oscillating, a normal fluid penetrating the space between the disks would induce a drag torque. In contrast a superfluid, having zero viscosity, induces no drag.

At lower and lower temperatures, a given volume of liquid helium becomes composed of a larger fraction of superfluid as compared to normal fluid. Therefore the amount of drag torque becomes a measure of the ratio of normal fluid to super fluid. In turn the drag torque is reflected in the resonant frequency of the system. Andronikashvili's experiment measured this resonant frequency versus temperature.

A similar approach can be taken to the determination of properties of granular media such as fine powders. Figure 5.24 shows a forced torsion pendulum (D'Anna 2000) that ends in a probe, which is immersed in the granular medium.

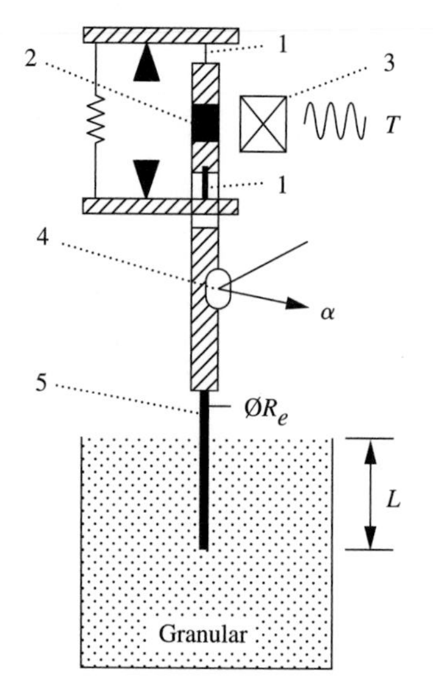

Fig. 5.24
Torsion pendulum immersed in a granular medium. Components are: suspension wires (1); permanent magnet (2); external coils (3); mirror (4); probe (5). Reprinted with permission from (D'anna 2000, p. 983). ©2000 by the American Physical Society.

By means of the small permanent magnet, a harmonic torque could be applied to the pendulum by sinusoidally exciting a pair of external coils. The natural frequency of the system (30 Hz) was higher than the forcing frequency (1 Hz). The experiment consisted of measuring the loss factor (by monitoring the time dependent angular displacement $\alpha(t)$ via the usual optical lever readout) as a function of the amplitude of the applied ac torque. Various granular materials were studied in this way; the list included large and small glass beads with diameters of 1.1 mm and 70 μm, sand with grain size 160 μm, and even snow.

5.5 Exercises

1. In a practical experiment the diameter d of the wire or fiber of the torsion pendulum is measured. Modify Eq. (5.6) to include the diameter rather than the radius. Now consider a pendulum of length 1 m with a moment of inertia of 500 gm cm^2. Using these data, data from the text, and the modified equation, fill in the blanks in the table.

Material	η	d	T_0
Cast iron	5×10^{11}	24 μm	24 μm
Bismuth	3.2×10^{11}		0.793 s
		1 mm	0.498 s
Quartz	3.0×10^{11}	0.1 mm	

2. Suppose that you wanted to design a wristwatch whose timing was based upon a small torsion pendulum. The pendulum can have a length of 2 mm and the suspended mass is a uniform wheel of radius 0.5 cm and mass of 5 gm. The moment of inertia of a uniform wheel is $I = 1/2(\mathrm{MR})^2$. The watch is to be made of super hard steel with a coefficient of rigidity $\eta = 10^{10}$ dynes/cm^2 and the period of the pendulum is to be 1 s. What must be the diameter for the steel wire of the suspension? Does such a pendulum seem to be mechanically viable?
3. A typical undergraduate physics laboratory exercise involving the torsion pendulum proceeds in the following way. The pendulum consists of a wire, fixed at one end (the suspension point) and connected at the other end to some configuration of unknown moment of inertia, I_1. The vibration period T_1 of this configuration is measured. Then a ring of known dimensions and therefore calculated moment of inertia I_2 is added to the oscillating part of the pendulum so that the total moment of inertia is now $I_1 + I_2$. The vibration period T_2 of this new configuration is measured. Measurements of the suspension wire length L, and the wire diameter d are made. The object of the experiment is to develop formulas (and therefore numerical values)—in terms of the known quantities T_1, T_2, I_2, d, and L—for the torsion constant τ, the modulus of rigidity η, and the unknown moment of inertia, I_1. Derive these formulas.
4. The logarithmic decrement, δ, measures the decay of the amplitude of an oscillating dissipative system. A typical definition is $\delta = \ln[A_i/(A_{i+1})]$ where A_i refers to the amplitude of the ith oscillation. Using Eq. (3.6) calculate δ for the underdamped pendulum. The definition can be generalized as follows. Let $\delta_n = \ln[A_i/(A_{i+n})]$, so that the logarithmic decrement now depends upon n. Plot a graph of δn versus n. The slope of this graph is given by our first definition of δ. Note that the graph is a straight line. How could measurement of the logarithmic decrement be used to test whether the amplitude is decaying exponentially?
5. A typical undergraduate experiment[15] might yield the following results for a Cavendish experiment: $M = 1$ kg, $m = 15$ gm, $r = 4.8$ cm, $L = 6.6$ cm, $T = 251$ s, and $\theta = 0.0014$ rad. Calculate the resulting value of the universal gravitational constant, G.

[15] The data in this problem are suggested by that found in the manual for the Cavendish balance manufactured by TEL-Atomic, 1223 Greenwood Ave., Jackson, Michigan, 49204, USA.

6. With reference to the previous exercise, suppose that the angle $\theta = 0.0014 \pm 0.0001$ rad is measured with an optical lever using a laser. The laser beam with the lever hits a scale where it has a cross-section of 3 mm. That is, suppose that the 3 mm beam corresponds to the 0.0001 rad uncertainty in the angular measurement. Approximately how far must the scale be from the mirror on the pendulum to achieve this precision? Note that even this small error will contribute a 7% error to the final result.
7. The velocity of a bullet can be measured with a ballistic galvanometer and a charged capacitor as follows. A capacitor C is placed in a circuit with a resistor R and a power supply of voltage V. Sufficient time is allowed so that the capacitor is fully charged. The circuit is constructed using, among other things, two pieces of aluminum foil. One strip of foil is placed so that a fired bullet will break it and thereby disconnect the circuit from the power supply. The capacitor now begins to discharge through the resistor. The second strip, distance d, is placed so that the circuit involving the capacitor and resistor is broken when the bullet breaks that strip. The capacitor stops discharging and now has a remaining fixed charge of Q that can be measured with the ballistic galvanometer. Assume that the galvanometer accurately measures the charge and derive an expression for the velocity of the bullet.
8. This problem is based upon a dynamic method of measuring the universal gravitational constant G using a Cavendish pendulum. Suppose that the structure around the pendulum is modified so that the position of the large balls can be easily switched between positions 1 and 2 as labeled in Fig. 5.13. Initially—and just for a moment—the large balls are placed in position 1 on one side of the small balls slightly pulling the pendulum. (The pendulum has a long period compared to the motion of the large balls.) Then the axis of the large balls is rotated away from the pendulum to a position of about 90° relative to the pendulum from which the large balls will have negligible influence. Gradually the pendulum will swing the other way. At the point of maximum angular displacement, the large balls are rotated to position 2 in Fig. 5.13, thereby pulling the small balls on the pendulum toward greater angular deviation. In this way a gravitational pull is exerted twice during each period of the pendulum's motion. The situation is identical to that of a swing being push twice each cycle and the resulting angular deviation is similar to that shown in Fig. 3.10. Somewhat analogous to the Foucault pendulum, the cumulative effect is much greater than that which is observed with a static method. Let us do the analysis using the methods developed for the swing in Chapter 3. (a) The equation of motion is

$$\ddot{\theta} + 2\gamma\dot{\theta} + \omega_0^2\theta = A\left[\delta(t) + \delta\left(t - \frac{T}{2}\right) + \delta(t - T) + \delta\left(t - \frac{3T}{2}\right)\right.$$
$$\left. + \delta(t - 2T) + \cdots\right],$$

where each of the pulses is approximated by a Dirac delta function. Using the Laplace transform methods of Chapter 3, solve this equation of motion. Assume that the damping is light. (b) Follow the outline of problem 12 in Chapter 3 and write the solution in terms of a finite geometric series. (c) For long times what is the equation of the limiting steady state solution? The gravitational constant G is embedded in the constant A and the damping constant γ is determined from observation of the decay of oscillations of the nonforced pendulum.
9. A cylinder of radius 5 cm and mass 1 kg is suspended with axis vertical by a long fine thread of torsional constant τ and is immersed to a depth of 10 cm in a liquid of viscosity 1.50 in CGS units contained in a coaxial cylinder of radius 5.10 cm. Neglecting forces on the base of the suspended cylinder, find the value of the torsion constant such that the cylinder, when twisted through a small angle, returns to its equilibrium position without oscillation.

The chaotic pendulum

6

6.1 Introduction and history

Much of the history of physics has been characterized by the effort to understand, in great detail, increasingly smaller pieces of nature; beginning with classical particles and waves, and progressing to molecules, atoms, nuclei, and elementary particles. This trend became especially pronounced in the twentieth century with the development of sophisticated experimental apparatus capable of probing deeply into nature's innermost parts. Aside from the sense that one is closer to reality at the deeper levels of nature, it is plausible to assume that a clear understanding of the small pieces of nature will lead to a clear view of the large picture. The whole is presumed equal to the sum of its parts. This approach is sometimes call *reductionism*.

More recently, in certain areas of physics, the opposite methodology has proved fruitful. New structure and organization may become evident when there is complexity, large numbers of parts, several degrees of freedom, or even just sufficient energy to make a discrete change in the system. Indeed, sometimes the whole is more than the sum of its parts. This creation of new richness of behavior often occurs in the study of processes that are pushed well beyond their equilibrium configurations. Researchers find new levels of organization, new complexity that does not seem to be obvious from a consideration of the individual parts of the process (Prigogine 1980). For example, if reactants are forced rapidly into certain chemical reactions, the resultant products may show spatial or temporal ordering (Zhabotinskii 1991). Or convective systems with large temperature gradients may exhibit new structural or dynamic organization of fluid motion. Even the motion of the humble pendulum achieves a new level of complexity if it is driven energetically at nonresonant frequencies. Why should this be? Part of the answer lies in the fact that when systems are well beyond equilibrium, they are often unstable, and instability can lead to new complex states. The entire field of chaotic dynamics is an important manifestation of a new order being achieved through instability. It is remarkable that the damped driven pendulum is an archetypical example of chaotic dynamics. The chaotic pendulum is therefore our next story of the pendulum.

What do we mean by chaotic dynamics? What are its characteristics? Chaotic systems are unstable deterministic systems whose trajectories are extremely *sensitive to the initial condition* of the system. The French mathematician Henri Poincaré is generally credited as being the first to articulate the behavior of such systems. In Chapter 1, we repeated his often quoted words that suggest how a deterministic system might behave in a probabilistic way. Let us give a larger version of that quotation.

A very small cause that escapes our notice determines a considerable effect that we cannot fail to see, and then we say that the effect is due to chance. If we knew exactly the laws of nature and the situation of the universe at the initial moment, we could predict exactly the situation of that same universe at a succeeding moment. But even if it were the case that the natural laws had no longer any secret for us, we could still only know the initial situation *approximately*. If that enable us to predict the succeeding situation with the same approximation, that is all we require, and we should say that the phenomenon had been predicted, that it is governed by laws. *But it is not always so:* it may happen that small differences in the initial conditions produce very great ones in the final phenomena. A small error in the former will produce an enormous error in the latter. Prediction becomes impossible, and we have the fortuitous phenomenon.[1]

In this statement, Poincaré makes the distinction between two kinds of deterministic systems. He notes that we would consider a system to be deterministic if our prediction of its final state were accurate to approximately the same degree as our knowledge of its original state. This is our common experience. In science education we perform laboratory exercises in which variables are slightly uncertain. Yet we are satisfied that our calculations and observations have proved some physical law if the results give *approximately* the "right" numbers. That is, uncertainty is not amplified. (Technically, we allow uncertainty to increase only *linearly* with time if the process is time dependent. If the initial uncertainty is Δx_0 then after a time t the uncertainty has grown according to $\Delta x = \Delta x_0 \lambda t$, where λ is some constant growth rate.) On the other hand, Poincaré suggests that with certain systems the initial uncertainty increases quite rapidly and therefore the final outcome looks as if it is the result of a random process or probability. Systems that behave in this latter manner are *unstable*. Unstable systems have the ability to be affected by very small changes in their initial conditions. This property is referred to as *sensitivity to initial conditions* (SIC) and it is the hallmark of chaotic dynamics. (Technically, sensitivity to initial conditions occurs when an initial uncertainty grows *exponentially* in time. The uncertainty grows according to $\Delta x = \Delta x_0 e^{\lambda t}$.) Of course, if a system is deterministic and if the initial conditions are somehow known to an *infinite* degree of precision, then the final state of the system is completely predictable. In practice however, no real system is completely determined, and therefore it is the instability of the system coupled with the initial uncertainty that leads to the possibility of widely varying outcomes. Thus, the deterministic unstable system, the chaotic system, gives the appearance of being probabilistic.

[1] See (Poincaré 1913, p. 397).

Chaotic dynamics, or Chaos as it is popularly named, is not to be confused with the primordial chaos suggested by various mythologies to exist at the moment of creation, or with the chaos that might ensue in the wake of a parade of small children and active dogs being turned loose in a confined space. In fact, the term "chaos" was coined by James Yorke and Tien-Yien Li of the dynamics group at the University of Maryland. In 1975 Li and Yorke published a paper with the title "Period Three implies Chaos" (Li and Yorke 1975) and the name stuck. Thus, we now define chaotic systems as deterministic systems that are unstable enough to exhibit a probability-like nature through their sensitivity to initial conditions. While confusion of the technical meaning with the common meaning is unfortunate, *chaos* is now firmly entrenched in the scientific literature as the short form for *deterministic chaotic dynamics*.

The history of chaos might be arbitrarily divided into two segments: first, from the time of Poincaré's large work on dynamics in 1892 to Edward Lorenz's 1963 computer simulations of unstable convection and second, from the early 1960s to the present. This division provides an approximate distinction between the scattered efforts of individual researchers, and the later epoch in which chaos came to be an accepted field complete with journals and conferences dedicated to the subject. We briefly mention a few of these contributions from the first epoch. In his studies of celestial motions (Peterson 1993), Poincaré realized that analysis had limitations especially when it came to nonlinear dynamics and revived the use of geometric methods to study stability in such systems. The American mathematician George David Birkhoff (1884–1944) continued the geometric tradition of Poincaré and helped to develop the general aspects of the connection between unstable systems and probability. In 1932 he published the first chaotic attractor to appear in the literature. The Russian mathematician Aleksandr Mikhailovich Liapounov (1857–1918)[2] studied the question of stability in dynamical systems and his name is now associated with the exponent λ that is the rate of exponential growth of uncertainty. In 1944 Levinson suggested that a system with three degrees of freedom and forcing (like a driven pendulum) could produce a chaotic attractor. Other important pioneers include the English mathematician Dame Mary Cartwright (1900–1998) who, in collaboration with John Littlewood (1885–1977), analyzed the van der Pol nonlinear oscillator during World War II (Tattersall and McMurran 2001), the Russian mathematicians Kolmogorov, well known for his work in probability, and Arnol'd and Moser, the last three for whom the famous KAM theory of nonlinear systems is named. Stephen Smale produced a series of seminal theoretical papers in the 1960s that are important for more advanced treatment of the chaotic pendulum. On the experimental side Balthasar van der Pol (1889–1959) used (*c.*1927) an LRC electrical circuit with a nonlinear resistance to produce chaotic behavior. Georg Duffing (1861–1944) created mechanical devices, specifically oscillators, with spring

[2] The spelling "Liapounov" is slightly different from that typically used to describe the "Liapounov" exponent.

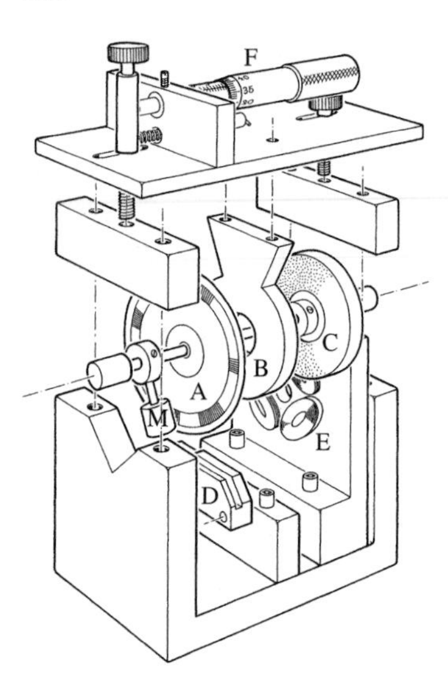

Fig. 6.1
Exploded view of an experimental forced, damped pendulum, designed by JAB and John Smith.

Fig. 6.2
Photograph of the Blackburn–Smith pendulum.

constants that depended upon position, to study nonlinear vibrations (*c.*1908). Finally we note the work of the mathematician Chihiro Hayashi (1911–1987) of Kyoto University in whose research group then (in 1961) graduate student Yoshisuke Ueda (born 1936) used an analog (as opposed to digital) computer to produce a strange attractor simulation of a nonlinear oscillator (Ueda 2000).

From the 1960s chaotic dynamics become a much more universal and active field. It is quite impractical to give anything like a complete history. We just note that many of the techniques and characterizations that we will discuss for the chaotic pendulum were developed in this period. The digital computer became a primary tool and an initial series of computer simulations of the chaotic pendulum were published by Gwinn and Westervelt from Harvard University during the mid-1980s (Gwinn and Westervelt 1985; 1986). Other researchers also saw the possibilities in a study of the pendulum. (See, for example (Bryant and Miles 1990).) The first physical model of the chaotic pendulum was built by one of us (JAB) in collaboration with John Smith of the University of Waterloo (Blackburn et al. 1989a). An exploded view of this design and a photograph of the actual system are shown in Figs. 6.1 and 6.2. Through the optical encoder, data giving the angular displacement of the pendulum is sent to a computer where it can be processed for further analysis. The fit of such data to the simple nonlinear dynamical model described below is remarkably good (see Fig. 6.3).

Variations on the basic chaotic pendulum have also been constructed and analyzed (See, for example, Starrett and Tagg (1995) and Shinbrot et al. (1992).) In 1990, one of us (GLB) with coauthor Jerry Gollub of Haverford College and the University of Pennsylvania gathered together many aspects of the behavior of the pendulum in an accessible undergraduate text. (Baker and Gollub 1996). Many of the figures in this chapter are from that book and further details of much of what we describe here may be found there.

Chapters 2 and 3 of this book provide a great many technical details about the mechanics of the pendulum. In order to introduce concepts gradually we used the linearized version of the pendulum to develop some sense of the pendulum's motions. The mathematics for the linearized pendulum is solvable by straightforward analytic techniques. However, the full nonlinear pendulum is more complex. As we discovered in Chapter 3, even an analysis of the natural frequency of the nonlinear pendulum becomes an involved calculation. While the linearized pendulum can be used to illustrate much of the expected behavior of the pendulum, only the nonlinear pendulum is capable of chaotic motion. This requirement of nonlinearity is actually a necessary condition for chaotic motion in any physical system. It is the nonlinearity which leads to instability and the possibility of an abrupt transition, a *bifurcation*, to a new kind of motion. A classic example of nonlinearity is the straw that breaks the camel's back. The camel may be loaded with increasing weights with only a slight proportional buckling of its legs. Yet there may come a point where the addition of just one small thing—a piece of straw—may be sufficient to "break" the camel's back. The camel would then be completely prostrate and unable to carry any weight at all. Thus there is an abrupt transition from one "state" to a completely different state. A slightly more scientific illustration is the example of a weight suspended at the end of thin wire. As more weight is added the wire stretches proportionately (linearly)—the behavior predicted by Hooke's law. But at some point a small extra weight causes a disproportionately large stretch or even breakage in the wire. Again, this behavior reflects a nonlinear response by the wire. Similarly, nonlinearity allows the pendulum to exhibit corresponding transitions. For small amplitudes of forcing the driven pendulum exhibits the usual periodic behavior that characterizes the linearized pendulum's motion, but for larger amounts of forcing where nonlinearity becomes important, the picture may change completely. The motion may become chaotic.

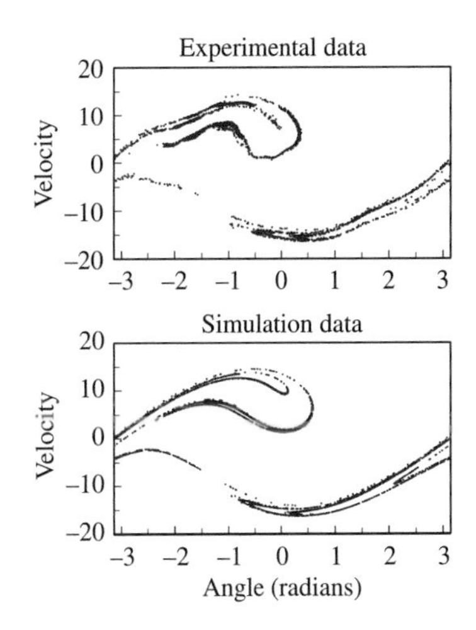

Fig. 6.3
Comparison of data from the experimental pendulum with simulated data computed from a mathematical model of the pendulum.

6.2 The dimensionless equation of motion

For a rigid compound pendulum of moment of inertia I, we may write the complete equation of motion for the sinusoidally driven, damped, nonlinear pendulum as

$$I\frac{d^2\theta}{dt^2} + b\frac{d\theta}{dt} + mgd\sin\theta = \Upsilon\cos\omega_F t, \tag{6.1}$$

where b is the friction parameter, mgd is the critical gravitational restoring torque, d is the distance from the pivot to the center of mass, Υ is the amplitude of the drive and ω_F is the angular driving frequency. The natural angular frequency of the corresponding linearized pendulum was given previously as $\omega_0 = \sqrt{mgd/I}$. The study of Eq. (6.1) can be simplified by reducing the number of adjustable parameters. This is achieved through the introduction of dimensionless parameters—those which have no physical units. Toward this end, we first introduce a dimensionless time parameter, $t' = \omega_0 t$. By appropriate use of the chain rule and the following conversions

$$\begin{aligned} Q &= \omega_0 I/b \\ A &= \Upsilon/(\omega_0^2 I), \\ \omega_D &= \omega_F/\omega_0 \end{aligned} \tag{6.2}$$

the nondimensional equation becomes:

$$\frac{d^2\theta}{dt^2} + \frac{1}{Q}\frac{d\theta}{dt} + \sin\theta = A\cos\omega_D t. \tag{6.3}$$

(For simplicity, the primes on the nondimensional time t have been omitted.) Instead of the seven parameters of the original equation of motion, this dimensionless equation has only three adjustable parameters, Q the inverse of the strength of the damping, A the strength of the forcing, and $\omega_D$the drive frequency relative to the natural frequency. Adjustments in any of the parameters can radically change the pendulum's motion. Thus, the value of each member of this minimal set of parameters will determine the dynamics.

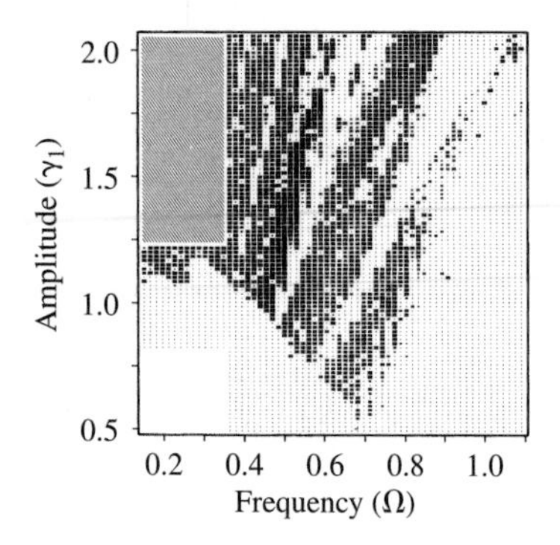

Fig. 6.4
Behavior of an experimental pendulum with fixed damping but variable normalized forcing amplitude (γ_1) and variable normalized forcing frequency (Ω). Filled squares represent chaotic motion and small dots represent periodic motion. Each point is derived from a single experiment and there are a total of 5250 points. Because of the time constraints, the large shaded rectangle was not covered experimentally, but the behavior in that region is mainly chaotic. The damping was fixed at Q = 4.2. See Blackburn et al. (1989*b*). The symbols used for the frequency and amplitude of the forcing are different from those found in this text.

The motion of the pendulum will either be periodic or chaotic. With periodic motion the pendulum repeats its behavior at regular intervals. In the simplest case, the period of the motion will coincide with the period of the forcing. But there are other possibilities. Periodic motion may also be more complex than a simple back and forth oscillation yet still be characteristically repetitive. To say that the motion of a driven pendulum is complex is an understatement. Much of the rest of this chapter describes tools that are used to display and characterize the wide variety of possible motions. Readers wishing to gain a quick impression of the possible complexity may look ahead to Figs. 6.9–6.11 in which the motions are schematically portrayed as the parameters A, ω_D, and Q, respectively, are varied. Ideally, one would like to study the pendulum's motion for every set of parameter values. That is, every state of the pendulum would be represented by a point with coordinates (A, ω_D, Q) and the actual motion might be indicated by the symbol placed at the point. Such a three-dimensional diagram would be very time consuming to produce and probably confusing to interpret. However, a two-dimensional version for which one parameter is held fixed is helpful. Such a diagram is shown here in Fig. 6.4. Unlike computer simulations, this figure was created with data obtained from a real driven pendulum.[3]

6.3 Geometric representations

There are a variety of ways to represent the pendulum's motion geometrically. Some of these we have already seen in other parts of the book. In Fig. 6.5 representative periodic motions are illustrated by sketches of the actual paths of the pendulum bob over several cycles. Situated beside the sketch of each motion is the corresponding phase plane diagram, first described in Chapter 2. The phase diagram or phase portrait is a graph of angular velocity versus angle. Each phase diagram or phase portrait is traced through many drive cycles and yet, in Fig. 6.5, the motion, although somewhat complex, remains periodic and therefore the phase portrait is fairly simple. As the amount of forcing increases complexity generally increases. For example, the period of the motion is sometimes twice or even four times that of the forcing motion. This phenomenon, known as *period doubling*, can be a mechanism that leads to chaotic motion. Thus, the understanding of chaotic motion is facilitated by using a variety of geometric representations.

[3] A small pendulum will have a period of perhaps one-half second. In Fig. 6.4, each data point is the result of a single experiment carried out under the following protocol: (a) the ac drive was switched off and the pendulum was allowed to come to rest (b) the drive amplitude and frequency were set electronically to the desired new values (c) the drive was turned on and the pendulum motion was allowed to stabilize (d) the pendulum motion was observed over a measurement interval using a data acquisition system. The information from this final stage was then interpreted to form a conclusion as to the type of motion that had occured. From start to finish, the protocol took several minutes. Obviously, more than 5000 points in the figure would require weeks of experiments.

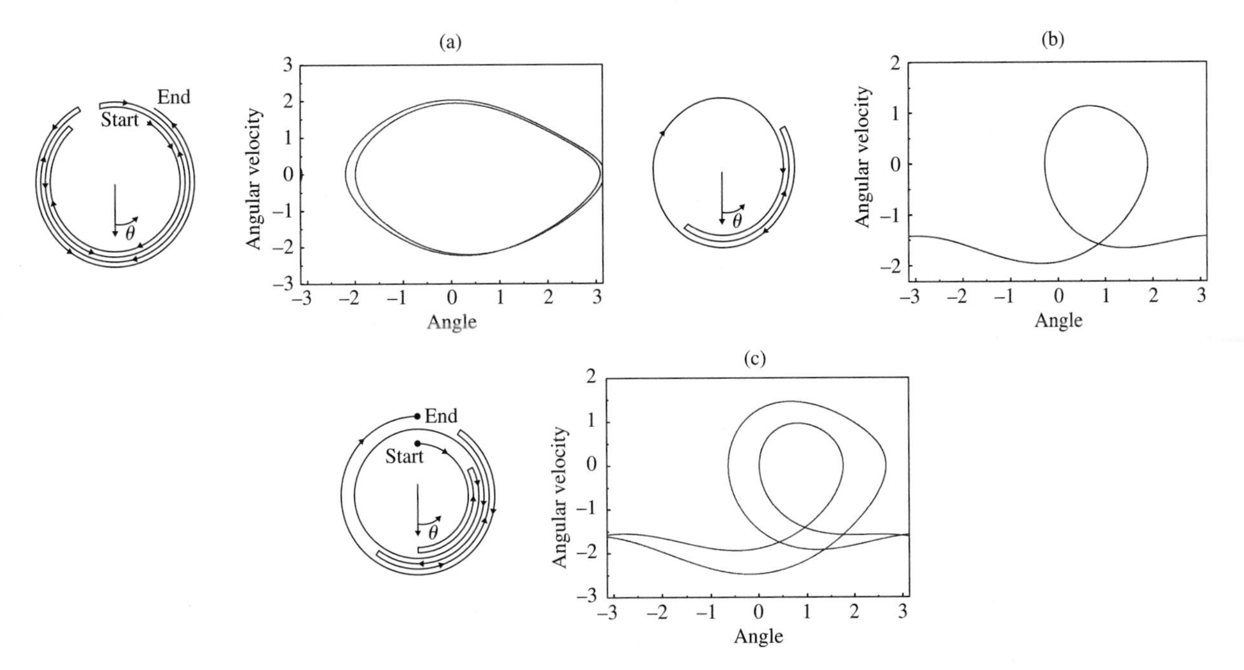

Fig. 6.5
Three examples of complex, yet periodic, motions of the driven pendulum. The figures on the left of each pair portray the pendulum motion in real space. The figures on the right are the corresponding phase plane diagrams. The boundary condition on the angle is periodic and therefore occasional apparent discontinuities appear. (Reprinted from Baker and Gollub (1996, p. 22), with permission of Cambridge University

6.3.1 Time series, phase portraits, and Poincaré sections

Poincaré introduced another geometric representation that will be of great value when phase portraits become more complex. If one imagines the motion occurring in a darkened room with only a strobe light for illumination, then the phase coordinates are only available when the light flashes on. If the strobe frequency is made equal to the forcing frequency the observer would see only those phase coordinates that are synchronized to the forcing. Thus in Fig. 6.5 (b) with the motion whose period is the same as the forcing, there would only be one point. However, in Fig. 6.5 (a) and 6.5 (c), a complete motion takes two forcing periods, (period doubling) so there are two points. Such a diagram, whereby time is strobed or sectioned according to the forcing cycle, is called a *Poincaré section*.

Time series, either of the angular displacement or the angular velocity as functions of time, are familiar geometrical devices and provide another picture of the motion. With the same parameter sets as in Fig. 6.5, corresponding time series for angular velocity, are shown in Fig. 6.6. While the motions are all periodic, unlike the linearized pendulum, the time series are not simple sinusoids. In Chapter 2 we introduced the idea that any periodic function could be decomposed into its constituent sinusoids using the machinery of the Fourier series. Certainly periodic motions of Fig. 6.6 contain several Fourier components. Chaotic motion contains an infinite number of components and therefore *Fourier analysis* is also an important tool.

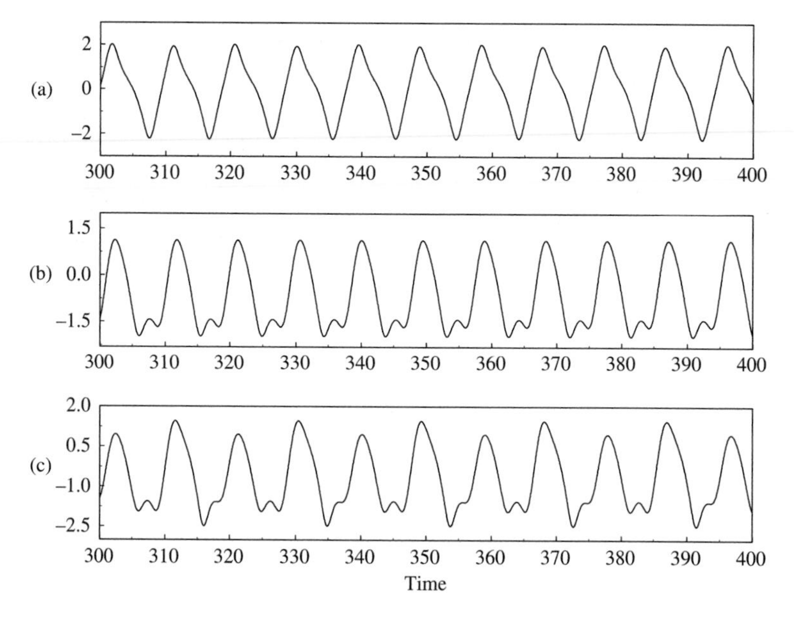

Fig. 6.6
Time series for the pendulum's angular velocity corresponding to parameters used in Fig. 6.5.

Thus we have several ways to represent the motion of the pendulum, time series, phase portraits, Poincaré sections, and Fourier analysis—although further elements of Fourier analysis or spectral analysis are needed for the discussion of chaotic motion.

Chaotic motion appears when the pattern of periodic motion is broken through instability. The motion then becomes nonrepetitive or non-periodic. Another way of describing the motion is to say that it contains *all* periodic frequencies, but that none of them last very long. More technically, it is said that chaotic motion contains an infinite number of unstable periodic orbits. As the phase trajectory wanders around the phase space, it comes very close to lots of periodic orbits but it is unable to maintain itself for very long on any one of these orbits. Thus, time series seem to have no discernible pattern or periodicity, the phase plane is very congested, and even the Poincare section becomes complex. These features are illustrated for the chaotic pendulum in Fig. 6.7.

The simplification caused by the strobe effect of the Poincaré section now becomes apparent. Whereas the phase portrait is quite congested, the Poincaré section has an appearance which lends itself to characterization. We note first that (for a dissipative system) the Poincaré section is confined to a particular complex region in space. No matter what the initial conditions the points will eventually all lie in this region which has a special characteristic appearance. Thus, the creation of the Poincaré section requires that the system be allowed to run until any initial aberrations are no longer evident, and then data are saved for the diagram. In essence the pendulum's motion is *attracted* to the set of points that form the Poincaré section. The Poincaré section and the corresponding phase portrait are

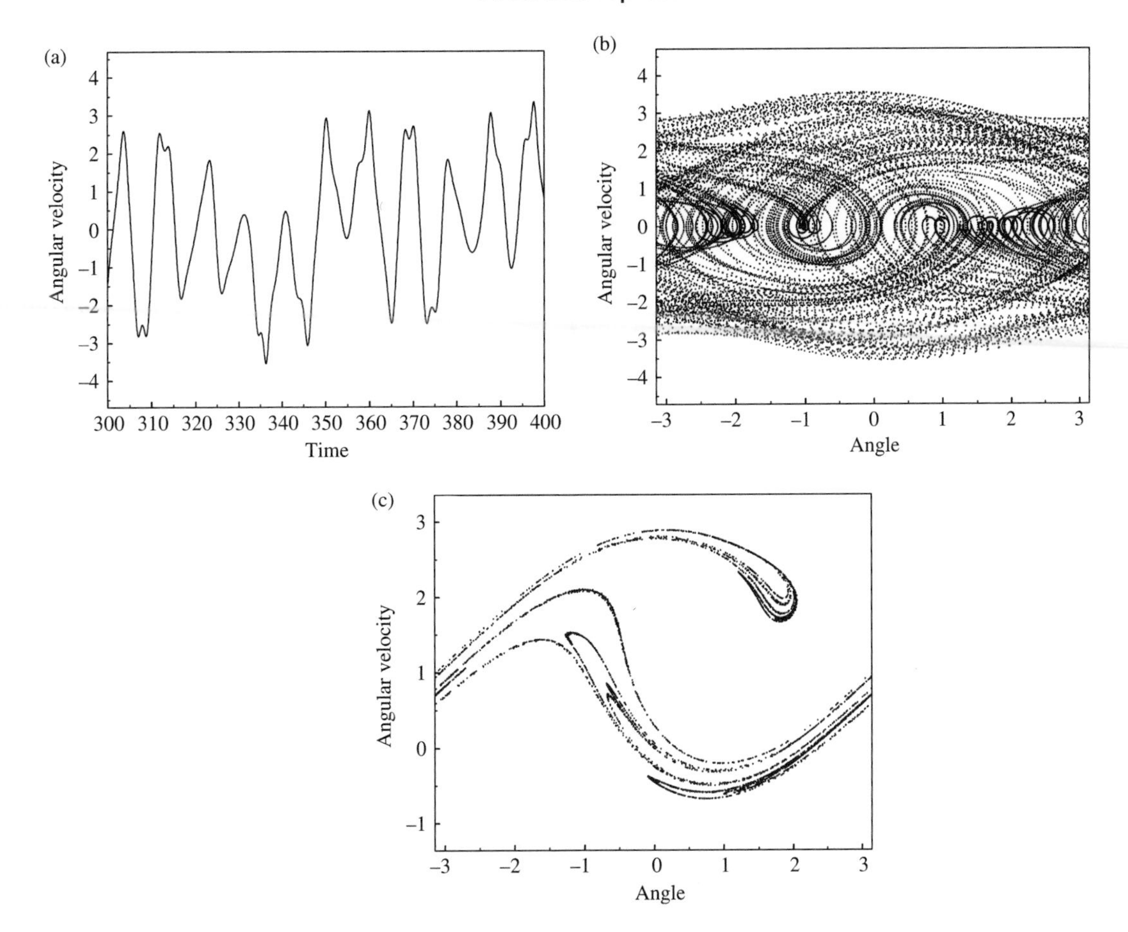

Fig. 6.7
Three geometric representations of the chaotic pendulum; (a) Time series of the angular velocity, (b) Phase plane diagram, and (c) Poincaré section.

said to be *strange attractors*. For nonchaotic motions, the attracting phase space set for periodic motion was referred to as an *attractor*. The word "strange" is related to the particular geometry of the attractor for chaotic motion. For chaotic systems in general and the chaotic pendulum in particular, the attractor is a *fractal*, about which more will be said later. Therefore, an attractor that is a fractal is called a "strange attractor".

Obviously chaotic motion is complex. This complexity suggests that nonlinear equations of motion are difficult and usually impossible to solve by analytic methods. The equation for the nonlinear driven pendulum has no analytic solution. Solutions are generated by numerical methods using computers to do repetitive calculations. It is not surprising that the bulk of the work in chaotic dynamics has occurred after the computer revolution. Edward Lorenz of MIT is generally credited with being the first to observe chaotic time series on a computer. Lorenz recounted the story of this bit of serendipity for a television program on chaos. During observation of the computer solution of his model equations for convective air flow in 1963, he had occasion to restart his computer. In doing this, he failed to put in the precise numbers used to start a previous computer simulation and the computer gave him quite a different result for the system trajectory. This was the first recorded instance of sensitivity to initial conditions, and led to the actualization of Poincaré's oft quoted prediction for the unstable behavior of a set of deterministic differential equations.

The computer is now the primary tool in the theoretical study of chaotic systems. In an ideal application of the scientific method experimental apparatus should be constructed whose behavior approximates that of the computer simulation. The experimental apparatus may then be used to test the theoretical model. But while chaotic systems are believed to be ubiquitous in nature, they are difficult to construct as controlled systems suitable for laboratory experimentation. Fortunately, the chaotic pendulum is one of the few exceptions to this chronic difficulty and, as we have seen, can be realized experimentally.[4] Thus, aside from the pendulum and a few other experimental systems, the computer is essential for the bulk of the research in chaotic dynamics. We therefore ask, "Given the instability and sensitivity of chaotic systems, does the computer accurately represent the motion of chaotic systems?". The answer is both "yes" and "no." Because of sensitivity to initial conditions and the finite precision of computer calculations, the actual trajectory of a system may, after a relatively short time not be the same as that which would be provided by an "infinitely" precise calculation using infinitesimally small time steps for the calculation. But fortunately chaotic attractors possess a characteristic that mitigates the deleterious effect of the approximate nature of the computer simulation. Figure 6.7(b) shows that chaotic trajectories are densely packed in the sense that their orbits in phase space are close together. Therefore the computer tends to produce a phase trajectory that "shadows" a "true" trajectory for some time. Then, as the computer orbit drifts away from the true orbit, it comes close to another "true" trajectory and shadows the true motion again and so forth (Grebogi et al. 1990; Fryska and Zohdy 1992). In this way, the computer does not necessarily provide a true trajectory but rather a series of shadowings of true trajectories. On the other hand, all true trajectories that are being shadowed by the computer calculations lie on the "true" strange attractor and consequently the shadow trajectories also lie on or extremely close to the strange attractor. Hence the computer does provide a true Poincaré section or phase portrait if data is taken over a long time. Therefore we accept the computer as a primary tool, partly from necessity and partly because it really does give true results in many circumstances.

6.3.2 Spectral analysis

But let us return to geometric representations of the pendulum's motion. What about spectral analysis? Since chaotic motion is not periodic, the

[4] While the Blackburn/Smith pendulum seems to have been the first experimental chaotic pendulum, there are other configurations. See, for example, (DeSerio 2003; Peters 1999).

Fourier decomposition into components as described in Chapter 2 no longer holds. There is no fundamental frequency of motion on which to base the Fourier series expansion. However, there is a generalization of the process used to obtain Fourier series that works for nonperiodic functions. This process requires a mathematical transformation known as the *Fourier transform*. In developing this transform, we start with the Fourier series of Eq. (2.11)

$$f(t) = \sum_{n=-\infty}^{n=\infty} a_n e^{in\omega_0 t}, \text{ where } a_n = \frac{\omega_0}{2\pi}\int_{-\pi/\omega_0}^{\pi/\omega_0} f(t)e^{-in\omega_0 t}dt. \tag{6.4}$$

in complex form. If the time series is not periodic as is the case with chaotic motion, then the nonperiodic motion can be thought of as part of a single wave whose period is *infinitely* long. The fundamental period of a possible Fourier series now becomes infinitely long or, conversely, the fundamental frequency becomes infinitely short. Creation of the Fourier transform then becomes a limit process for several of the variables as shown by the following transformations:

$$T \Longrightarrow \infty$$
$$n\omega_0 \Longrightarrow \omega,$$

where ω is a continuos variable, and

$$a_n \Longrightarrow a(\omega)d\omega \tag{6.5}$$

$$f(t) = \sum_{n=-\infty}^{n=\infty} a_n e^{in\omega_0 t} \text{ becomes } f(t) = \int_{-\infty}^{\infty} a(\omega)\, d\omega e^{i\omega t}, \text{ and}$$

$$a_n = \frac{\omega_0}{2\pi}\int_{-\pi/\omega_0}^{\pi/\omega_0} f(t)e^{-in\omega_0 t}\, dt \text{ becomes } a(\omega)d\omega = \frac{d\omega}{2\pi}\int_{-\infty}^{\infty} f(t)e^{-i\omega t}\, dt.$$

From these transformations the *amplitude*, $a(\omega)$, of each of the uncountably infinite number of frequency components is found using the *Fourier transform*, given by the equation

$$a(\omega) = \frac{1}{2\pi}\int_{-\infty}^{\infty} f(t)e^{-i\omega t}\, dt. \tag{6.6}$$

While the graph of the spectral components, a_n, of a Fourier series is a set of *discrete* lines at harmonics of the fundamental frequency ω_0, a graph of the Fourier transform $a(\omega)$, or frequency density, is a continuous function of the *continuum* of frequencies present in nonperiodic motion. Similarly, the original function $f(t)$ is now expressed in terms of an integral with each member of the continuum of frequencies being weighted by the frequency density according to the formula

$$f(t) = \int_{-\infty}^{\infty} a(\omega)e^{i\omega t}\, d\omega. \tag{6.7}$$

The Fourier transform of the time series for the chaotic pendulum shown in Fig. 6.7(a) is illustrated in Fig. 6.8.

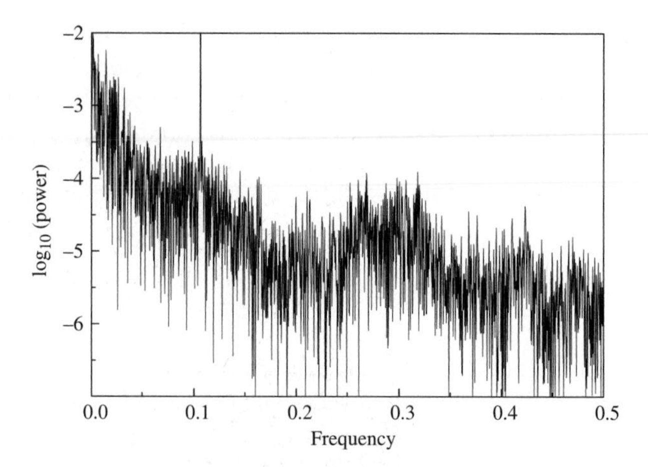

Fig. 6.8
Fourier spectrum for a chaotic pendulum. Aside from the strong component at the forcing frequency, the spectrum power diminishes inversely with increasing frequency.

It seems fairly clear that the idea for the Fourier transform stems from Fourier's 1811 paper (Kline 1972) although other mathematicians were working on similar ideas. But for the effective implementation of the transform in chaotic dynamics we rely on a modern approach. Toward this end we note that our development of the Fourier transform implies the use of a continuous function, $f(t)$. But the computer generates time series as a noncontinuous, discrete set of points and therefore chaotic times series generated by computers will be Fourier transformed, again by numerical integration. Furthermore, the discretization of the points requires that the appropriate Fourier transform is slightly modified and renamed as the *Discrete* Fourier transform (DFT). The process is essentially that of the continuous transform but the many integrations done in the transform process are performed by discrete numerical integration. (We also note in passing that the discrete nature of the sampling in the time series can cause subtle problems, but that is another story.) At any rate, for much of spectral analysis up until the mid-1960s the discrete transform was standard procedure. But in 1964, J. W. Culley and J. W. Tukey, at the urging of R. L. Garwin at the IBM research center, created a better method or "algorithm".[5] One way to assess an algorithm is by the number of computational steps required. A "better" algorithm therefore has fewer steps and computes faster. Thus was born the *Fast Fourier Transform* (FFT). This ingenious technique took advantage of certain symmetry properties of trigonometric functions at their points of valuation. The increase in speed over the discrete transform method is substantial. If the time series contains N points then the DFT requires a number of calculations that is of the order of N^2, whereas the FFT computation requires a number of calculations on the order of $N \log_2 N$. For $N = 1000$ points the efficiency is increased by a factor of more than 100 (Press et al. 1986). The spectrum shown in Fig. 6.8 was generated using the FFT algorithm. Let us now return to the application of spectral analysis to the chaotic pendulum.

[5] The FFT seems to have been discovered independently at different times, possibly as early as 1942, and implemented on hand calculators.

Periodic motions in time typically yield a small number of discretely spaced frequency components. Furthermore, even for periodic motions with very sharp corners that yield spectra with a large number of components or harmonics, the components are still spaced apart by a frequency interval equal to the fundamental frequency. But, as shown in Fig. 6.8, chaotic motion gives rise to an *infinite* number of *densely* packed frequency components. This observation reinforces the statement made earlier that chaotic motion contains an infinite number of unstable periodic orbits. The spectrum also shows a very strong component at the frequency of the pendulum forcing, together with a couple of harmonics, that characterize the forcing motion. By ignoring these components we see that, for the bulk of the components, the amplitude of the components is approximately proportional to the inverse of the frequency, $1/\omega$. This behavior points to much of the pendulum's motion as being in (unstable) periodic orbits of very long periods and is consistent with the notion that one route to chaos is through increasing complexity of the periodic orbits which, in turn, implies increasingly longer periodic orbits. A spectrum that is inversely proportional to frequency is also characteristic of a certain type of electrical noise, called "$1/f$" noise. This noise is above that which is inherent in an electrical system due to simple thermal agitation. "$1/f$" noise can be caused, for example, by fluctuations in the electrical resistance in a circuit. These fluctuations are quite prominent at low frequencies. In vacuum tubes—now largely a museum artifact—current was transmitted between the cathode and anode. The electron emission from the cathode is not strictly uniform and again, this generated noise whose spectrum exhibits $1/f$ behavior. The analogy between $1/f$ noise and a $1/\omega$ spectrum for chaotic time series is important. The processes that generate electrical $1/f$ noise are random. On the other hand, the governing rules for chaotic systems are deterministic. Therefore we once again witness a manifestation of Poincaré's statement that deterministic processes can sometimes appear to be probabilistic; that is, according to spectral analysis, the chaotic pendulum appears to behave randomly.

6.3.3 Bifurcation diagrams

Previously described geometric representations of chaotic motion all picture the development of a particular pendulum being driven with a particular forcing strength and frequency. We can also provide a more global picture, in the sense that the motion of a variety of pendulums can be represented on a single diagram. That is, we observe some aspect of the motion of the pendulum for a particular configuration, but then change slightly the configuration and repeat the process. This is done many times for incremental changes in one of the pendulum's parameters, (A, ω_D, Q). In this way, the motion is observed over a range of pendulum configurations. The bifurcation diagram shown in Fig. 6.9 is a typical example.

The vertical axis coordinate is the pendulum's angular velocity $d\theta/dt$ as determined at a single point during a drive cycle. It is essentially the vertical coordinate of the Poincaré section. The horizontal axis coordinate is one of the parameters of the system. In this example, it is the strength of the

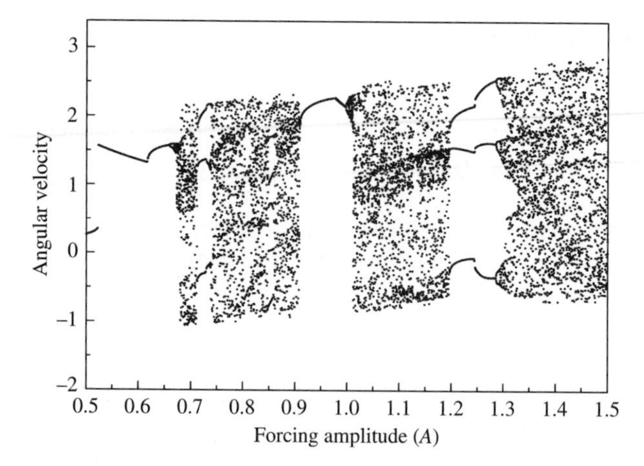

Fig. 6.9
Bifurcation diagram. A plot of the angular velocity, taken many times at a fixed point in the forcing cycle, versus the forcing amplitude.

forcing, A. In essence, each vertical segment of the diagram requires data equivalent to that of a Poincaré section and therefore the construction of the entire diagram is computationally intensive as it requires the creation of a Poincaré section for each value of A. The advantage of this sort of presentation is that it provides a summary of the possible types of motion over some range of the system parameter. Points where the pendulum's motion changes to a new kind of motion are called *bifurcations* and since this diagram is characterized by many such points, it is called a *bifurcation diagram*. At the left side, where the forcing amplitude is small, the motion has a period equal to the forcing period. Further to the right on the diagram the motion, still periodic, becomes more complex through successive *period doublings*. After a few such doublings the strobed angular velocity takes on many values and the motion is now chaotic. However, the motion does not remain uniformly chaotic but reverts to *windows* of periodicity throughout the range of forcing amplitudes. We see that there is often a certain sameness to the progression via period doubling from periodic motion to chaotic motion over the range of amplitudes. By looking at increasingly smaller ranges over which the system parameter is varied we magnify the bifurcation diagram and observe another phenomenon; namely, the number of periodic windows increases as the bifurcation diagrams are increasingly magnified. This property of increasing complexity under magnification is also found in phase portraits and Poincaré sections.

Similar bifurcation diagrams may also be created through variation of each of the other two parameters, ω_D the forcing frequency, and Q, the friction parameter. Examples of these are shown in Figs. 6.10 and 6.11.

The appearance of these figures is similar to that of Fig. 6.9. We noted earlier that, although impossible to draw, one might try to imagine a parameter space with coordinates (A, Q, ω_D) in which the various types of motion could be exhibited and thereby gain a comprehensive overview of the pendulum's behavior. Such a diagram would somehow contain in one figure all the information found in the two-dimensional representations of Figs. 6.9–6.11. Thus one could identify parameter sets of periodic (stable) and chaotic behavior (unstable).

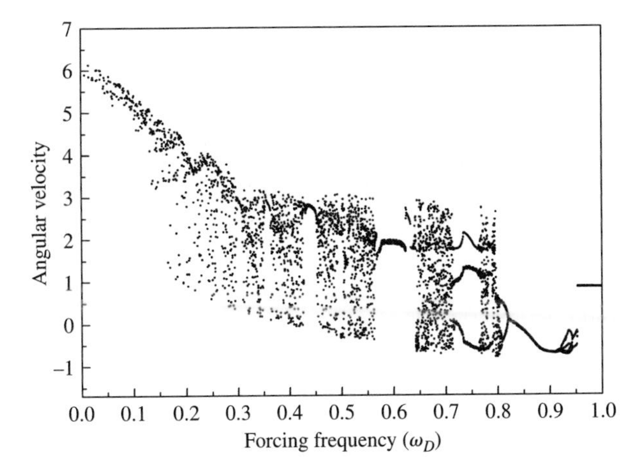

Fig. 6.10
A bifurcation diagram using the forcing frequency as the independent parameter.

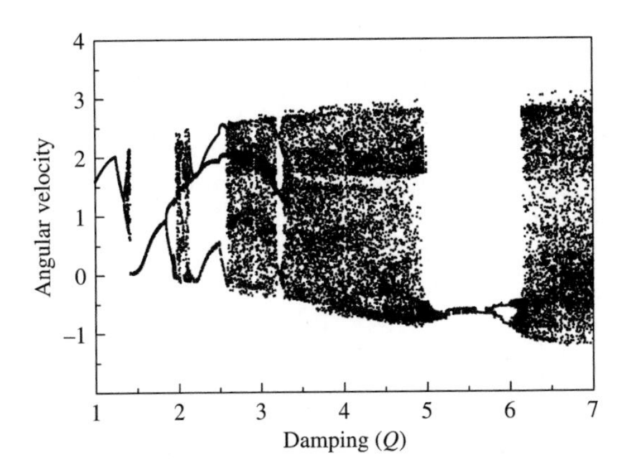

Fig. 6.11
Bifurcation diagram with the damping Q as the independent parameter.

6.4 Characterization of chaos

6.4.1 Fractals

The geometric representation of chaotic motion by, for example, the strange attractor of the Poincaré section, has an almost artistic quality. This representation is also a fruitful subject for mathematical analysis as an example of *fractal* geometry. As we will see, the quantitative description of the Poincaré section fractal is intimately related to the pendulum's dynamics. But let us first give some idea of the general properties of a fractal. The term "fractal" was coined by the mathematician Benoit Mandelbrot (Mandelbrot 1977). It comes from the Latin adjective "fractus" which means "broken". If we imagine the coastline of, say, Norway, and look at that country's map at ever increasing magnification, we would see that there is no characteristic length that can be applied. Any apparently straight lines are, upon further magnification, seen to break into smaller line segments. Hence any characteristic length is "broken". Fractals then are roughly defined as geometric sets that have no characteristic length. Fractals also exhibit a property known as *self-similarity*. For example, viewed at one level the coast of Norway appears to be fairly jagged. If a portion of that coast is now magnified to the same size as the original view, it will have an appearance similar to that of the original stretch of coastline. This sameness under observation at increasing magnifications is called self-similarity. In summary, fractals typically have (a) no characteristic length and (b) some degree of self-similarity.

Yet long before the term "fractal" was invented, several mathematicians studied geometric configurations that later came to be known as fractals. Some of the mathematicians that are associated with these fractal precursors are the German mathematicians Georg Cantor (1845–1918) and David Hilbert (1862–1943) who proposed *space-filling* curves, the Italian Mathematician Guiseppe Peano (1858–1932), Helge von Koch, the Swedish mathematician who, in 1904, published the *Koch snowflake*, the Polish mathematician Waclaw Sierpinski (1882–1969) who published the *Sierpinski gasket* fractal in 1916, the French mathematician Gaston Julia (1893–1978) who gave us the family of fractals known as *Julia sets*, and the German mathematician Felix Hausdorff (1868–1942) with whom we associate the idea of dimension of a fractal. (During World War II Hausdorff and his wife, both Jewish, committed suicide in anticipation that they were within a week of being sent to a concentration camp.) Mandelbrot, born in 1924, entered the scene much later. Mandelbrot's uncle recommended Julia's 1918 paper to him around 1945. For various reasons Mandlebrot did not like the paper and he found his own path to Julia's work in about 1977. With the aid of the computer Mandelbrot greatly expanded Julia's work and showed the potential for creating beautiful pictures. The *Cantor* set (1883) is simple but popular, and its construction is indicated in Fig. 6.12.

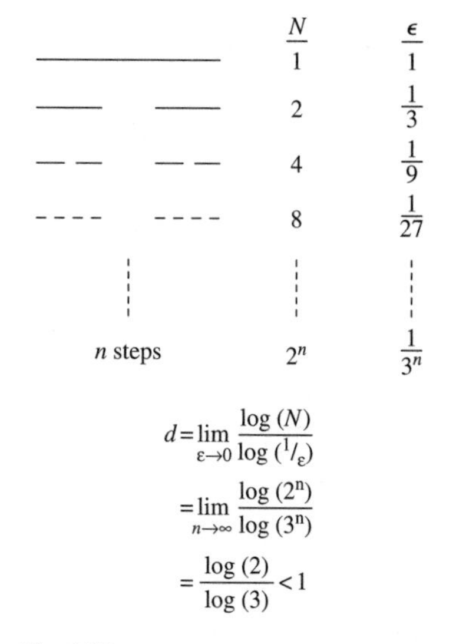

Fig. 6.12
Calculation of the fractal dimension for the Cantor set. (Reprinted from Baker and Gollub (1996, p. 113), with the permission of Cambridge University Press.)

The set is constructed by consecutively removing the middle third out of each line segment. After an infinite number of steps there are an infinite number of points but the pieces do not form a connected line. Clearly the Cantor set is self-similar—same appearance at varying magnifications—and has no characteristic length. The Poincaré section for the pendulum also has these properties in some measure. Figure 6.13 shows the Poincaré section under increasing magnification, and there is a sense that further structure is revealed with increasing magnification. This is especially the case in looking "across" the Poincaré sections.

We are quite used to the idea that objects have integer dimensionality—either one, two, or three dimensions. A single point or perhaps a finite collection of points have zero dimension. Lines are one-dimensional and consist of an infinite set of closely spaced points. Yet, one of the distinguishing features of most fractals is that they have noninteger values of dimension. This notion seems strange; but we see that while the Cantor set consists of an infinite number of points it has no parts that can be considered to be line segments. By some sort of intuitive interpolation, the

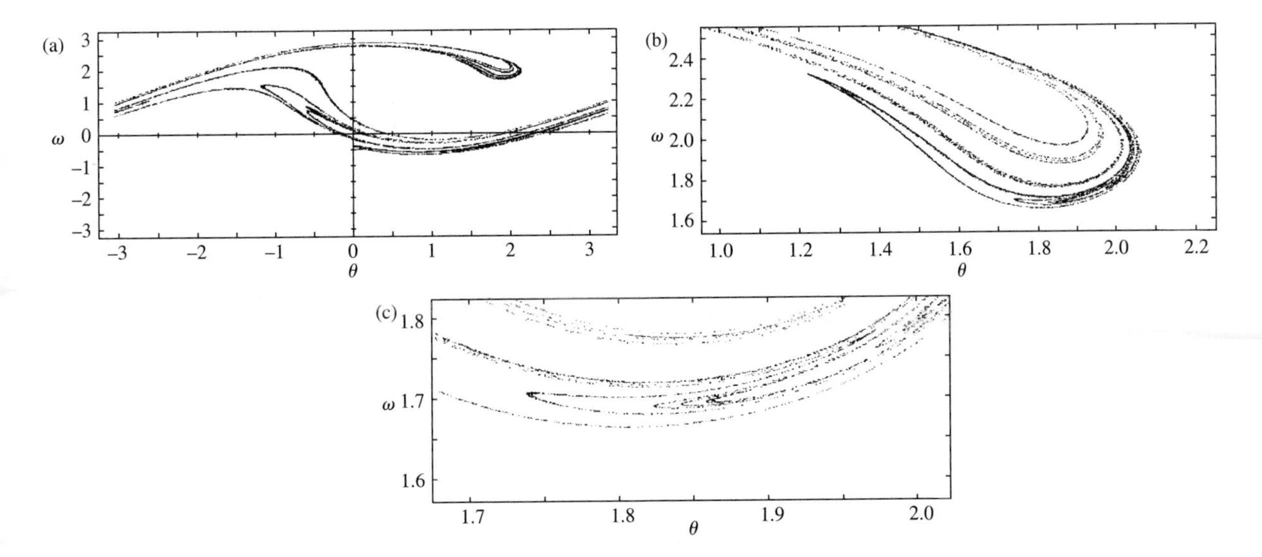

Fig. 6.13
Increasing magnifications of the Poincaré section of the chaotic pendulum that reveals the fractal structure of the attractor. (Reprinted from Baker and Gollub (1996, 56,57 pp.), with the permission of Cambridge University Press.)

Cantor set should have dimension somewhere between zero and one. To find the precise number requires a methodology for the calculation of dimension.

Perhaps surprisingly, there are several definitions of dimension. We will use a method that gives the *capacity dimension* d_C, introduced by Kolmogorov in 1959 (Kolmogorov 1959). As shown in Fig. 6.14, a line may be "covered" by a set of one-dimensional "boxes" of ever decreasing size, ϵ. If the length of the line is L then the number of boxes $N(\epsilon)$ is clearly equal to $L(1/\epsilon)$. Similarly the two-dimensional square in Fig. 6.14 can be covered by $N(\epsilon) = L^2(1/\epsilon)^2$ boxes. For a three-dimensional cube the exponent would be equal to 3. Therefore we define dimension as the exponent d in the expression

$$N(\epsilon) = L^d(1/\epsilon)^d \text{or, taking logs,}$$
$$d = \frac{\log N(\epsilon)}{\log L + \log(1/\epsilon)}. \quad (6.8)$$

In the limit as N gets very large and ϵ becomes very small the fixed term $\log L$ becomes negligible and the expression for dimension is simplified to

$$d = \lim_{\epsilon \to 0} \frac{\log N(\epsilon)}{\log(1/\epsilon)}. \quad (6.9)$$

This expression can be used to calculate the dimension, the *fractal* dimension, of the Cantor set as shown in Fig. 6.12. The result is found to be (log 2/log 3) = 0.63093, a number between zero (for a point) and one (for a line).

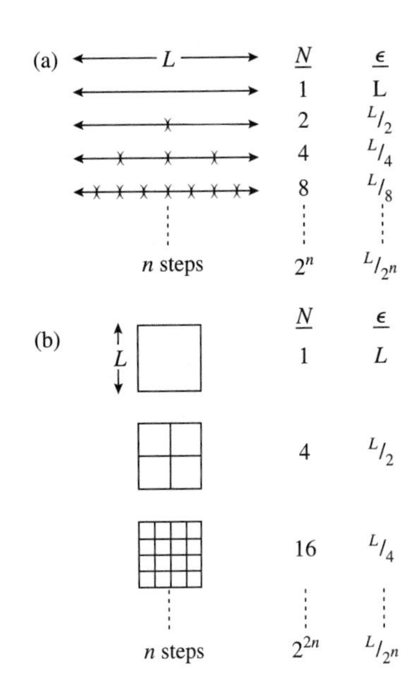

Fig. 6.14
Application of the box counting technique to calculate the capacity dimension. The scaling exponent gives the dimension. (Reprinted from Baker and Gollub (1996, p. 111), with the permission of Cambridge University Press.)

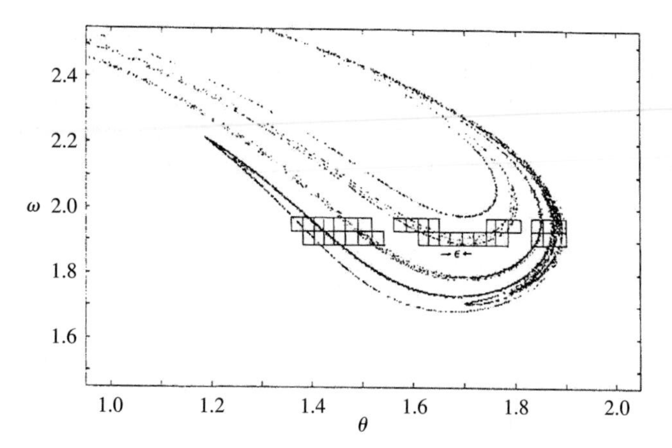

Fig. 6.15
A portion of the Poincaré section with some representative covering boxes. (Reprinted from Baker and Gollub (1996, p. 115), with the permission of Cambridge University Press.)

This procedure (or a similar one) may be used to calculate the fractal dimension for the Poincaré section of the chaotic pendulum. Figure 6.15 shows a set of boxes covering the Poincaré section.[6] Of course the process must now be done on a computer and is subject to some limitations. A close examination of the Poincaré section shows that its points do not cover an area, but are really a (possibly infinite) set of closely spaced lines. Therefore the Poincaré section is more than a line and less than an area. We expect its dimension to lie between one and two. Again, the methodology is indicated in Fig. 6.14.

A set of square boxes of side $1/\epsilon$ cover the Poincaré section. The logarithm of the number of boxes $N(\epsilon)$ is plotted against $\log(1/\epsilon)$ and the slope gives the fractal's dimension. For the parameter set $A = 1.5$, $Q = 4$, $\omega_D = 0.66$ the fractal dimension is about 1.3. The capacity or *box counting* dimension is the simplest measure of dimension. In this sort of work it is more common to use another, related but somewhat less intuitive measure of dimension, called the *correlation* dimension. For the details of this process as applied to the pendulum, see, for example, Baker and Gollub (1996).

The fact that the geometry of the Poincaré section has unusual fractal properties is itself an interesting and surprising feature. Even more striking is the fact that the fractal dimension is intimately and beautifully related to the dynamics of the pendulum through a measure of the pendulum's sensitivity to initial conditions. This measure is known as the *Lyapunov exponent*.

6.4.2 Lyapunov exponents

At the beginning of this chapter we introduced the notion of sensitivity to initial conditions and defined it in terms of the exponential separation in

[6] There exist many variations on the definition of dimension. For simple fractals there is a high degree of equivalency among them. But in some cases the more general definitions provide a higher degree of specificity of the fractal. For a concise summary see the table in Takayasu's book on page 150.

time of identical systems whose initial conditions are separated by an amount Δx_0. That is, the time dependence of this separation is given by $\Delta x(t) = \Delta x_0 e^{\lambda t}$. The parameter λ is called the Lyapunov exponent. By following the change in separation of an initial pair of coordinates one can determine the Lyapunov exponent. If the parameter is negative or zero then the separation either diminishes or stays constant. But if λ is positive then SIC holds and the system is chaotic.

For a one-dimensional system with the single variable $x(t)$, the interpretation of λ is relatively straightforward. However, the pendulum is a three-dimensional system, with angle, angular velocity, and time (or forcing phase) as the three variables. We need to think about a three-dimensional phase space and imagine following the growth or decay of a ball of coordinates (points) representing many identical pendulums in states that momentarily differ only slightly from each other. The evolution of this ball of coordinates is complicated. It may grow in some directions and shrink in other directions—becoming a kind of twisting and shape-changing football. In this process, the directions of growth and shrinkage are not constant but vary as the system evolves over the phase space in which the attractor is imbedded. If the initial volume of the ball is ΔV_0 then its volume a short time later is $\Delta V(t) = \Delta V(0)e^{(\lambda_1+\lambda_2+\lambda_3)t}$ where the three λ's—Lyapunov exponents—give the expansion or contraction rates in the three "directions" that define the football. (The time over which we follow these many points needs to be short because the attractor is limited in size and the evolving ball will fold back on itself for longer times. SIC is exhibited between foldings.)

For the chaotic pendulum, time, and thus the forcing phase, increases at a constant rate, and therefore there is no spread of the ball in the phase direction. The Lyapunov exponent corresponding to the forcing phase or time coordinate is, say $\lambda_2 = 0$. This leaves the exponents for contraction and expansion. The directions of expansion and contraction change as the ball moves along its central phase space orbit and therefore, unlike the orbit direction, no constant direction can be associated with contraction or expansion of the ball. Expansion implies the existence of a positive Lyapunov exponent, a necessary condition for SIC. Let us call this exponent λ_1. What about contraction of the phase ball of the damped driven pendulum? If millions of pendulums were started each at one of the millions of possible starting points in the entire phase space, these points would tend to collect on the attractor, and therefore there must be overall shrinkage of phase volume ball. Thus there will be a negative Lyapunov exponent, say λ_3. Shrinkage of the phase volume requires, not only that there be the negative Lyapunov exponent, but that it must be sufficiently negative so that

$$\lambda_1 + \lambda_2 + \lambda_3 < 0. \tag{6.10}$$

A schematic diagram of this shrinkage process is shown in Fig. 6.16. Algorithms for computation of the Lyapunov exponents are somewhat complex and beyond the needs of this book. (See, for example, Wolf et al. (1985) and Eckmann et al. (1986).) A computer program for the computation of the exponents for the pendulum is given in Baker and Gollub (1996).

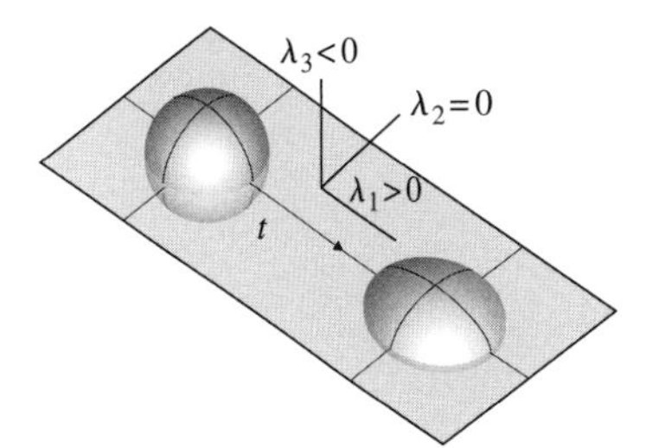

Fig. 6.16
Schematic of the process of shrinkage and stretch of an ensemble (ball) of phase coordinates for the chaotic pendulum.

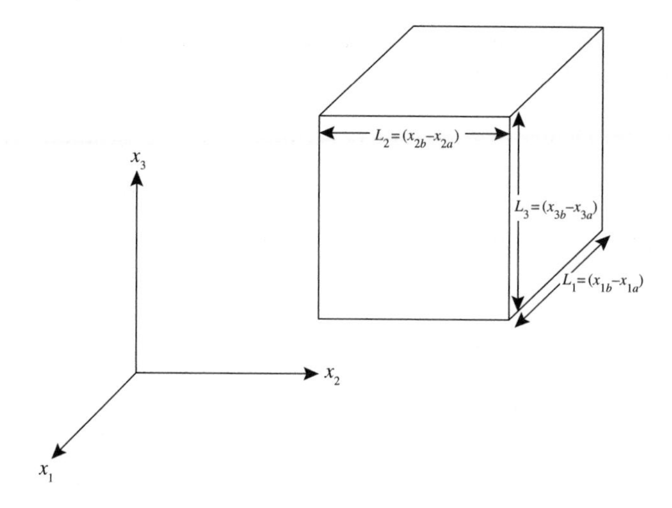

Fig. 6.17
Evolution of an initial volume of phase coordinates in phase space. (Reprinted from Baker and Gollub (1996, p. 13), with the permission of Cambridge University Press.)

What is the connection between Lyapunov exponents and the dynamics of the pendulum? The Lyapunov exponents measure the evolution of the phase space ball. Therefore we require some sort of analysis of how a phase volume changes in the general case. Consider the general three-dimensional phase space of Fig. 6.17 where the coordinates are the generic triplet of dynamical coordinates (x_1,x_2,x_3). The figure shows a rectangular box of sides

$$L_1 = x_{1b} - x_{1a}, \quad L_2 = x_{2b} - x_{2a}, \quad L_3 = x_{3b} - x_{3a}. \tag{6.11}$$

and the product of these sides gives the volume. Let us calculate the time rate of change of the volume. Using the product rule for differentiation we obtain

$$\frac{dV}{dt} = L_2 L_3(\dot{x}_{1b} - \dot{x}_{1a}) + L_1 L_3(\dot{x}_{2b} - \dot{x}_{2a}) + L_1 L_2(\dot{x}_{3b} - \dot{x}_{3a}), \tag{6.12}$$

where $\dot{x}$ means dx/dt. Now the dynamical coordinates each have a rate, $\dot{x}_i$, that depends on the other coordinates and the various system parameters. In general these dynamical rates may be expressed as

$$\begin{aligned} \frac{dx_1}{dt} &= F_1(x_1, x_2, x_3) \\ \frac{dx_2}{dt} &= F_2(x_1, x_2, x_3) \\ \frac{dx_3}{dt} &= F_3(x_1, x_2, x_3) \end{aligned} \tag{6.13}$$

and are simply the equations of motion for the system expressed as a set of first order differential equations.

Therefore the rate of change of phase volume may be written in terms of the dynamics, as represented by the functions (F_1, F_2, F_3), as

$$\begin{aligned} \frac{dV}{dt} &= L_2 L_3[F_1(x_{1b}, x_{2a}, x_{3a}) - F_1(x_{1a}, x_{2a}, x_{3a})] \\ &\quad + 2\,\text{similar terms}. \end{aligned} \tag{6.14}$$

For small evolution times with small change in phase volume we use a linear approximation by expanding the first term in a Taylor series as

$$F_1(x_{1b}, x_{2a}, x_{3a}) \approx F_1(x_{1a}, x_{2a}, x_{3a}) + \frac{\partial F_1(x_{1a}, x_{2a}, x_{3a})}{\partial x_1}(x_{1b} - x_{1a}) \quad (6.15)$$

and similarly with the other terms. This linearization means that our final result for the rate of change of volume is only valid for small periods of the system's evolution. Nevertheless, by averaging the changes over many small steps we obtain an average rate of shrinkage of the phase volume. Combining the above equations gives the rate

$$\frac{dV}{dt} = L_2 L_3 \frac{\partial F_1}{dx_1}(x_{1b} - x_{1a}) + L_1 L_3 \frac{\partial F_2}{dx_2}(x_{2b} - x_{2a}) + L_1 L_2 \frac{\partial F_3}{dx_3}(x_{3b} - x_{3a}), \quad (6.16)$$

which may be more compactly expressed as a logarithmic derivative

$$\frac{1}{V}\frac{dV}{dt} = \nabla \cdot \mathbf{F}. \quad (6.17)$$

Putting this expression together with the definition of the Lyapunov exponents leads to

$$\lambda_1 + \lambda_2 + \lambda_3 = \nabla \cdot \mathbf{F}. \quad (6.18)$$

Amazingly, the dynamics are now connected to the phase volume shrinkage.

Let us apply this general formulation to the pendulum. We simply transform the equation of motion for the pendulum into the form of a dynamical system—the collection of rate equations for each dynamical variable. We begin with the pendulum's equation of motion, Eq. (6.3)

$$\frac{d^2\theta}{dt^2} + \frac{1}{Q}\frac{d\theta}{dt} + \sin\theta = A\cos\omega_D t.$$

Rather than using just the single dependent variable we define new variables, ω for angular velocity, and ϕ for phase. These new variables, combined with Eq. (6.3) lead to the dynamical system

$$\begin{aligned} \frac{d\omega}{dt} &= -\frac{1}{Q}\omega - \sin\theta + A\cos\phi \\ \frac{d\theta}{dt} &= \omega \\ \frac{d\phi}{dt} &= \omega_D. \end{aligned} \quad (6.19)$$

In terms of Eq. (6.17) the logarithmic volume rate is $-1/Q = \lambda_1 + \lambda_2 + \lambda_3$. This equation makes the connection between the pendulum dynamics—in this case the rate of dissipation—and the phase space volume evolution as characterized by the Lyapunov exponents. Now that we know how phase space volumes evolve, we will make one more connection between geometry and dynamics. That is, we connect the fractal dimension of the attractor with the Lyapunov exponents.

6.4.3 Dynamics, Lyapunov exponents, and fractal dimension

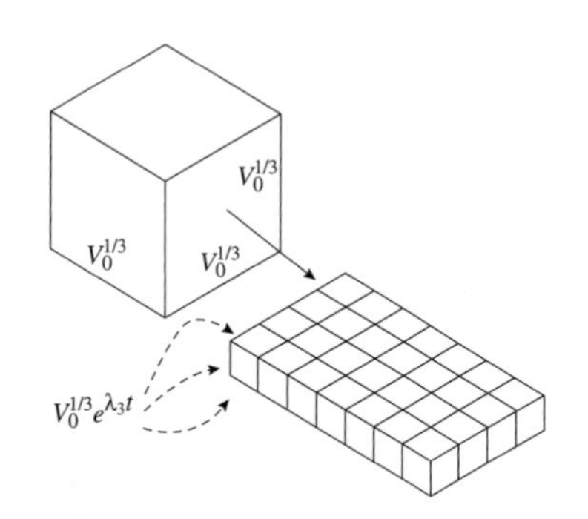

Fig. 6.18
Schematic illustrating the stretching and shrinkage of a small volume (ensemble) of phase points.

The relationship between Lyapunov exponents and fractal dimension was posited by Kaplan and Yorke in 1979 (Kaplan and Yorke 1979). In the following paragraphs, we provide an informal argument to make the conjecture plausible. Consider the schematic diagram, as shown in Fig. 6.18, of the simultaneous stretching and shrinking of a set of points that fills a small portion of a three-dimensional phase space. Like the pendulum this system is dissipative and therefore the volume occupied by the set shrinks in time.

Let the evolution of the unit volume be described by $V(t) = e^{(\lambda_1+\lambda_2+\lambda_3)t}$ where, as in Fig. 6.16, the stretching rate is described by the positive Lyapunov exponent $\lambda_1 > 0$, λ_2 may be positive or zero, and the shrinking rate is described by the negative Lyapunov exponent $\lambda_3 < 0$. Note that in the labeling the exponents decrease numerically as the index increases. Eq. (6.9) defines the fractal dimension in terms of $N(\epsilon)$, the number of boxes needed to cover the evolved volume, and $\epsilon(t)$, the time dependent length of one side of an elemental box. From the figure we see that

$$N(\epsilon(t)) = \frac{V(t)}{\epsilon^3(t)} = \frac{e^{(\lambda_1+\lambda_2+\lambda_3)t}}{e^{3\lambda_3 t}} = e^{(\lambda_1+\lambda_2-2\lambda_3)t} \tag{6.20}$$

and therefore

$$d_L = \frac{\log(N(\epsilon(t)))}{\log(1/\epsilon(t))} = \frac{(\lambda_1+\lambda_2-2\lambda_3)t}{-\lambda_3 t} = 2 + \frac{\lambda_1+\lambda_2}{|\lambda_3|}. \tag{6.21}$$

Dimension that is calculated in this way is sometimes called the *Lyapunov dimension* and in certain simple cases it will equal the box counting dimension. A discussion of the conditions under which the Kaplan-Yorke conjecture is true has been given by Grassberger and Procaccia (1983).

(The Kaplan-Yorke conjecture may also be used for the higher dimensional phase space of more complex systems. In these cases the formula becomes

$$d_L = j + \frac{\lambda_1+\lambda_2+\cdots+\lambda_j}{|\lambda_{j+1}|}, \tag{6.22}$$

where the λ_i are ordered from largest to smallest with λ_j being the smallest nonnegative exponent.)

Let us put all the pieces together. We are now thinking of the pendulum as a dynamical system that may be represented in a three-dimensional phase space with coordinates (ω, θ, t). The full attractor for the pendulum in this space is shown in Fig. 6.19. Side and end views are shown in Figs. 6.20 and 6.21.

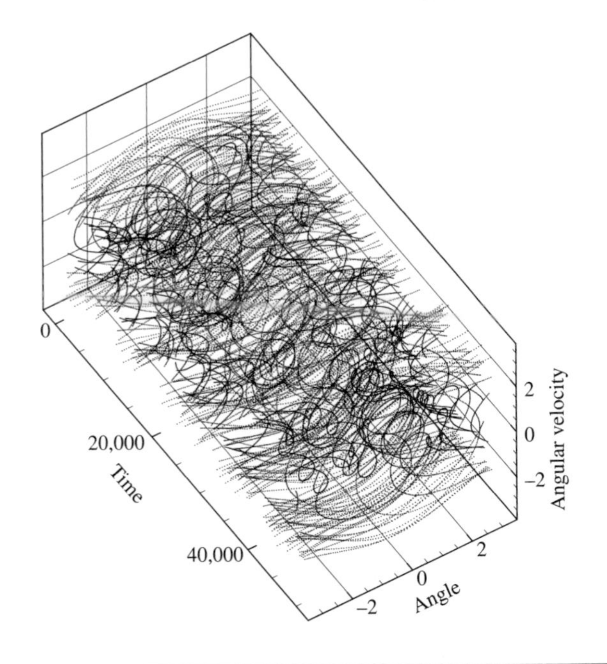

Fig. 6.19
Full three-dimensional attractor for the chaotic pendulum. The angle coordinate θ is constrained by periodic boundary conditions at $\pm\pi$.

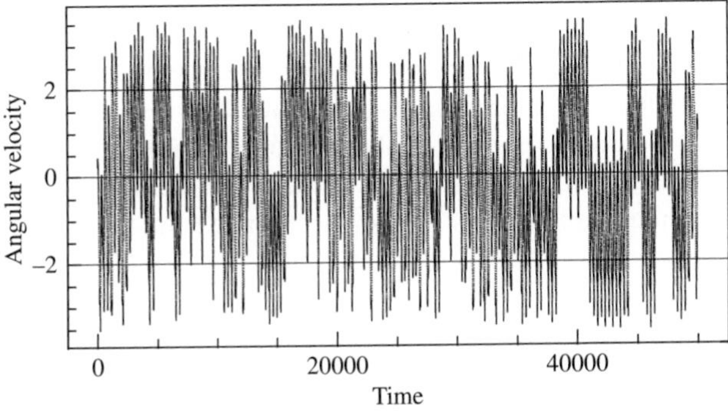

Fig. 6.20
Side view of the 3D attractor showing a time series similar to that of Fig 6.7(a).

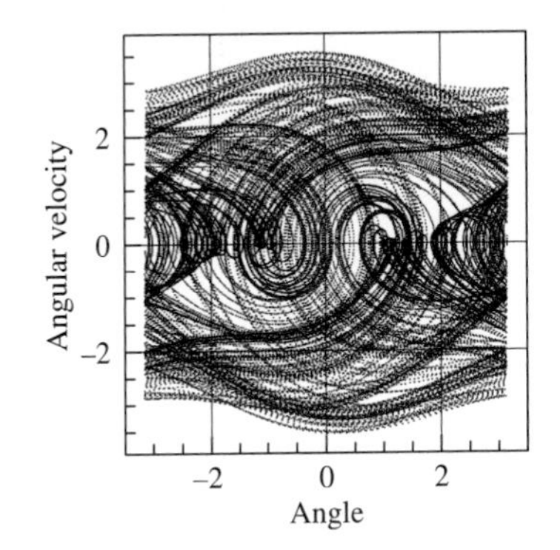

Fig. 6.21
End-on view of the 3D attractor (looking down the time axis) showing a phase plane diagram similar to that of Fig. 6.7(b)

The attractor is now embedded in three-dimensional space but is not space filling. Therefore we expect its dimension to be less than three. In this particular case it is computed to be about 2.4. The time coordinate t is a measure of the phase ϕ of the forcing function. Furthermore, motion along the time (phase) axis is represented by the second Lyapunov exponent λ_2. We noted that since time grows linearly and there is no change in phase volume due to time, then λ_2 is zero. The connections between Lyapunov exponents, fractal dimension, and dynamics, may now be summarized in two equations:

$$d_L = 2 + \frac{\lambda_1 + \lambda_2}{|\lambda_3|} \quad \text{and} \quad -1/Q = \lambda_1 + \lambda_2 + \lambda_3. \tag{6.23}$$

Baker and Gollub (1996) computed the various pertinent quantities for several values of the damping factor, Q. The results of that work are summarized in Table 6.1. Remember that $\lambda_2 = 0$. (We previously

Table 6.1 Lyapunov exponents and dimension for several values of damping

Q	λ_1	λ_3	$\sum_{i=1}^{3} \lambda_i$	$-1/Q$	$d_L = 2 + \frac{\lambda_1+\lambda_2}{\lvert\lambda_3\rvert}$
4.0	0.16	−0.42	−0.26	−0.25	2.38
3.7	0.16	−0.43	−0.27	−0.27	2.37
3.0	0.11	−0.44	−0.33	−0.33	2.25
2.8	0.09	−0.45	−0.36	−0.36	2.2
2.0	0.12	−0.58	−0.46	−0.5	2.2

calculated dimension using the box counting technique for the two-dimensional attractor—see Fig. 6.14—and found that $d_c \approx 1.4$. For the three-dimensional attractor the corresponding dimension would be $d_c \approx 2.4$, a value that is similar to those shown in Table 6.1)

As Q increases the damping diminishes and the pendulum has freer motion in both real space and in phase space. Therefore the attractor is more open and fuller, and its dimension increases with less damping. We have seen that Lyapunov exponents, fractal dimension, and pendulum dynamics are all intimately connected. While our focus is on the pendulum it is useful to note that other nonlinear dynamical systems exhibit these same SIC properties and lead to fractal geometries that are susceptible to similar analyses. On a more speculative note, we might ask whether fractal geometry leads to SIC. Does the appearance of fractal geometry—such as in tree leaves, spiral galaxies, and coastlines—suggest a mechanism of nonlinear chaotic dynamics as the functional source of this geometry? The relationship of the pendulum's damping to geometry may indicate the operation of similar dissipative mechanisms in these other diverse phenomena.

However, we return to the pendulum and look at further connections between geometric properties and dynamics.

6.4.4 Information and prediction

The notion of information is related to the thermodynamic concept of entropy. Entropy is the measure of disorder or uncertainty in a closed system. Information, or more precisely *missing information*, is also a measure of uncertainty in a system. For example, suppose there is a system which may be in one of two states, A and B. If one knows that the system is in state B, then there is no missing information. On the other hand, if the state of the system is completely unknown, and one must assume that A or B is equally likely, then we say that there is 1 bit of missing information. Intermediate between these two possibilities might be a condition where we know that A is twice as likely as B. Then the associated probabilities are $p_A = 2/3$ and $p_B = 1/3$. In this case the missing information is equal to 0.918 bits. Again, if $p_A = 9/10$ and $p_B = 1/10$ then the missing information is equal to 0.465. All of these results may be calculated from the formula

$$I = -(p_A \log_2 p_A + p_B \log_2 p_B). \tag{6.24}$$

The use of the base 2 for the logarithm is typical of information theory. However, in nonlinear dynamics the base e is more commonly used. We follow that practice here and define the missing information as

$$I(t) = -\sum_{i=1}^{n} p_i(t) \ln p_i(t). \tag{6.25}$$

For a dynamical system, the probabilities for the system's configuration may change in time. Consider three cases. If the system is *deterministic* and *nonchaotic*, then if the system is known initially, it will be known for all time. For example, if the system's starting configuration is confined to a small number of states, W then its starting information is $\ln W$. As time goes on the state of the system is exactly predictable and therefore the time evolution of the small number of initial states is known precisely. Thus the entropy starts at a constant value and remains equal to that value, and therefore

$$I(t) = I(0) = \text{constant}. \tag{6.26}$$

The other limiting case is the *random* system. Again the system can be started in a specific state. But after some transient epoch, the system will randomly travel through all possible states. After these initial moments, the system's specific configuration always remains unknown. Therefore the missing information becomes and remains infinite for all time:

$$I(t) = I(0) + \infty. \tag{6.27}$$

The intermediate *deterministic* case is the *chaotic* system. As in the previous deterministic case, the initial state is known and the initial information is constant. This condition may be represented by a set of points in phase space all being located in a small region. However, SIC now causes a gradual increase in uncertainty. One can imagine the points in phase space beginning to spread and a measure of this spread is the system's positive Lyapunov exponent. Thus the spreading into other states after a time t is proportional to $e^{\lambda_+ t}$, where λ_+ is the positive Lyapunov exponent. (More generally the increase in possible (but unspecified) states would be proportional to $e^{\Sigma \lambda_{+i} t}$ where the sum is taken over all positive Lyapunov exponents.) If the "size" in phase space of the initial state, or equivalently the number of states, is taken as ϵ then after time t the number of states is $\epsilon e^{\lambda_+ t}$. Assuming all states are equally probable we see that the information develops in time according to

$$I(t) = \ln(1/\epsilon) + \lambda_+ t. \tag{6.28}$$

In general, it has been shown for many chaotic systems that the information change is linear (Atmanspacher and Scheingraber 1987)

$$I(t) = I_0 + Kt, \tag{6.29}$$

where K is called the *Kolmogorov entropy*. (Note that the Kolmogorov entropy is really a rate of change of entropy.) Under many circumstances (Grassberger and Procaccia 1983) it is reasonable to suppose that $K \approx \sum \lambda_+$

and therefore knowledge of the positive Lyapunov exponents can give some estimate of the time variation of the information content of a system. If $K = 0$ then the system is deterministic and nonchaotic, if $K =$ constant, then the system is deterministic and chaotic, and if $K = \infty$, then the system is random. For the chaotic pendulum, one typical set of pendulum parameters gave $\lambda_+ \approx 0.16$, thus providing some measure of the growth of uncertainty.

Growth in uncertainty implies a limitation on the ability to predict the future behavior of the chaotic pendulum. Such uncertainty is very familiar to all those who follow weather reports. In such reports we often see the initial uncertainty in the knowledge of present weather conditions amplified into a much larger uncertainty in what seems to be a relatively short time. While we probably do not know the values of the positive Lyapunov exponents for weather systems, we can make some predictions for simple systems like the pendulum.

Calculation of Lyapunov exponents for growth and decay of a phase volume are made over times that are short compared to the time required to traverse the phase space. Once the ball of initial conditions has stretched to a size comparable to that of the attractor, the ball typically folds over, and the stretching and shrinking process begins again. Thus the prediction time T is approximately the time taken for a set of points in small volume of phase space to diverge to the boundaries of the phase space of size, say L. In one dimension, $L \approx \Delta e^{\lambda_+ T}$ where Δ is the initial linear dimension of the phase volume. Therefore the prediction time, T is given by

$$T = \frac{\ln(L/\Delta)}{\lambda_+} = \frac{\ln(L/\Delta)}{K}. \tag{6.30}$$

We can obtain some idea of this prediction time for the pendulum. Suppose that the initial linear dimension, or uncertainty in the dynamical coordinates, is about one, one hundred millionth of the attractor size. With $\lambda_+ = 0.16$ we obtain a prediction time of about 115 in the dimensionless time units used in Eq. (6.3), which is about 12 forcing cycles. This prediction is tested by the simulation in Fig. 6.22 where two pendulums are started with the above, very small, difference in one of their initial coordinates. The figure shows that after about a dozen forcing cycles the two time series start to come apart, as predicted.

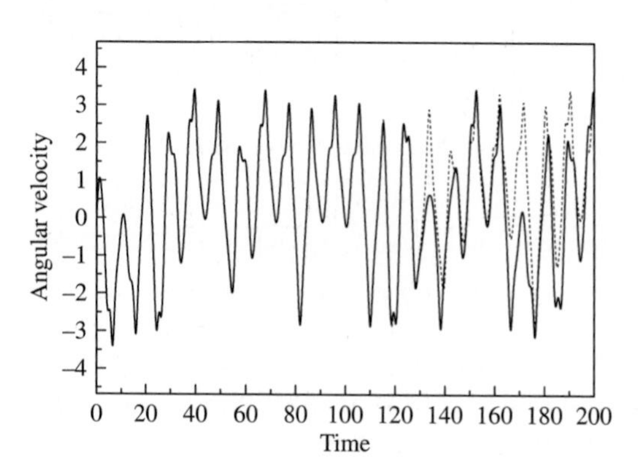

Fig. 6.22
Two angular velocity time series (one solid, one dotted) with a difference between their initial angles of one part in 10^8.

One goal of science is to develop models of reality that will predict the future. Some theories seem to work amazingly well. For example, astronomers can predict the motions of the planets hundreds of years into the future. On the other hand meteorologists seem unable to achieve anything more than approximate short term weather forecasts and very approximate long term forecasts. Similarly, accurate long term prediction of the future of chaotic systems is not possible. But within this constraint, scientists have built up a series of methods in their efforts to enhance the prediction of chaotic systems. There are two aspects to this problem. In our treatment of prediction of the pendulum's motion we have used the model provided by the deterministic differential equation. The accuracy of any prediction then depends strongly on the degree of specificity of the initial state, as indicated by Eq. (6.30). Such specificity can theoretically go well beyond that which is possible from experimental data. A more realistic approach is to consider the problem of prediction from a set of experimental data. This data may be such that there is no corresponding mathematical model or, if there is such a model, it may be approximate at best. This latter problem has led to a great deal of interesting work which is really beyond the scope of this book. (The interested reader is referred to the paper by Abarbenal et al. (1990) and the book by Weigend and Gershenfeld (1994).) For the pendulum, two of the simpler methods are discussed in Baker and Gollub (1996) and applied to data from the Blackburn pendulum. Ingenious as some of these methods are, the fundamental limitation described by Eq. (6.30) is still the best that can be done. Thus, despite the fact that chaotic systems are deterministic as to the causes of their motions, we have yet another confirmation that they are unpredictable.

6.4.5 Inverting chaos

Model equations, such as that for the pendulum, are often used to simulate chaotic data. But let us turn the process around and consider the more challenging task of fitting of model equations to chaotic data, especially real, experimental chaotic data. In this final section of this chapter, we briefly describe this "inverse" problem.

A real pendulum, such as the design of Blackburn and Smith, produces data which roughly approximates simulated data from a computerized solution of the model differential equation. Figure 6.3 shows simulated and experimental data for the pendulum. In order to obtain experimental data, the physical characteristics of the real pendulum are first calibrated according to the parameters (Q, A, ω_D) used in the model equation. Then the settings are adjusted so that the parameters of the real pendulum are as close as possible to those used in the model equation, (Q, A, ω_D). The real pendulum and the simulated pendulum then produce data sets for comparison. Minor discrepancies between the two pictures may be due to a variety of sources: perhaps the model equation does not exactly represent the dynamics of a real pendulum, or perhaps there are effects that result from flaws in the manufacture of the pendulum, or perhaps there are limitations on the quality of the parts. Nevertheless, the agreement between the simple nonlinear simulation and the data produced by a complex mechanical and electronic pendulum is quite good.

Now let us suppose that one does *not* have a calibration of the real pendulum. Is there some way that the parameters of the model pendulum, given by the differential equation, could be determined? This question is part of a larger question in science. Can experimental data help to determine a theoretical model of some phenomenon? In chaotic dynamics, much effort has gone into such work. Sometimes the equations of motion are not known and one is starting only with the data (Brown et al. 1994). In other cases, information about the model is sufficient to suggest equations of motion but the values of the parameters of the equation set are not known (Parlitz 1996). This latter situation is the case for the driven pendulum. A good set of equations of motion is known. Thus one can test an inversion method, a method to obtain the equation parameters, by using data sets from a real chaotic pendulum. The statistical method of *least squares fitting* is the mathematical basis for the study of the pendulum inversion problem (Baker et al. 1996).

As we will see, the set of *three* parameters (Q, A, ω_D) is not sufficient for this work and therefore we use a modified dimensionless equation,

$$\omega' = -\alpha\omega - \beta \sin\theta + \gamma \sin(\delta t + \phi). \tag{6.31}$$

with *five* unknown parameters $(\alpha, \beta, \gamma, \delta, \phi)$. The two extra unknown parameters account for our lack of knowledge (a) of the pendulum's natural frequency and (b) of the initial phase of the pendulum. In the test of this inversion process, only the range of possible experimental values is known, as shown in Table 6.2

Table 6.2 Range of possible experimental parameters

$2 < \alpha < 10$
$60 < \beta < 100$
$80 < \gamma < 200$
$0 < \delta < 12$
$0 < \phi < 2\pi$

The real pendulum produces a set of data points $\{\omega_i, \theta_i\}$ each data pair being separated by a time of $\Delta t = 0.007$ s. The "experimental" angular acceleration ω_i' is calculated by a standard finite difference approximation defined as

$$\omega'_{i\text{Expt}} = \frac{-\omega_{i+2} + 8\omega_{i+1} - 8\omega_{i-1} + \omega_{i-2}}{12\Delta t} \tag{6.32}$$

Least squares fitting is a technique whereby one minimizes the sum of all the differences between the "experimental" value of $\omega'_{i\text{Expt}}$ and the value of ω_i' using the right side of Eq. (6.31). That is, the sum

$$S = \sum_{i=1}^{n} \left[\omega'_{i\text{Expt}} - (-\alpha\omega_i - \beta \sin\theta_i + \gamma \sin(\delta t_i + \phi))\right]^2 \tag{6.33}$$

is minimized with respect to each parameter by setting

$$\frac{\partial S}{\partial \alpha} = 0, \frac{\partial S}{\partial \beta} = 0, \text{ etc.} \tag{6.34}$$

Differentiation with respect to the set $(\alpha\ \beta\ \gamma)$ leads to linear equations in these parameters which may be solved to give complex but closed form

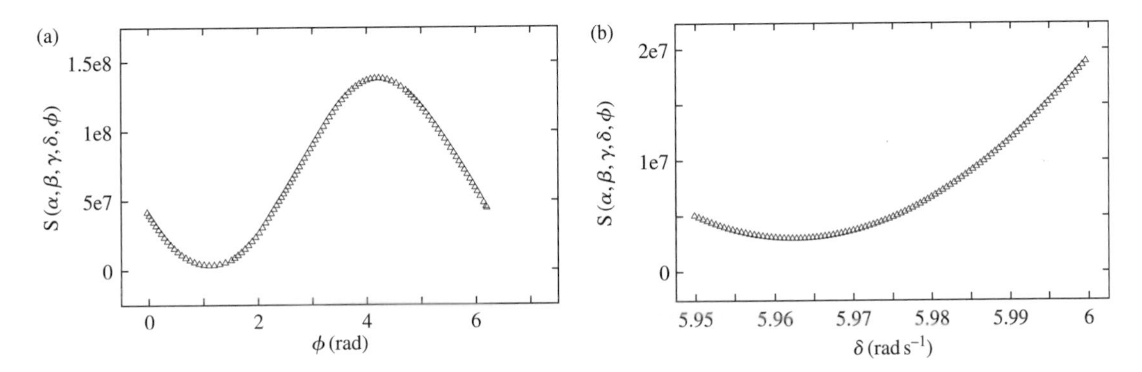

Fig. 6.23
Minima of the graphs are used to determine the best values for δ and ϕ. (Reprinted with permission from Baker et al. (1996, p. 531). ©1996, by the American Institute of Physics.)

expressions for these parameters in terms of the data sets $\{\omega_i, \theta_i\}$. On the other hand, equations involving the parameters δ and ϕ are nonlinear and it is not possible to find a closed form expression for these parameters. Therefore the process of determining all five parameters requires several steps.

As a first step the experimental time series is subjected to a fast Fourier transformation. The transform gives the spectrum of all frequencies present in the data and the experimental data has an appearance similar to the spectrum shown in Fig. 6.8. The forcing frequency δ is a strong component and therefore a first estimate of its value is provided by the spectrum. The other problematic parameter ϕ, the phase of the time series is not especially sensitive to values of α, β, and γ. Using the newly acquired estimate of δ and almost any reasonable values of (α, β, γ) it is possible to look for a minimum in S as a function of the phase ϕ. The estimates of δ and ϕ are then used together with the linear equations for (α, β, γ) to determine a first estimate for all five parameters. Now all parameter values can be further refined by iteration of the process of minimizing S as a function of each parameter. Figure 6.23 shows minimization graphs for δ and ϕ.

This procedure was followed by the authors in an effort to find parameters that would best fit real experimental data. The final results from the successive iterations are given in Table 6.3. The experimental values were known only to one of us (JAB) while the analysis was being carried out by the other (GLB).

The agreement between the two parameter sets is quite good although not perfect. In a previous section we discussed the Lyapunov exponents and fractal dimension of an attractor. Data from the real pendulum and data from a simulated pendulum with the "fitted" parameters both provide attactors. The corresponding Lyapunov exponents and dimensions of the attractors should match if the simulation accurately models the experiment. Table 6.4 provides a comparison of these geometric "invariants," and the agreement is very good. How well the fitting process works depends upon the number of points and upon the amount of noise on the data. Various tests involving both factors were conducted and, within certain

Table 6.3 Comparison of experimental and fitted parameters

Parameter	Exp. value	Fitted (from exp. data)
α(rad s^{-1})	2.24±0.1	2.12
β(rad s^{-2})	80.6±0.1	76.1
γ(rad s^{-2})	121±6	117
δ(rad s^{-1})	5.98±0.02	5.96
ϕ(rad)	unknown	1.05

Table 6.4 Comparison of experimental and simulated values of dimension and positive Lyapunov exponent

	Exp. data	Simulated data
Dimension	2.2±0.15	2.1±0.1
Positive Lyap. Exponent	0.9±0.1	0.9±0.1

limits, the fitting process is found to fairly robust against noise and small data sets. Thus the combination of Least Squares, FFT, and iteration led to a successful "inversion process." The data successfully "predicted" the model equations—or at least the parameter values.

The chaotic pendulum exhibits a variety of unusual and surprising phenomena, including beautiful strange attractors and felicitous connections between geometry and dynamics. Further effects appear when pendulums are coupled together as in the next chapter.

6.5 Exercises

1. Use the transformations of Eq. (6.2) to convert Eq. (6.1) to Eq. (6.3).
2. Periodic motion often changes to chaotic motion because of increased forcing of the dynamical system. In many cases, subharmonics of the original frequency of the forcing or even the natural frequency of the system become more prominent. With increased forcing, more and more subharmonics are created until the system destabilizes and becomes chaotic. The musical scale can also provide examples of subharmonics. For example, period doubling of middle C whose frequency is 256 Hz (scientific scale) leads to the first subharmonic of 128 Hz that is the C one octave below middle C. For 256/3 Hz there is period tripling and the musical note is the "F" below the lower C. Calculate the frequencies of all the subharmonics down to 256/16. Refer to the *Handbook of Physics and Chemistry* to determine which of the subharmonics correspond to real musical notes. Use the "scientific" or "just" scale.
3. Find the Fourier transform $a(\omega)$ of the function

$$\begin{aligned} f(t) &= 0, \quad t < -b \\ &= c, \quad -b \leq t \leq b \\ &= 0, \quad t > b. \end{aligned}$$

   Sketch the graphs of $f(t)$ and $a(\omega)$. What is the width Δt of the graph of $f(t)$? Define the width $\Delta\omega$ of $a(\omega)$ as the distance between the two zeros on either side of the line, $\omega = 0$. Calculate $\Delta\omega$ and find the product of the two widths. Note that the product is independent of either separate width. This observation is generally true. This product illustrates the mathematical basis of the uncertainty principle of quantum physics.
4. *Fixed points* in phase space are important in the study of dynamics. These are points in phase space where the time derivatives of the variables vanish. Suppose we write the dimensionless equation for the damped, but not driven, pendulum as

$$\begin{aligned} \frac{d\dot{\theta}}{dt} &= -\frac{1}{Q}\dot{\theta} - \sin\theta \\ \frac{d\theta}{dt} &= \dot{\theta} \end{aligned}$$

   and consider the phase portrait which, for this system is given in Chapter 3 as Fig. 3.21. From that diagram or a consideration of the equations find the fixed points. Note that there are two kinds of points. One kind corresponds to the pendulum bob being at the top of its motion and therefore unstable, and the other kind corresponds to the pendulum bob being at the bottom of its motion and therefore stable. We noted in Chapter 3 that this latter point is called an *attractor* and the former point is called a *saddle point*.
5. A *Bifurcation diagram* shows that as the amplitude of the driving force for the pendulum is increased, more and more subharmonics occur. The transition

points are known as *bifurcation points*. While location of these bifurcation points are often impossible to predict, the following simple (nonpendulum) model illustrates their basic characteristics. Consider the curve in phase space

$$\dot{\theta} = f(\theta) = \theta^2 - h.$$

(a) Graph $f(\theta)$ in phase space. (b) For $h > 0$ what are the roots of $f(\theta)$? Note that these roots are the fixed points of the equation. (c) The behavior of $f(\theta)$ near the fixed points (roots) may be approximated by a linear Taylor series. For the ith root θ_i this approximation is $f(\theta) \approx f'(\theta_i)(\theta - \theta_i)$. Using your answers to (b) for θ_1 and θ_2, write the two linearized equations for $f(\theta)$. (d) Now let us examine the stability of the fixed points. First, near a fixed point, we define a new variable $\eta = \theta - \theta_i$ and look at whether this small deviation from the fixed point increases with time. If the deviation η does increase then the fixed point is unstable. If the deviation decreases then the fixed point is stable. Thus $d\eta/dt = d(\theta - \theta_i)/dt = \dot{\eta} \approx f'(\theta_i)\eta$. The sign of $f'(\theta_i)$ determines whether the system will regress toward or away from the fixed point. Find and solve the linearized differential equations for η near each of the fixed points. Determine the stability of each fixed point. (e) Now consider the changes in the model curve $f(\theta)$ and, particularly, in the fixed points as the parameter h tends toward zero. The parabola gradually moves upward, its roots (the fixed points) approach each other and the origin along the θ axis. As the roots coalesce when $h = 0$, the fixed points disappear and the system has undergone a distinct change or bifurcation. Therefore $h = 0$ specifies a *bifurcation point* (See, for example, Strogatz, (1994).)

6. Calculate the fractal dimension of a Cantor set where the middle *one-quarter* is removed rather than the middle third. Follow the process shown in the text. Prove that the length of the set is zero.
7. Construct a two-dimensional Cantor set as follows. Draw a square, one unit on a side. Inside the square remove a square of sides $1/3$ from the center of the original square. The result is a square that contains 8 squares of size $1/3 \times 1/3$. Now, in each of those remaining squares, remove from the center a square of size $1/9 \times 1/3$, and so on. The look will be somewhat like Swiss cheese. Sketch a few iterations of this process. (a) Find the dimension by following the limiting process in the text. Note that ε is the length of the side of the appropriate covering squares. (b) Prove that the area of the resulting set is zero.
8. The first computer generated attractor is due to the pioneering work of Lorenz (1963) who modeled the atmosphere according to the equations

$$\dot{x} = -\sigma x + \sigma y$$
$$\dot{y} = -xz + rx - y$$
$$\dot{z} = xy - bz,$$

where σ, r, and b are positive constants. Calculate the logarithmic rate of change of the phase volume, $(1/V)dV/dt$ using the formula that connects Lyapunov exponents, volume change, and the equations of motion. If the result is negative then the system is dissipative and has an attractor. The attractor in question is shaped somewhat like a butterfly and is ubiquitous in the popular literature of chaos.

9. Consider a physical system that is capable of being in one of the n states. The missing information for the system is given by $I(t) = -\sum_{i=1}^{n} p_i(t) \ln p_i(t)$. Prove that if all states are equally likely then $I(t) = \ln n$.
10. Suppose a physical system can exist in one of two states whose probabilities are $p_1 = e^{-Kt}$ and $p_2 = 1 - e^{-Kt}$. At $t = 0$, what are the values of the probabilities? At $t = \infty$, what are the values of the probabilities? What are the values of the missing information at these two times? Write the missing information as a function of time. Using a plotting program, sketch the function. For what value

of t is the missing information a maximum? Prove this result in two ways: (a) by using calculus, and (b) using the result of problem 9.

11. If the positive Lyapunov exponent for a chaotic system is $\lambda_+ = 0.1$, how big would an initial relative uncertainty of one billionth grow to be in a time of 200 (dimensionless) units?
12. Suppose a dissipative system has Lyapunov exponents of $\lambda_1 = 0.10, \lambda_2 = 0.0$, and $\lambda_3 = -0.13$. (a) Calculate the relative change in phase volume, $(V_I - V_F)/V_I$, after a time T. (b) Using the Kaplan–Yorke conjecture, calculate the dimension of the system's attractor.
13. "Inverting" chaos in the case of the Lorenz equations (given in exercise 8) is relatively straightforward since all equations for the unknown parameters, σ, r, and b are readily solvable. Use the least squares method described in the text to develop equations for the parameters in terms of sums involving the data sets $[x_i, y_i, z_i]$ and $[\dot{x}_i, \dot{y}_i, \dot{z}_i]$. Show that $\sigma = (\sum \dot{x}_i y_i - \sum \dot{x}_i x_i)/(\sum x_i^2 - 2\sum x_i y_i + \sum y_i^2)$, $r = (\sum x_i \dot{y}_i + \sum x_i^2 z_i + \sum x_i y_i)/\sum x_i^2$, and $b = (-\sum z_i \dot{z}_i + \sum x_i y_i z_i)/\sum z_i^2$.
14. The Kaplan–Yorke conjecture may also be demonstrated using the time evolution of an initial sphere of radius r as it becomes an ellipsoid (see Fig. 6.16) with semi-axes, $a = re^{\lambda_1 t}, b = re^{\lambda_2 t}, c = re^{\lambda_3 t}$, where $\lambda_1 > 0, \lambda_2 = 0, \lambda_3 < 0$, as with the pendulum. Start from Eq. (6.20) and let the elemental volume be a sphere of volume $4\pi c^3/3$. Then follow the corresponding derivation in the text to arrive at the Kaplan–Yorke conjecture. (Hint: the volume of an ellipsoid is $V = 4\pi abc/3$.

Coupled pendulums

7

7.1 Introduction

It has been asked, "You know the sound of two hands clapping; what is the sound of one hand clapping?" Of course one hand does not clap. But there are answers to a related question, "You know the sound of two hands clapping; what is the sound of many hands clapping?" Zolten Neda and his physicist colleagues in Romania have studied this question extensively in concert halls, theaters and opera houses (Neda et al. 2000). One of their conclusions is that at some point during the applause the aggregate noise from hand clapping will be periodic. Probably, most of us have experienced this phenomenon. (Being physicists, Neda and his colleagues also learn a lot of other things that the casual observer might miss.) This periodicity of the applause is an example of *synchronization*. Each component system—each pair of hands—acts in concert with all the other component systems to create a uniform pattern of clapping. But the applause may not necessarily stay periodic. If some audience members are more enthusiastic than others and break the pattern by clapping more frequently, then the applause can again become chaotic. Thus the degree of synchronization seems to depend upon the audience members being mutually coupled by a common level of enthusiasm. As an example, Neda et al. note, historically, that clapping at now-defunct communist party rallies typically stayed in the rhythmic state. One might conclude that the level of enthusiasm was fairly low at such gatherings. But let us discuss pendulums.

Possibly the first and certainly one of the most celebrated instances of synchronization of pendulums (or anything for that matter) was described by Christiaan Huygens in a letter to his father, written in 1665. Here is a partial quote.

While I was forced to stay in bed for a few days and made observations on my two clocks of the new workshop, I noticed a wonderful effect that nobody could have thought of before. The two clocks, while hanging [on the wall] side by side with a distance of one or two feet between, kept in pace relative to each other with a precision so high that the two pendulums always swung together, and never varied. While I admired this for some time, I finally found that this happened due to a sort of sympathy: when I made the pendulums swing at differing paces, I found that half

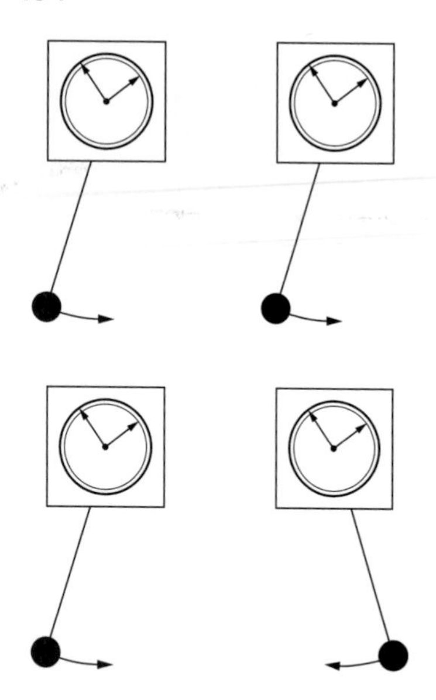

Fig. 7.1
Two sets of synchronized pendulum clocks. The upper set is locked *in phase* and the lower set is locked but in *antiphase*.

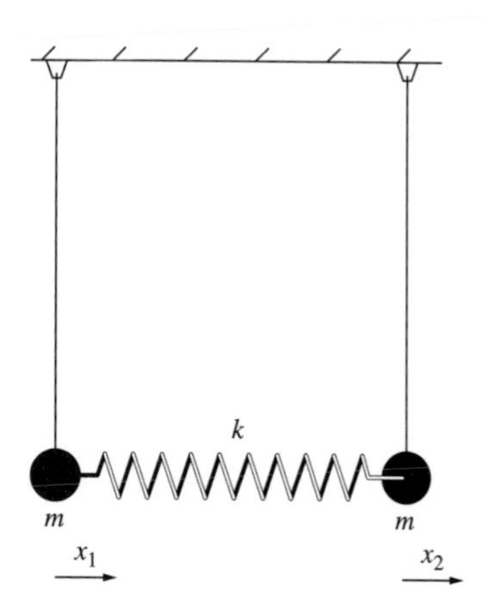

Fig. 7.2
Two pendulums coupled by a spring.

an hour later, they always returned to synchronism and kept it constantly afterwards, as long as I let them go. (Quoted in Pikovsky et al. (2001, p. 357).)

Huygens goes on at great length, describing his observations over several days and his attempts to test the degree of synchronization by moving the clocks apart or by placing obstacles between them. Huygens also describes this synchronization effect as acting between two pendulum clocks attached to a common beam. Figure 7.1 shows two possible types of synchronization, in-phase and out-of-phase. As it happened, the physical configurations of Huygens clocks were such that he only observed the out of phase synchronization.

Let us examine a very simple model of coupled pendulums (Feynman 1965) that will show us how the in-phase and out-of-phase motions arise. Figure 7.2 shows a pair of equal pendulums coupled together by a spring. We consider only small amplitude motion and can therefore revert to a linearized model of the type described in Chapter 2. With the coordinates as shown in Fig. 7.2 the equations for each pendulum become

$$\begin{aligned} m\frac{d^2x_1}{dt^2} &= -m\omega_0^2 x_1 - k(x_1 - x_2) \\ m\frac{d^2x_2}{dt^2} &= -m\omega_0^2 x_2 - k(x_2 - x_1), \end{aligned} \tag{7.1}$$

where $\omega_0 = \sqrt{g/L}$ and k is the force constant of the connecting spring.
We substitute the trial solutions

$$x_1 = A_1 e^{i\omega t} \text{ and } x_2 = A_2 e^{i\omega t} \tag{7.2}$$

into Eq. (7.1) and the subsequent elimination of common factors leads to a pair of equations for the amplitudes A_1 and A_2;

$$\begin{aligned} \left(\omega^2 - \omega_0^2 - \frac{k}{m}\right)A_1 &= -\frac{k}{m}A_2 \\ \left(\omega^2 - \omega_0^2 - \frac{k}{m}\right)A_2 &= -\frac{k}{m}A_1. \end{aligned} \tag{7.3}$$

We had hoped to obtain two equations that would uniquely determine the amplitude of the motions, but this thought is overly optimistic. Only the *ratio* of the amplitudes can be found. Yet the ratio must be consistent with both equations. The trick is to multiply both equations together,

$$\left(\omega^2 - \omega_0^2 - \frac{k}{m}\right)^2 A_1A_2 = \left(\frac{k}{m}\right)^2 A_2A_1 \tag{7.4}$$

and then solve for the unknown frequency, ω. We find a quadratic expression for ω and thereby obtain two frequencies

$$\omega_1 = \omega_0 \text{ and } \omega_2 = \omega_0\sqrt{1 + \frac{2k}{m\omega_0^2}} \tag{7.5}$$

for the two kinds of motions. The oscillatory mode represented by ω_1 is the in-phase motion and the connecting spring is not stretched at all. The other mode, represented by ω_2, is the out of phase motion whereby the spring is periodically stretched and shrunk and therefore the spring constant k contributes to the expression for the frequency, ω_2. As a check on these conclusion we note that substitution of these frequency values back into Eq. (7.3) leads to ratios of the amplitudes corresponding to each frequency; that is

$$A_1(\omega_1) = A_2(\omega_1) \text{ and } A_1(\omega_2) = -A_2(\omega_2). \tag{7.6}$$

The motion characterized by frequency ω_1 is shown in the upper part of Fig. 7.1 and the frequency is just the frequency of a single pendulum. On the other hand, the out-of-phase motion characterized by frequency ω_2 is shown in the lower part of Fig. 7.1. Huygens observed the out-of-phase motion for his pendulum clocks. However, the coupling of his two clocks through the wall or a beam must have been quite weak as he did not apparently observe a noticeable change in frequency for the motion of the coupled clocks as compared to individual clocks.

We might ask how Huygens came to have arranged two clocks near to each other. Was it just serendipity? Perhaps. In 1660 Huygens mounted a concerted effort to solve the problem of calculating longitude at sea. Determination of position at sea requires knowledge of both latitude and longitude. Latitude may be calculated by measuring the angle between the horizon and some reference celestial body such as the north star (in northern latitudes). But longitude is another matter. Without it ships were forced to use crude estimates of speed or stick to certain known channels, thus making them prey to enemy attack. Accurate determination of longitude would not only make for shorter voyages but would also go far toward eliminating the loss of ships at sea. Seafaring countries promised financial incentives to those who could find a solution. At the time, there were two possible solutions (a) complicated analysis of astronomical data, or (b) the construction of a reliable clock. Both Huygens and Galileo saw the clock, regulated by pendular motion, as a good solution to the problem. Huygens built the first practical pendulum clock and during 1662–1665 these clocks underwent sea trials supervised by the British Royal Society (Bennett et al. 2002). Now what about the two clocks? During these trials, it was important to have two clocks going because if one happened to quit, the other would continue to keep time during the interval needed to restart the stopped clock. Thus, synchronization of clocks was a quite probable event. (We note, however, that the most robust solution to the longitude problem was the Harrison clock—a spring driven clock, and not the pendulum clock (Sobel 1996).)

The synchronization of Huygens's pendulum clocks is of sufficient interest that the phenomenon has been both mathematically analyzed and physically reproduced (more or less) several times in recent decades. The latest reproduction of Huygens' work and a new analysis has been provided by Schatz and coworkers at Georgia Institute of Technology

Fig. 7.3
Magnetically coupled damped, driven pendulums. (Photo by JAB.)

(Bennett et al. 2002). This group confirms that the out of phase mode is the only one observed (or indeed possible) with the clocks that Huygens constructed. (For a discussion of other work see the references contained in Bennett et al. (2002).)

The complete analysis of Huygen's coupled clocks is fairly complex—see (Bennett et al. 2002)—and would carry us too far from our main topic. Therefore we look at a simpler realization of coupled pendulums. Figure 7.3 is a photograph of two pendulums that are magnetically coupled (Smith et al. 1999). The pendulums are of the same type as that shown in Figs. 6.1 and 6.2 except that extending from the shaft of each pendulum is a disk. On one pendulum the disk is a magnet whereas on the other pendulum the disk is made of copper. The coupling is through the disks rather than through a wall or beam as was the case for Huygen's clocks.

Motion of the magnetic disk relative to the copper disk causes electrical eddy currents to develop in the copper disk. The eddy currents then produce magnetic fields that oppose the magnetic fields in the magnetic disk. Therefore motion of the magnetic disk relative to the copper disk causes dissipation of the motion of the magnetic disk. Now when the motions of the two pendulums are synchronized, there are no eddy currents. But when the pendulums are not aligned then the coupling between the pendulums is dissipative and therefore the loss of energy is velocity dependent. With these two ideas, that eddy currents depend on differences in velocity and require energy, we see that the coupling, affecting one pendulum, is of the general form,

$$c\left(\frac{d\theta_1}{dt} - \frac{d\theta_2}{dt}\right) \tag{7.7}$$

and a similar term for the other pendulum with the subscripts reversed. The equations of motion for this system are symmetric and the coupling is equally shared between the two pendulums. Such coupling is said to be *bidirectional*. Both pendulums are equal partners. The addition of coupling to the pendulum equations of motion leads to:

$$\frac{d^2\theta_1}{dt^2} + b\frac{d\theta_1}{dt} + \sin\theta_1 + c\left(\frac{d\theta_1}{dt} - \frac{d\theta_2}{dt}\right) = A\cos\omega_D t \tag{7.8}$$

and

$$\frac{d^2\theta_2}{dt^2} + b\frac{d\theta_2}{dt} + \sin\theta_2 + c\left(\frac{d\theta_2}{dt} - \frac{d\theta_1}{dt}\right) = A\cos\omega_D t. \tag{7.9}$$

Let us note two consequences of the form of these equations. First, the coupling of the two equations and hence the possibility of synchronization rests upon there being difference terms such as $(d\theta_1/dt - d\theta_2/dt)$. This means that if the two pendulums have identical angular velocity coordinates then they are no longer coupled. It seems strange that if the coupling creates a situation of complete synchronization then that state removes the coupling between the pendulums. In such a state, the pendulums behave identically but do not, at that instant, communicate with each other. Second, we ask under what circumstances such a state of complete synchronization might be possible. Let us first consider real pendulums in a laboratory. Given slight differences in the manufacture of the pendulums, the possibility of other small physical effects acting on the pendulums, the existence of even miniscule amounts of thermal noise, it seems unlikely that real pendulums would match exactly for more than a moment. Therefore the feedback provided by a nonzero coupling term is preserved. The two pendulums are, in some sense, always communicating through the coupling. They are always *slightly* unsynchronized. Now let us imagine a pair of completely matched *theoretical* pendulums. The solutions to the equations—the trajectories of the pendulums—are now expressed as real number coordinates of the phase space moving in time. Since real numbers, theoretically, have an infinite number of digits, the phase space consists of an infinite number of points. Therefore the probability that even theoretical pendulums would match exactly is vanishingly small. Again, less than perfect synchronization is achieved. The pendulums remain in communication via coupling of infinitesimal coordinate differences. Finally, there is the intermediate case of computer simulations of the pendulums. Computer simulation is a popular tool for the study of synchronization in a host of examples. Computers do calculations with numbers having finite precision and therefore each number is expressed by a finite number of digits. The phase space numbers become a grid of cells or "course grained." For example, if only 10 digits are used then the computer will truncate numbers such as 1.243567882035 and 1.243567882179 and 1.243567882243, all as 1.243567882. Thus there are now only a finite number of cells in phase space and they are of finite size. Therefore, it is quite possible that the two pendulums may find themselves with identical coordinates—to this level of precision—in the same cell. Synchronization has been achieved and the computer algorithm that generates the orbits will now make the pendulum orbits identical for all future times. This is a state of *false* synchronization.

Therefore it is important to add a tiny bit of noise to each step of the orbit calculation in order that any state of synchronization represents a real physical process, and not a computer artifact. After all, physics is about real objects. With these thoughts in mind, let us return to the mathematical analysis.

We introduce new sum and difference coordinates to help our analysis. These new coordinates are

$$\theta_- = \frac{\theta_1 - \theta_2}{2} \text{ and } \theta_+ = \frac{\theta_1 + \theta_2}{2} \tag{7.10}$$

and substitution of these coordinates into Eqs. (7.8) and (7.9) lead to new equations of motion in the sum angle θ_+ and the difference angle coordinate θ_-. We are especially interested in the equation for the difference coordinate,

$$\frac{d^2\theta_-}{dt^2} + (b + 2c)\frac{d\theta_-}{dt} + \cos\theta_+ \sin\theta_- = 0, \tag{7.11}$$

since the difference angle coordinate goes to zero when the pendulums are synchronized. We had hoped to achieve a complete separation of coordinates such that the resulting equation would have only θ_- or θ_+. However, the desired separation of coordinates is not realized here. Nevertheless we can look at some special cases and perhaps gain some insight. For example, if at some point $\theta_+ = \pi/2$ then the differential equation contracts to

$$\frac{d^2\theta_-}{dt^2} + (b + 2c)\frac{d\theta_-}{dt} = 0 \tag{7.12}$$

whose solution is

$$\theta_-(t) = \theta_A + \theta_B e^{-(b+2c)t}, \tag{7.13}$$

where $(\theta_A + \theta_B)$ would be the instantaneous value of θ_- at the moment that $\theta_+ = \pi/2$. Therefore the effect of this solution would be to push θ_- toward the constant θ_A rather than zero. Thus the original two pendulums would be synchronized as to their angular velocities, or the phase of the motion, but their angles would differ by a constant angle. This state of affairs is often called *phase synchronization*.

More generally, we notice that Eq. (7.11) is, aside from the coefficient containing θ_+, the equation of motion for a damped pendulum. Near synchronization of the two original pendulums, this equivalent pendulum has a small amplitude and therefore we may, for the moment, continue our analysis using a linear approximation,

$$\frac{d^2\theta_-}{dt^2} + (b + 2c)\frac{d\theta_-}{dt} + (\cos\theta_+)\theta_- = 0. \tag{7.14}$$

In Chapter 3 we studied this sort of linearized equation of motion and found a variety of possible behaviors. However, this equation is slightly different in that the $\cos\theta_+$ term may be positive or negative. Let us assume, for the moment that it is positive and therefore it acts like the restoring force term found in springs or low amplitude pendulums. Thus $\cos\theta_+$ acts as an attractive force and the damping term $(b+2c)$ ensures that θ_- goes toward the potential minimum at $\theta_- = 0$. Therefore, for positive $\cos\theta_+$, the two pendulums will synchronize. How fast the pendulums synchronize depends on the relative size of the various coefficients. As shown in Chapter 3, the damped spring exhibits, damped, underdamped, and overdamped motion. For relatively light damping, b, and small coupling c, the tendency toward synchronization is likely to be underdamped motion. Figure 7.4 shows a typical decay of the "difference" pendulum angle θ_- toward zero. The upper graph shows the periodic variation of $\cos\theta_+$ when the pendulums are synchronized. The lower graph shows the oscillatory decay of the difference angle θ_- toward zero and again indicates that the pendulums synchronize. Why should this be a dominant behavior? The reason lies with the controlling factor that is the discriminant of the characteristic equation for Eq. (7.14),

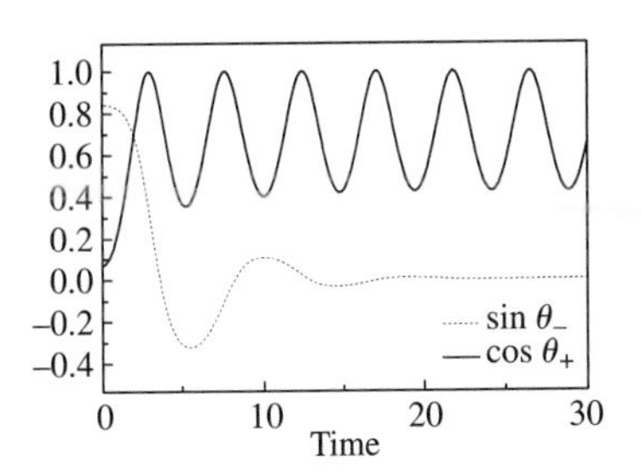

Fig. 7.4
The upper trace shows the cosine of θ_+ whereas the lower trace shows the corresponding values of the sine of θ_-.

$$\sqrt{\frac{(b+2c)^2}{4} - \cos\theta_+}. \tag{7.15}$$

For small amplitudes of oscillation of real pendulums, both angles θ_1 and θ_2 are relatively small and therefore the sum angle θ_+ is also likely to be small which means that $\cos\theta_+ \approx 1$. Therefore, in the likely event that $(b+2c) < 2$, there will be an oscillatory decay of the difference angle, θ_-. One other thing to note is that even in the absence of a coupling term, c, there will still be a decay toward synchronization but it will not be as fast. (This eventual synchronization is due to the coupling implicitly present through the use of a common forcing term.)

Other behaviors are also possible. For example, when $\cos\theta_+ < 0$, then the discriminant will be positive and $\theta_-(t)$ will diverge toward large values. Therefore such behavior would be destabilizing and not promote synchronization. But such behavior is unlikely for small amplitude motions, and therefore we defer that discussion until we come to the situation where the swings are large and chaotic. Another possible behavior is that for larger values of the coupling the decay toward synchronization will be monotonic rather than oscillatory. Such is the case when $\cos\theta_+ < (b+2c)^2/4$. This case is illustrated in Fig. 7.5 where the coupling is made significantly larger than was the case for Fig. 7.4. Finally there is the unlikely event that $\cos\theta_+ = 0$, where the pendulums would synchronize but would maintain a constant phase difference such that $\theta_+ = \theta_1 + \theta_2 = \pi/2$. We summarize the four possibilities for the case of small forcing in Table 7.1.

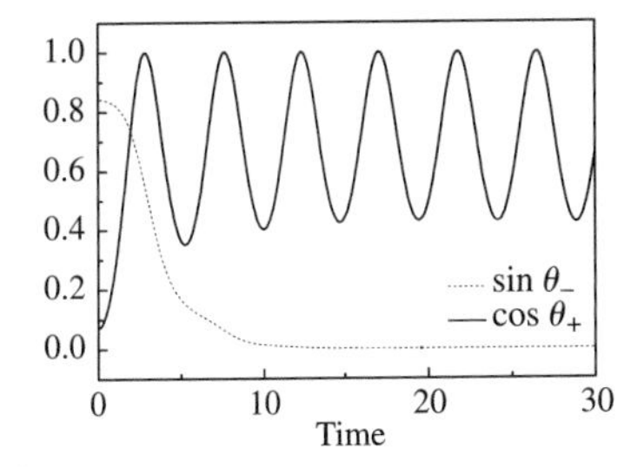

Fig. 7.5
The upper trace is the cosine of θ_+ and the lower trace is the sine of θ_-.

Thus Eq. (7.14) allows us to achieve some understanding of how the two pendulums achieve synchronization at least when the angles are almost equal.

When the pendulums are further from synchronization two things happen. Case number four becomes more likely and the case four scenario for the linear system holds less and less as $\sin\theta_-$ deviates increasingly from its linear approximation, θ_-. Nevertheless the above considerations

Table 7.1 Possible synchronization behaviors for small forcing

Condition on $\cos\theta_+$	Behavior of linearized (small forcing) system
$\cos\theta_+ = 0$	Pendulums go to phase-only synchronization (unlikely)
$(b+2c)^2/4 < \cos\theta_+ \leq 1$	Pendulums oscillate toward synchronization (likely)
$0 < \cos\theta_+ < (b+2c)^2/4$	Pendulums go monotonically to synchronization (likely)
$\cos\theta_+ < 0$	Pendulums are forced away from synchronization(unlikely)

provide us with some sense of how pendulums that execute small amplitude motion can be synchronized. To recap, the system as a whole behaves like a damped oscillator. The damping is enhanced by the coupling and the harmonic potential is provided by the $\cos\theta_+$ term. If the latter is positive it will promote the tendency toward synchronization. (Incidentally, the adjacent pendulum clocks that Huygen's observed oscillated with relatively low amplitudes. At first glance we might think that the analysis would therefore be similar to the simple linear model given above. However, clocks have a nonlinear feature, the escapement, which gives a periodic pulse to the pendulum that provides the energy to overcome friction and thereby keep the clock ticking. This feature and other factors of the coupling between the clocks necessitate a detailed analysis (Bennett et al. 2002).)

The situation is more complex for large and unpredictable deviations from synchronization. For strongly chaotic pendulums with large angular displacements, $\cos\theta_+$ spends about equal amounts of time in both positive and negative regions, even though the pendulums are approximately synchronized for much of the duration. This behavior is illustrated in Fig. 7.6. After the initial transient somewhat erratic variations, the two pendulums approximately synchronize for many drive cycles as shown by the somewhat erratic line near $\theta_- = 0$. The synchronization is much stronger than would be suggested by the behavior of $\cos\theta_+$ extrapolated from the linear case. Once again nonlinearity reveals behavioristic features that are not found in the linear approximation.

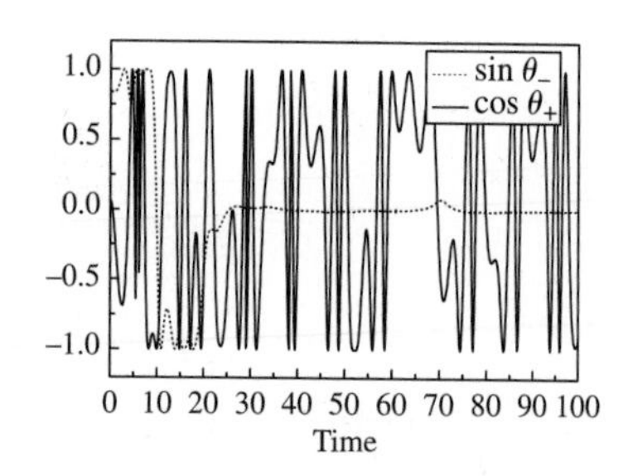

Fig. 7.6
The highly variable trace is a time series of $\cos\theta_+$ and the more quiescent trace is a time series of $\sin\theta_-$.

Smith et al. (1999) studied the synchronization of the driven pendulums shown in Fig. 7.3. Figure 7.7 shows time series for θ_- at various levels of mutual coupling between the pendulums. From the series of graphs in Fig. 7.7 it becomes clear that as the coupling increases the difference coordinate θ_- spends more time near zero or a state of synchronization for the two pendulums. These states are often referred to as *laminar* states in contrast to the "turbulent" states when the pendulums are not synchronized. (The coupling increases with each time series lower on the page.)

There is another effect of increased coupling. When the coupling is nonexistent, each pendulum generates its own Poincaré section. Because the pendulums are quite similar, the Poincaré sections, if plotted on the same set of axes, should be the same and simply overlap as shown in Fig. 7.8(a). But once there is coupling, the pendulums are now joined to form one large system. Poincaré sections graphed from data acquired from the separate pendulums, as in Fig. 7.7(b), now begin to "thicken" each

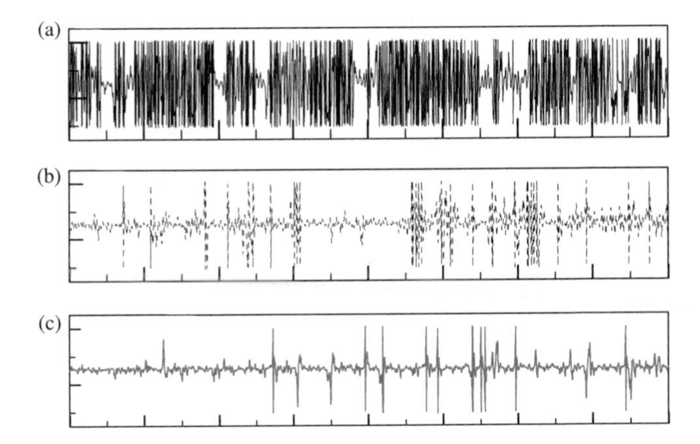

Fig. 7.7
Experimential time series of θ_-. Coupling between the pendulums increases for the successive traces.

attractor. Then as the coupling increases, and the two pendulums act more and more as one unit, the thickening goes away and we are again left with the attractor of the uncoupled pendulums as in Fig. 7.7(c). That is, the pendulums are now synchronized and act like a single pendulum.

Let us now try to achieve some understanding of the mechanism of synchronization with chaotic pendulums.

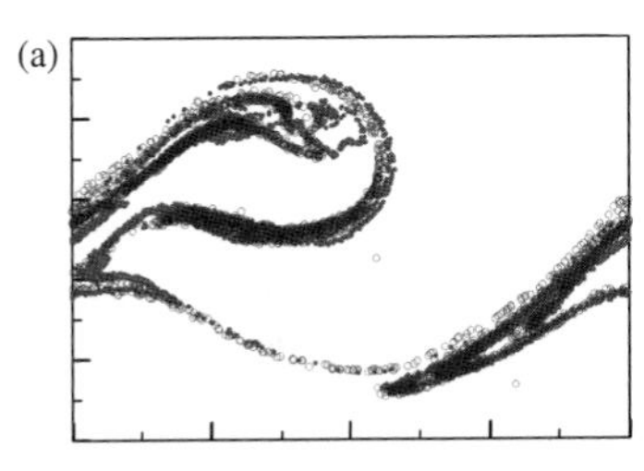

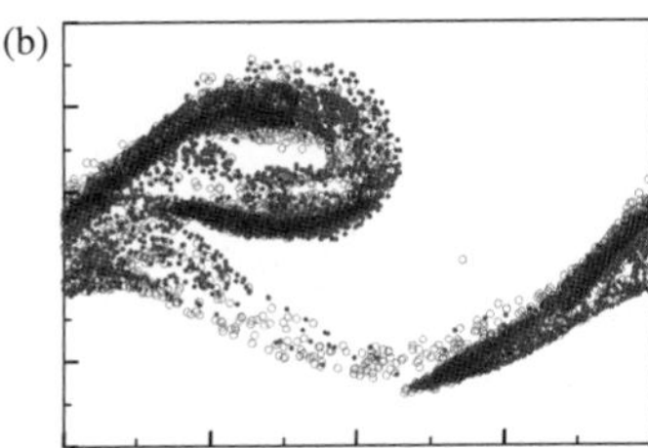

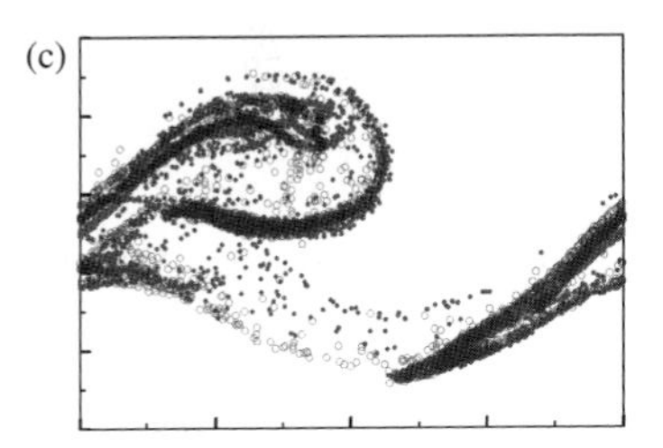

Fig. 7.8
Poincaré plots for the coupled pendulums that correspond to the time series of Fig. 7.7.

7.2 Chaotic coupled pendulums

What happens when two pendulums synchronize and, are there various kinds of synchronization? In the case of Huygen's clocks there was some sort of synchronization of the clocks' pendulums. Yet the pendulums oscillated out of phase and therefore the pendulum's angles, θ_1 and θ_2, were identical only twice momentarily (at the bottom of the swing) during each cycle of the motion. Similarly the pendulums' velocities $\dot{\theta}_1$ and $\dot{\theta}_2$ were opposed for all but two moments (at the extremes of the swing) during the cycle. Yet we saw, from our simple model of pendulums coupled with springs, how this state could arise. And it was a kind of synchronization in that there was a dynamic *phase* relationship between the angles of the pendulums such that $\theta_1 = -\theta_2$.

With *chaotic* systems there are a variety of possibilities as to the type of synchronization. Sometimes each subsystem may have some associated frequency that gets in synchronization with all the other subsystems, but other variables may not be in synchronization. There is the famous example of a certain type of firefly which, in large groups, tend to synchronize the timing of their flashes[1]. Yet while the flashing may be synchronized, the firefly is a complex structure with many degrees of freedom, and individual fireflies may be totally out of synchronization with regard to other bodily functions.

For our chaotic pendulums we choose synchronization to mean—unlike Huygen's clocks—that all coordinate pairs line up: $\theta_1 = \theta_2$ and $\dot{\theta}_1 = \dot{\theta}_2$.

[1] See (Strogatz 1994, p. 103).

The phase space geometry of the pair of pendulums is five-dimensional, $\theta_1, \theta_2, \dot{\theta}_1, \dot{\theta}_2$, and φ. (We take the phases φ of the drives for both systems to be identical during the motion.) This state is sometimes called *general* synchronization. When synchronization occurs, the two pairs become identical and the dimension of the resulting new phase space contracts to three dimensions $(\theta, \dot{\theta}, \varphi)$, a subset of the original five-dimensional phase space. Thus synchronization forces the instantaneous orbit of the phase point of the original system onto a subset of the entire phase space where $\theta_1 = \theta_2$ and $\dot{\theta}_1 = \dot{\theta}_2$. This subset space is called the *synchronization manifold.* If the pendulums become somewhat unsynchronized then the instantaneous phase point jumps off the synchronization manifold into the original five-dimensional phase space. Abstractly, we can think of the direction "perpendicular" to the synchronization manifold as the direction that the system takes when it loses synchronization. On the other hand, the system can stay synchronized by maintaining its motion "along" the surface that represents the synchronization manifold. The motion, whether synchronized or slightly nonsynchronized, is characterized by the spreading and shrinking that we associate with the pendulum's chaotic dynamics. When the pendulums are synchronized, they act as a single chaotic pendulum—locked to the other—and the stretching and shrinking of sets of phase points is described by the Lyapunov exponents ($\lambda_+ > 0, \lambda = 0$, and $\lambda_- < 0$) of a single pendulum, *on the synchronization manifold.* See, for example, Table 6.1. When the system is not synchronized and only slightly coupled, there are *five* Lyapunov exponents. However, when the system is close to synchronization, it is convenient to think of there being the three Lyapunov exponents of a single system, plus a new *transverse Lyapunov exponent*, $\lambda_\perp$, that describes whether the system is tending into or out of synchronization, onto or off the synchronization manifold. If $\lambda_\perp < 0$ then the pendulums tend to synchronize; if $\lambda_\perp > 0$ then the synchronous state is unstable. When $\lambda_\perp = 0$, then the coupling coefficient is said to take its critical value: $c = c_{\text{critical}}$.

Now, some of our thinking about the dynamics of synchronization of chaotic systems rests with the notion of chaotic systems acting as if they are random. While the transverse Lyapunov exponent, $\lambda_\perp$, is a fixed quantity for a certain level of coupling between pendulums, the nature of the active chaos is such that in addition to $\lambda_\perp$ the Lyapunov exponent has a random part, which will be a function of chaotic coordinates and therefore time. Over relatively long times this fluctuating part averages to zero, but in the short term it strongly affects the degree of synchronization. This short term fluctuation is called the *local* or *finite time Lyapunov exponent* and may be denoted as $\Lambda(t)$. The effect of this fluctuation is to cause the phase point near synchronization (near the manifold) to randomly move away from or toward the manifold and that effect is added onto the constant drift—toward or away from the manifold—caused by the action of $\lambda_\perp$. The degree of synchronization depends upon the dynamics of the pendulums and on the strength of the coupling between them. Specifically, the transverse exponent $\lambda_\perp(c)$ is a static function of the coupling, and the finite time exponent $\Lambda(t)$ contains the time dependence caused by the unstable dynamics of a chaotic system. In a physical experiment, noise can also be a factor in determining the degree of synchronization. The analysis provided below is based upon computer simulations and therefore noise is not a factor. However, when we examine data from simulations and data from experiment, differences due to noise—and the inexact correspondence between the simulation model and experimental reality—will be evident.

We showed earlier that absolutely complete synchronization is an experimental and theoretical impossibility. But for practical purposes we make a somewhat arbitrary distinction between the synchronized and unsynchronized state in the following way. Let us define a synchronization variable

$$\eta(t) = \sqrt{(\theta_1 - \theta_2)^2 + (\dot{\theta}_1 - \dot{\theta}_2)^2} \tag{7.16}$$

and say that coupled pendulums are synchronized when η is less than ϵ, where ϵ is some arbitrary small value. The variable η measures the "distance" off the synchronization manifold. We call ϵ the *threshold* of synchronization. The choice of threshold is somewhat flexible and is usually some number that is much smaller than the size of the phase space. As we shall see, certain behaviors are not very sensitive to the size of the threshold. But we also note that the arbitrariness of any quantity in a scientific analysis often encourages a search for criterion on which to more firmly base its definition. Thresholds are no exception and suggestions have been made as to other possible criteria for synchronization. See, for example, Blackburn et al. (2000). These concepts are illustrated schematically in Fig. 7.9.

Chaotic systems can *hard-lock*—be pretty much synchronized—or the synchronization may be *intermittent*. In the latter case the coupled system acts "locked" for long periods of time and then occasionally burst into a state of desynchronization. With the type of coupling described by Eqs. (7.8) and (7.9), chaotic pendulums seem to synchronize intermittently. Figure 7.7 shows actual experimental time series from two magnetically coupled pendulums (Smith et al. 1999) and the synchronization is clearly intermittent.

A common method for studying synchronization of systems is to look at the statistics of the synchronization. For example, we might ask what percentage of the time is a system synchronized. A more complex measure of synchronization would be the probability distribution of the various durations of synchronization or laminar intervals. Such measures of synchronization lend themselves to analysis by statistical methods.

We may model the process in a variety of ways. For example, we may think of the process as a simple two-state random process. Either the pendulums are locked such that $|\eta| < \epsilon$, or the pendulums are unlocked with $|\eta| > \epsilon$. This model explains some of the behavior of the system and is a good first conceptual approach. We give the details of this model in the next section. But in passing, we note that in the technical literature of synchronization a more sophisticated model is used. After a detailed

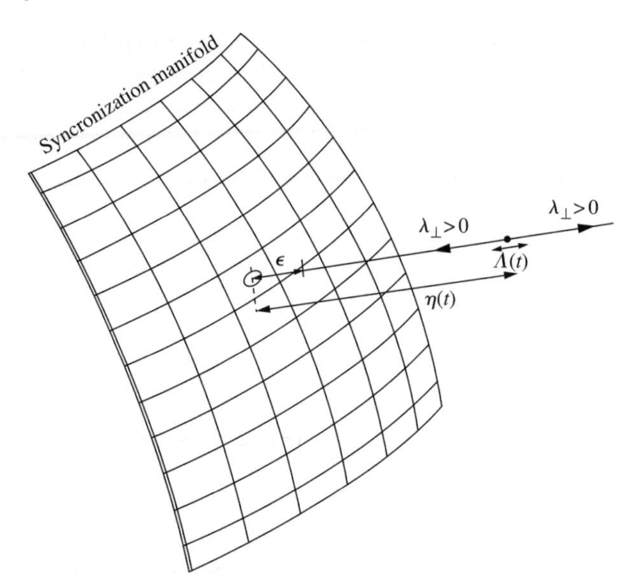

Fig. 7.9
Schematic diagram to illustrate the main features of synchronization as described in the text.

discussion of the two-state model we provide a brief summary of the more complex model.

7.2.1 Two-state model (all or nothing)

The two state model (Baker et al. 1998) is an example of a *Markov* process, one for which the present state of the system only depends on the state at the previous time step. Suppose that at time t the state of the system of coupled pendulums is either synchronized (S) with probability $p_S(t)$ or unsynchronized (U) with probability $p_U(t)$. Furthermore let us define the conditional probability $P(S|U)$, sometimes called a transition probability, as the probability that the system goes from an unsynchronized state to a synchronized state. We also define three other similar conditional probabilities, $P(U|S)$, $P(S|S)$, $P(U|U)$ for the corresponding changes of state. A Markov process then requires that $p_i(t+\Delta t)$ depends on all the probabilities $p_j(t)$ through the conditional probabilities. This dependence is expressed by the *total probability theorem:*

$$\begin{aligned} p_S(t+\Delta t) &= P(S|S)p_S(t) + P(S|U)p_U(t) \\ p_U(t+\Delta t) &= P(U|S)p_S(t) + P(U|U)p_U(t). \end{aligned} \tag{7.17}$$

We assume that the time step Δt is small and that, in this approximation, the transition probabilities are linearly proportional to the time steps according to

$$P(S|U) = \beta\Delta t \quad \text{and} \quad P(U|S) = \alpha\Delta t. \tag{7.18}$$

The transition rates β and α are, respectively, the so-called "birth" and "death" rates for synchronization, and are respectively the rates at which a system becomes synchronized or becomes unsynchronized. We assume

α and β to be constants for a given set of pendulum parameters. Furthermore, the transition probability pairs $P(S|S)+P(S|U)$ and $P(U|S)+P(U|U)$ must each sum to unity, which means that the other transition probabilities are readily found to be

$$P(S|S) = 1 - \beta\Delta t \quad \text{and} \quad P(U|U) = 1 - \alpha\Delta t. \tag{7.19}$$

This two state model is represented schematically in Fig. 7.10.

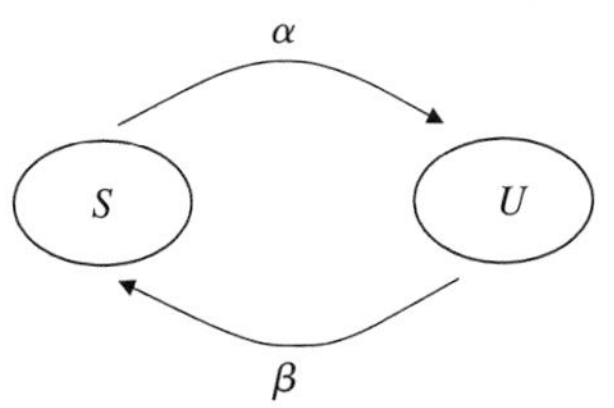

Fig. 7.10
A state diagram for the "all or nothing" model of synchronization. The arrows indicate transitions to the opposite state.

It is important to understand that the seemingly innocuous probability rates α and β contain much of the physics. The rate α describes the tendency away from synchronization whereas the rate β describes the tendency toward synchronization. As the coupling c increases, α will decrease and β will increase. Implicit in these rates is the action of the transverse Lyapunov exponent $\lambda_\perp$ and the noise effect of the finite time Lyapunov exponent $\Lambda(t)$. Figure 7.11 gives the dependence of the fraction of time, that the system is synchronized, on the coupling constant c. Remarkably the simple two state model can account for the shape of this curve.

Let us continue with the mathematics of the model. The results of Eqs. (7.18) and (7.19) can be inserted into the total probability theorem and then we take the limit as Δt approaches zero. This procedure results in the following pair of differential equations;

$$\begin{aligned} \frac{dp_S(t)}{dt} &= -\alpha p_S(t) + \beta p_U(t) \\ \frac{dp_U(t)}{dt} &= +\alpha p_S(t) - \beta p_U(t). \end{aligned} \tag{7.20}$$

First we consider the behavior of the probabilities over a long time period. This is the equilibrium state determined by setting the derivatives in the rate equations equal to zero, yielding the very simple result;

$$p_S^e = \frac{\beta}{\beta+\alpha} \quad \text{and} \quad p_U^e = \frac{\alpha}{\beta+\alpha}. \tag{7.21}$$

Bearing in mind that β increases and α decreases with increasing coupling, these equations verify that, on average, the probability of synchronization,

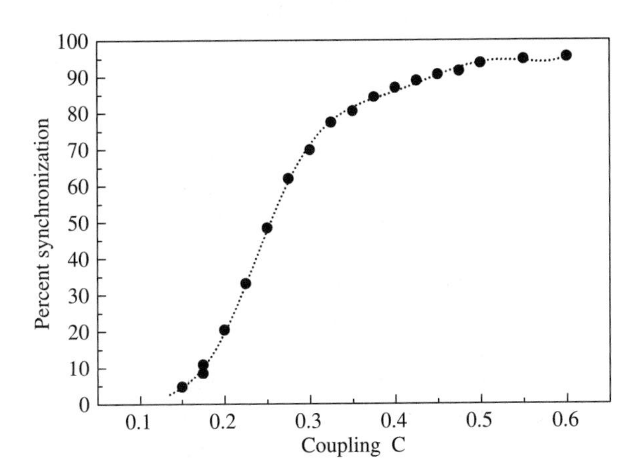

Fig. 7.11
The percentage of synchronization as a function of coupling strength.

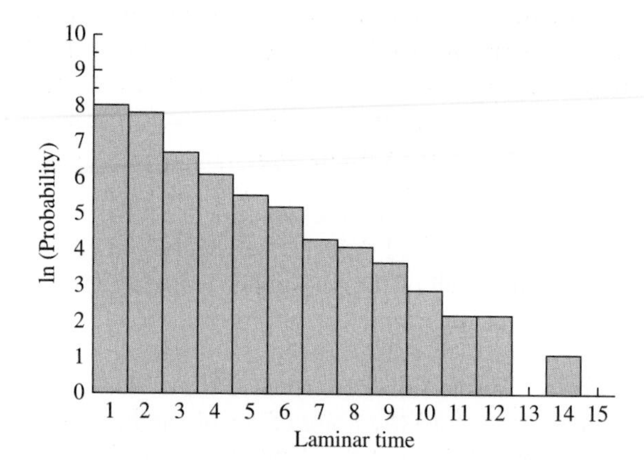

Fig. 7.12
Semilogarithm plot of the probability distribution for the duration of laminar times versus laminar time, as determined from simulation of the coupled pendulums.

p_S^e, increases with increasing coupling. Figure 7.11 is just a plot of p_S^e versus coupling. This probability is equivalent to one of the statistics we mentioned earlier; namely, the percentage of time that the system is synchronized.

Now let us consider the time dependent behavior, and focus on a single value of the coupling. Figure 7.12 shows the probability distribution for the duration of the laminar times as determined by computer simulation of the coupled pendulums. (This is the other statistical measure that we described as a characterization of synchronization studies.) The distribution has an exponential behavior for all but the shortest times. Therefore we ignore, for the moment, the nonexponential behavior for short times. For most of the laminar times the data may be characterized by the equation

$$\text{Probability } (t < \text{laminar time} < t + \Delta t)\ \alpha\ e^{-t/\tau}, \tag{7.22}$$

where τ is the mean laminar time. From individual probability distributions of laminar times for each value of coupling one can determine the corresponding mean laminar times, using the straight line portion of the semilog plots as in Fig. 7.12. For comparison, data from real magnetically coupled pendulums is also given (Smith et al. 1999) in Fig. 7.13.

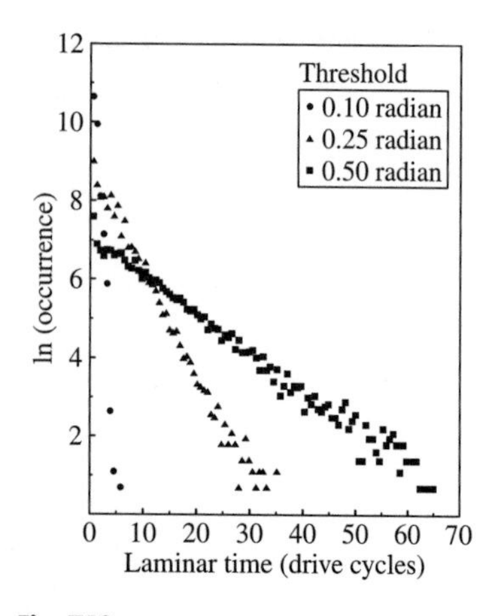

Fig. 7.13
Semilogarithm plot of the frequency of long laminar times versus laminar time for different values of thresholds of laminar behavior. (Reprinted from Smith et al. (1999, p. 1914), with permission from WSPC, NJ)

Next, let us see how the two state model predicts this exponential behavior. We first develop a differential equation for the probability that the system remains laminar at any time $t + \Delta t$ *given* that it is in the laminar state at time t. Since the system is specified to be in the synchronized state at time t, $p_U(t) = 0$ and the total probability equation simplifies to

$$p_S(t + \Delta t) = P(S|S)p_S(t) = (1 - \alpha\Delta t)p_S(t). \tag{7.23}$$

Passing to the limit as Δt approaches zero leads to

$$\frac{dp_S(t)}{dt} = -\alpha p_S(t) \text{ with solution } p_S(t) = e^{-\alpha t}, \tag{7.24}$$

where we have used the initial condition $p_S(0) = 1$.

The second step in the process is to determine the conditional probability density function for the system becoming unsynchronized. That is, given that the system is in a synchronized state at time t, what is the probability that the system desynchronizes during the next time interval Δt? Put into symbols, what is $p_U(t + \Delta t)$, given that $p_U(t) = 0$? Again we use Eq. (7.17) with the previous solution $p_S(t) = e^{-\alpha t}$ as follows. First

$$p_U(t + \Delta t) = \alpha \Delta t p_S(t) \tag{7.25}$$

and, using the fact that $p_U(t) = 0$, we obtain

$$\frac{p_U(t + \Delta t) - p_U(t)}{\Delta t} = \alpha e^{-\alpha t} \tag{7.26}$$

leading to the probability density function $\alpha e^{-\alpha t}$. It is this density function that allows us to calculate the average laminar time as

$$\tau = \int_0^\infty t \alpha e^{-\alpha t}\, dt = \alpha^{-1}. \tag{7.27}$$

In this way our model makes a connection with the data from the simulation. The transition rates α and β may now be determined from the data. Figure 7.14 shows a plot of the logarithm of α versus the coupling, c. The approximate straight line behavior suggests that α can be fit to a curve of the form $\alpha = Be^{-bc}$.

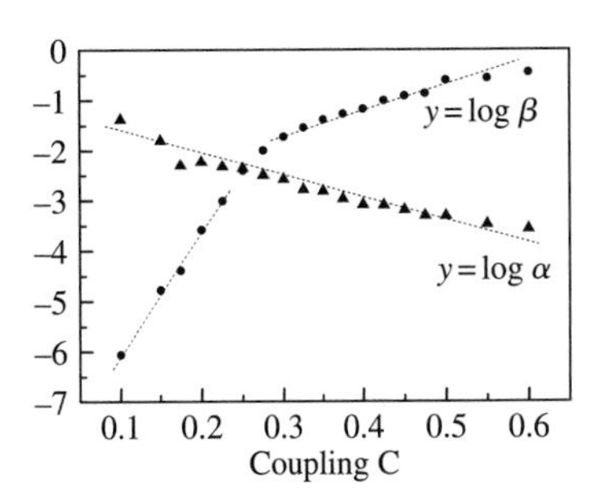

Fig. 7.14
Semilogarithmic plot of the transition probabilities α and β as functions of coupling.

We are now ready to complete the connection between the data and the model parameters. Note that the simulation data provides the average fraction of time that the system is synchronized, p_S^e, and the mean laminar time or, according to the model, the mean desynchronization rate, α. The last model parameter is the synchronization rate, β, and this may determined using Eq. (7.21) as

$$\beta = \frac{\alpha p_S^e}{(1 - p_S^e)}. \tag{7.28}$$

Figure 7.14 also includes a plot of the logarithm of β versus coupling. Unlike the graph of α, the graph of β does not seem to be characterized by a single exponential behavior throughout the range of couplings. However, the 50% synchronized coupling of about $c(50\%) \approx 0.252$ provides a natural breakpoint between two exponential behaviors with different rates. That is, if $\beta = Ae^{ac}$ then there are two pairs of parameters; (A_1, a_1) below $c(50\%)$, and (A_2, a_2) above $c(50\%)$.

With the fitted expressions, $\alpha = Be^{-bc}$ and $\beta = Ae^{ac}$, we go back to Eq. (7.21) and produce an expression for the equilibrium synchronization fraction as a function of coupling and overlay the results on Fig. 7.11. The required expression is given by

$$p_S^e = \frac{1}{1 + \frac{B}{A} e^{-(b+a)c}}, \tag{7.29}$$

where the parameters are found from least squares data fits: $A_1 = 2.2723 \times 10^{-4}$, $A_2 = 5.4291 \times 10^{-2}$, $a_1 = 23.6687$, $a_2 = 4.3412$, $B = 0.26205$, $b = 4.0267$.

Given the simplicity of the model the fit seems fairly good. Let us note some features. We saw that the coupling, $c(50\%) \approx 0.252$, provides a breakpoint about which the possible symmetry above and below this coupling seems to break down. The lack of symmetry is slightly noticeable in Fig. 7.11 and more so in Fig. 7.14. Similar asymmetry has been observed in other forms of pendulum coupling (see, for example, Baker et al. (1998)). Thus the processes by which the pendulums synchronize and desynchronize do not seem entirely symmetric. (The lack of perfect symmetry may also indicate a limitation of the two-state model.)

The phenomenon of bursts of desynchronization and subsequent stretches of synchronization is sometimes called *on–off synchronization* or *intermittency*, and is often characterized by some degree of symmetry between the bursting and laminar stretches. (Cenys et al. 1997). The fact that the data in Fig. 7.11 is almost symmetric suggests that intermittency is the primary phenomenon. In this circumstance, the transverse Lyapunov exponent $\lambda_\perp$ is slightly positive, causing the coupled pendulums to temporarily jump off the synchronization manifold. Conversely, the nonlinear dynamics and the variation in the finite time exponent $\Lambda(t)$ tend to bring the coupled pendulums back into synchronization. In a real system noise also affects the on–off nature of the synchronization.

7.2.2 Other models

While the simple two state model describes many features of intermittency, a more sophisticated analysis is typically used in the technical literature (Heagy et al. 1994; Pecseli 2000; Pikovsky et al. 2001), but because of the mathematical complexity we provide only a brief description. Rather than just consider a two state model where the pendulums are either absolutely synchronized or not, the more complex models postulate a large number of states. We assume that the system spends much of its time in the laminar (synchronized) phase. Therefore, for most of the probability states, the system is in the laminar state, but has bursts of desynchronization as it nears some sort of reflecting barrier. The barrier is reflecting because it acts to direct the temporarily unsynchronized system back toward approximate synchronization. The system then acts as a *random walker* moving between various states occasionally being redirected by the reflecting wall. The random walk model goes back to the early nineteenth century when small particles suspended in liquid were seen to undergo erratic, random movements. This phenomenon, first reported in 1837 by the botanist Robert Brown, and now appropriately known as *Brownian motion*, is also observed with smoke and dust particles in the air. In 1905 Einstein provided an analysis of the motion using the idea that even relatively large particles—like smoke and dust—have motion due to thermal agitation. The unpredictability of the direction of travel of the dust particle suggests the metaphor of a random walker, one that makes a random decision, at each time step Δt to go either right or left, for example. The random walk model is like the two state model in that it is Markovian, but is unlike the two state model in that the system must change state (position) at each time step and there are many states (positions). The walker travels back and forth randomly. His average position is zero because of the equal probability of going right or left. Yet after some time the mean square displacement will be nonzero and proportional to the square root of the time passed. (For a uniform velocity the displacement is directly proportional to the time passed.)

The random walker model describes a process of discrete steps taken at discrete times. If the time intervals and the space intervals are allowed to become relatively small, the random walk process becomes a diffusion process, a process involving continuous materials such as fluids. More concretely, if we can imagine that the steps of the walk are relatively small and numerous, then the walker's position may be modeled by some sort of continuous probability density that will spread or diffuse in time. This behavior is exemplified by the spreading of an ink blob that is released into water. Furthermore we might imagine that there is also a current in the water. The blob will then move in some direction as well as spread. In general, diffusion exhibits these two behaviors. Similarly, a biased random walker may choose left or right with some degree of favoritism. Thus he drifts more in one direction than the other.

The characteristics of intermittent synchronization, such as are found for coupled chaotic pendulums, are similar to a biased diffusion problem. We previously introduced the transverse Lyapunov exponent $\lambda_\perp$ that measures the rate at which the coupled system goes toward or away from the synchronization manifold. In the regime where $\lambda_\perp$ is just slightly positive and the system is almost, but not quite, completely synchronized there is intermittent synchronization. The system may then be mathematically represented by the so-called Fokker–Planck equation (Pikovsky et al. 2001):

$$\frac{\partial P(x,t)}{\partial t} = -\lambda_\perp \frac{\partial P(x,t)}{\partial x} + \frac{D}{2}\frac{\partial^2 P(x,t)}{\partial x^2}, \tag{7.30}$$

where $P(x, t)$ is the probability of the random walker being at position x at time t. The rightmost term in the equation determines the spreading or diffusion of the probability, and D is a measure of the diffusion rate and is called the "diffusion constant." It is also proportional to the variance of the Gaussian distribution. Larger values of D indicate faster rates of spreading. The next term in the equation and the value of $\lambda_\perp$ regulate the mean drift of the system, toward or away from synchronization. If there is no bias term then the distribution just spreads out with a normal (Gaussian) shape. But if there is a drift term, the mean of the distribution moves off zero and heads in the direction of the bias. Thus the distribution becomes a moving and broadening Gaussian shape.

Experimentally we measured the laminar time distribution for the coupled pendulums. In terms of this latest model, the laminar time between bursts of desynchronization is just the time it takes for the system to diffuse from one instance near the reflecting barrier to a subsequent recurrence of state near the reflecting barrier. Unlike the two-state model, the probability distribution of laminar times is no longer simply exponential but now has a power law component as well (Pecseli 2000);

$$P(t) \propto t^{-3/2} e^{-t/\tau}, \tag{7.31}$$

where $\tau = D(\lambda_\perp)^{-2}$. Note that the time τ in this model probably does not have the same meaning as time constant τ in the two state model, and numerically they are not found to be the same when the models are applied to data. The mean laminar time τ in the random walk/diffusion model gives a measure of the diffusion constant and the transverse Lyapunov exponent. We saw earlier in Fig. 7.13 that most of the data for the histogram of laminar time distributions followed the exponential decay predicted by the two state model. However, for small laminar times the 3/2 power law behavior of the random walker model provides a better fit. Thus a more graduated model of synchronization seems to better fit the data and seems more in keeping with the continuous nature of solutions to the equations of motion for the pendulums. The random walk model for coupled pendulums has been briefly discussed by Grassberger (1999).

The analysis of synchronized chaotic systems is complex. Our discussion has just given a small sampling of this complexity. Let us now discuss some applications of synchronization; a synchronization machine, secure communication, control of chaos, and coupling of many chaotic systems. Each area is treated in the context of the pendulum, but other mechanical and electrical systems may be substituted—often resulting in greater utility.

7.3 Applications

7.3.1 Synchronization machine

We might speculate as to whether a system that consists of two coupled pendulums could somehow be thought of as a single pendulum. If so, what sort of form would this "resultant" single pendulum take? Whatever the form, we will call such a device a "synchronization machine."

Our starting point is the system of two pendulums magnetically coupled through the difference in their angular velocities as described above. One approach to the building of a synchronization machine, that corresponds to this pair of coupled pendulums, is to consider a system whose primary variable is the *difference* between the angles (or the angular velocities) of two separate pendulums. In other words we make a single pendulum whose angular coordinate is proportional to the difference coordinate. Let us begin with the equations for the original coupled pendulums

$$\frac{d^2\theta_{1,2}}{dt^2} + b\frac{d\theta_{1,2}}{dt} + \sin\theta_{1,2} + c\left(\frac{d\theta_{1,2}}{dt} - \frac{d\theta_{2,1}}{dt}\right) = A\cos\omega_D t \tag{7.32}$$

and reintroduce the sum and difference coordinates

$$\theta_- = \frac{\theta_1 - \theta_2}{2} \quad \text{and} \quad \theta_+ = \frac{\theta_1 + \theta_2}{2}. \tag{7.33}$$

Substitution of these coordinate transformations into the previous equations leads to two differential equations. However, we will focus on the θ_- coordinate and find the following equation of motion;

$$I\frac{d^2\theta_-}{dt^2} + (b + 2C)\frac{d\theta_-}{dt} + [mgr]\cos\theta_+ \sin\theta_- = 0. \tag{7.34}$$

We have seen this equation before in the dimensionless form and used it to analyze coupled pendulums for nonchaotic motion. But now we take a different approach. Let us compare this equation to another equation of motion,

$$I_m\frac{d^2\phi}{dt^2} + b_m\frac{d\phi}{dt} + [ma(t)r]_m \sin\phi = 0. \tag{7.35}$$

The physical configuration to which this latter equation belongs is also a pendulum. But this new pendulum is tipped on its side such that the axis for the pendulum's motion is vertical rather than horizontal. Thus gravity is removed and the only force acting on the pendulum, aside from the usual friction term, is a time dependent acceleration $a(t)$ that jiggles the axis of the pendulum. A schematic diagram of this "synchronization machine" is shown in Fig. 7.15.

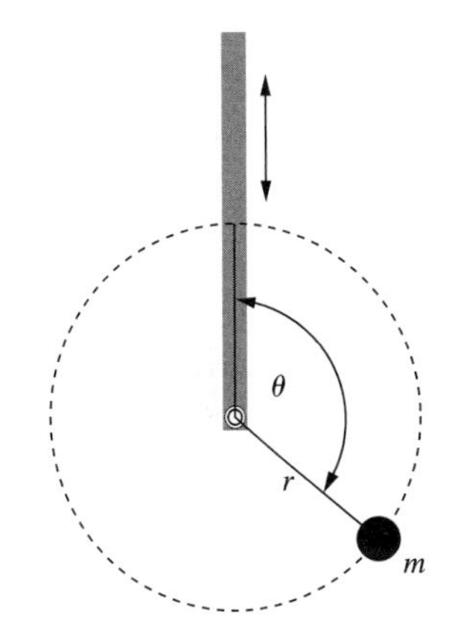

Fig. 7.15
Schematic diagram of the pendulum with an oscillating pivot.

A photograph of the physical realization of this device is shown in Fig. 7.16. The pendulum bob is a mass—partially hidden by the pendulum axis—mounted on a disc. The mass is somewhat spread out because it must be able to get underneath the structure that supports the axis of the pendulum. There is also a damping magnet that can be adjusted to vary the damping parameter b_m. Angular displacement data are transmitted to the computer via the optical encoder as in previous arrangements.

If this synchronization machine or equivalent parametric[2] pendulum truly corresponds to the action of two coupled pendulums then time series and laminar time distributions derived from the synchronization machine should reflect similar entities, shown above, from the coupled pendulums. For the parametric pendulum the angle is θ_- and when $\theta_- \approx 0$, the corresponding coupled pendulums are approximately synchronized. That is, synchronization implies that the parametric pendulum is not oscillating: the little wheel is stationary relative to the center. On the other hand the bursting out of synchronization for the coupled pendulums is registered by erratic oscillations of the wheel of the parametric pendulum. Figure 7.17 shows a time series of the angle for the parametric pendulum with quiescent and irregular bursts. This time series has the same general form as that shown for the difference angle of the coupled pendulums in Fig. 7.7. Similarly, for a given level of damping for the parametric pendulum, there is a distribution of laminar times somewhat corresponding to the distribution for the coupled pendulums at a certain level of coupling. Figure 7.18 shows such a distribution for the parametric pendulum and it may be

[2] The term parametric is applied here because the pivot point of the new pendulum is not stationary.

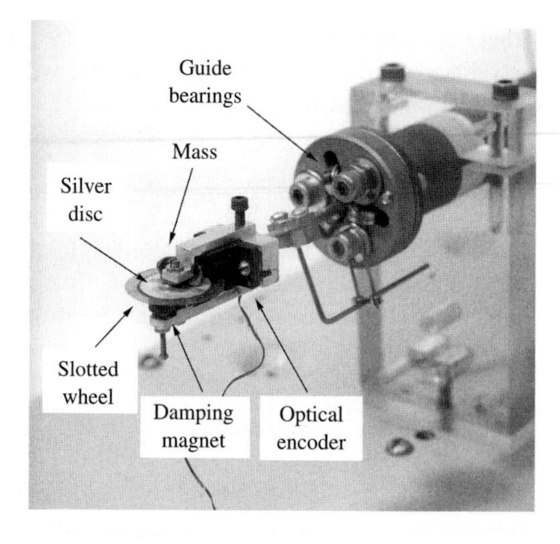

Fig. 7.16
Photograph of the "head" of the apparatus, including the pendulum "bob," the slotted wheel and the silver disk. (Reprinted from Smith et al. (2003, p. 11), with permission from WSPC, NJ.)

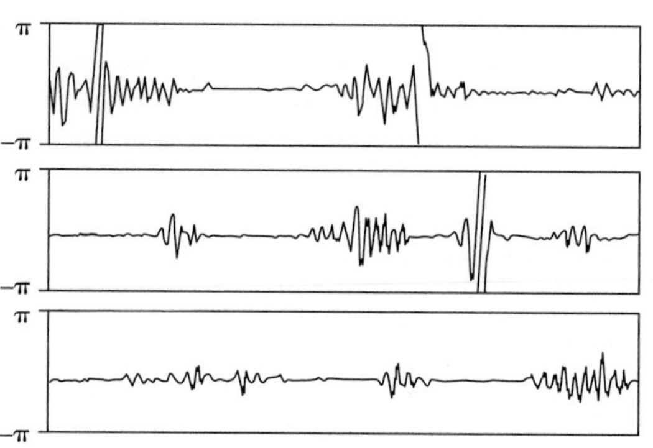

Fig. 7.17
Experimental time series from the synchronization machine. The three panels are time sequenced.

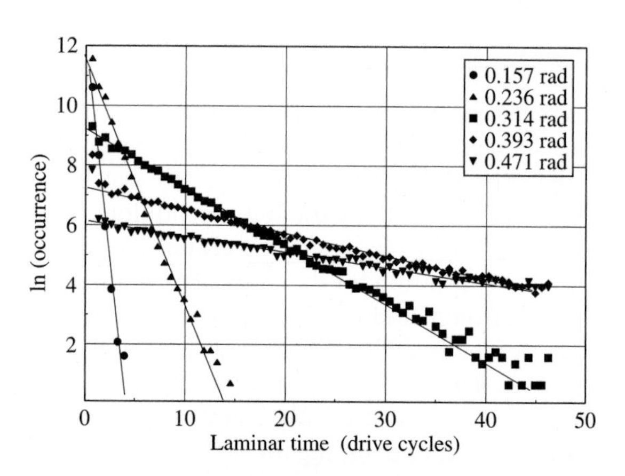

Fig. 7.18
Semilogarithmic plot of laminar time distributions as a function of laminar time for various threshold values. The machine parameters are the same as those in Fig. 7.17. (Reproduced from Smith et al. (2003, p. 15), with permission from WSPC, NJ.)

compared to the distribution shown in Fig. 7.13 for the real coupled pendulums or Fig. 7.12 for the simulated coupled pendulums.

The driving force for the synchronization machine is the acceleration applied to the pivot of the pendulum. The fact that $a(t)$ involves a $cos\theta_+$ term is interesting in that the θ_+ is just the average of the angles of the equivalent coupled pair of pendulums. Therefore, the action of the parametric pendulum is not really independent, but is the result of an earlier state of two pendulums. There is feedback of information. But, *in practice* the drive can be provided by a single chaotic pendulum. Even a *random* drive yields time series and laminar time distributions much like those in Fig. 7.18. Finally, Smith (Smith et al. 2003) has explored the states of the synchronization machine by letting the variable θ_+ be a *periodic* function. He finds that the parametric pendulum exhibits patterns of quiescence, period one, period two, period four, and chaotic states. Thus, like the single pendulum, the synchronization machine exhibits the range of dynamical states characteristic of a typical nonlinear system. The synchronization machine is the first mechanical contrivance that combines two chaotic synchronized devices into a single device and demonstrates the difference coordinate directly. It is remarkable that the result of combining two pendulums is also a pendulum, albeit a different type. We note that a similar idea seems possible in the optical regime using lasers. Based upon work at Georgia Tech and Universite Libre de Bruxelles, Rajarshi Roy and coworkers suggested that coupled lasers might act as a single modulated laser; again, a kind of synchronization machine (Thornburg et al. 1997).

7.3.2 Secure communication

It seems natural that the study of synchronized chaotic systems would result in the search for practical applications of such systems. One such application is the idea of using chaos to produce transmitting and receiving systems through which can be channelled messages that are secure against nonsystem decoding (Pecora and Carroll 1990; Cuomo and Oppenheim 1993). As with other chaotic systems, coupled chaotic pendulums could, at least in theory, also be used in this application. While we are not suggesting a system that is physically realizable, it is instructive to see how such a system might work. Our starting point is a modification of the magnetically coupled system represented by Eq. (7.32) described earlier. In our communications system one chaotic pendulum is the transmitter and the other chaotic pendulum is the receiver. Unlike the symmetric magnetic coupling, the transmitter pendulum has an effect on the receiver whereas the receiver pendulum has *no* obvious effect upon the transmitter. For such a coupling scheme the transmitter is sometimes called the *master* whereas the receiver is the *slave*. This coupling scheme contrasts with the bidirectional system described earlier in which the partners had equal roles. Thus we modify Eq. (7.32) so that (unphysically) the magnetic coupling is felt to go only from the transmitter pendulum to the receiver pendulum and not the reverse. The angular velocities are coupled such that the angular velocity of the receiver has a dose of the angular velocity of the transmitter,

$$\frac{d^2\theta_T}{dt^2} + b\frac{d\theta_T}{dt} + \sin\theta_T = A\cos\,\omega_D t$$

and

$$\frac{d^2\theta_R}{dt^2} + b\frac{d\theta_R}{dt} + \sin\,\theta_R + c\left(\frac{d\theta_R}{dt} - \frac{d\theta_T}{dt}\right) = A\cos\,\omega_D t. \qquad (7.36)$$

There is one feature about synchronized chaotic pendulums (or any chaotic systems) that we have not emphasized, but is actually quite remarkable. In the various coupling schemes described, only one pair of variables from each system—for example, ω_1 and ω_2—is explicitly coupled. That is, there are two sets of equations of motion, but their connection is only through a single variable. Yet if that pair of variables is synchronized, then the other pairs of corresponding variables are also synchronized. This seems surprising until we remember that within a given system (single pendulum) the equations of motion couple all the variables together. Thus, for synchronization of two chaotic systems, it is enough that a single pair be coupled and synchronized. But let us return to the description of communication between pendulums.

We can write the above equations as a set of first order differential equations for simulation on the computer. For the "transmitter" the equations of motion are

$$\begin{aligned}
\frac{d\theta_T}{dt} &= \omega_T \\
\frac{d\omega_T}{dt} &= -b\omega_T - \sin\theta_T + A\cos\varphi_T \\
\frac{d\varphi_T}{dt} &= \omega_D
\end{aligned} \qquad (7.37)$$

and let the "receiver" equations be

$$\begin{aligned}
\frac{d\theta_R}{dt} &= \omega_R \\
\frac{d\omega_R}{dt} &= -b\omega_R - \sin\theta_R + A\cos\varphi_R - c(\omega_R - \omega_T) \\
\frac{d\varphi_R}{dt} &= \omega_D.
\end{aligned} \qquad (7.38)$$

Having established the transmitter and receiver pendulums, we superpose a signal or message $m(t)$ on top of a chaotic signal from the transmitter. An unwanted agent that intercepts the transmitted signal would not be able to distinguish the message from the chaotic carrier signal. But the operator of the receiver will be able to make the distinction, because he or she has a system identical to and synchronized with the chaotic transmitter, and will be able to subtract away the chaotic part of the transmission and thereby recover the signal. The apparent randomness of the chaotic "carrier" ensures that the message has been securely transmitted, and the synchronization ensures that it is accurately received. Specifically, the message to be transmitted, $m(t)$, is added to the angular velocity $\omega_T(t)$ to produce a modified transmitter angular velocity, $\hat{\omega}_T(t)$ that is then sent to the receiver,

$$\overset{\wedge}{\omega}_T(t) = \omega_T(t) + m(t).$$

When the transmitter and receiver are approximately synchronized, subtraction of $\omega_R(t)$ from the transmission leads to recovery of the message,

$$\hat{m}(t) = \overset{\wedge}{\omega}_T(t) - \omega_R(t) \approx m(t).$$

The synchronization between the systems must be fairly good in order that there is faithful transmission of messages, although the method does seem to be robust against small quantities of systemic noise. For the case of mostly synchronized pendulums, the results of such a transmission are shown in Fig. 7.19. The upper trace shows the original message. In the middle part of the diagram the signal is modulated by the chaotic transmitter output and therefore appears to be quite meaningless. The lower trace shows the message—a sine wave—after the receiver has demodulated the signal. The process seems fairly robust in that the synchronization does not have to be perfect for faithful signal reception, nor do small amounts of noise in the system substantially degrade the received signal. In the case of weak coupling between transmitter and receiver, the received message is soon degraded as shown in Fig. 7.20. In fact, the only reason that the

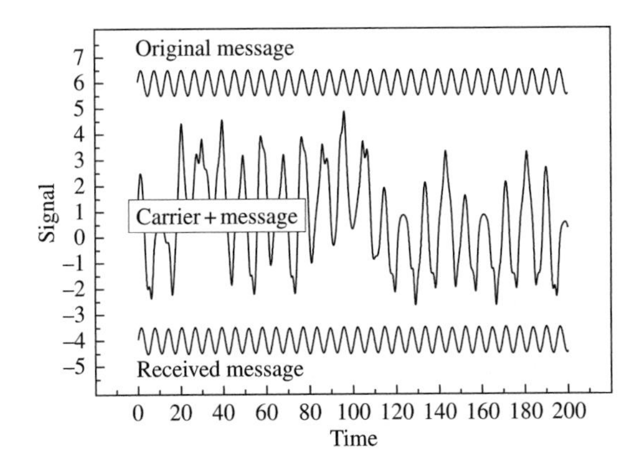

Fig. 7.19
Time series of the message, its transmission, and its reception for a strongly coupled transmitter and receiver.

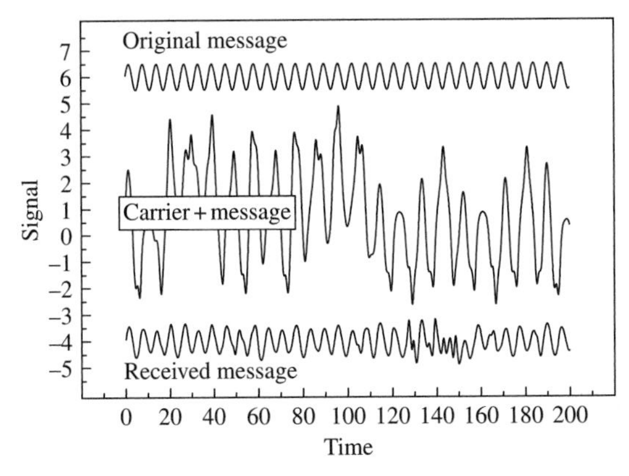

Fig. 7.20
Time series of the message, its transmission, and its reception for a weakly coupled transmitter and receiver.

systems seem temporarily synchronized, resulting in a few cycles of the faithful reception, is that the motions of both pendulums were initiated from the same dynamical coordinates.

The use of pendulums for secure communication may seem a little bizarre or at least impractical. However, this general method of secure communication can be applied to a variety of chaotic mechanical and electronic configurations including lasers. The key requirements are that the transmitter and receiver must be similar systems (although this condition can sometimes be relaxed) and the mode of operation must be chaotic. Cuomo and Oppenheim (1993) demonstrated this type of communication using voice transmission. Electronic circuits have advantages over mechanical systems in that there is much possible variation in parameters and in regimes of operation. The coupling between the chaotic systems would be via some sort of wire or wireless radio communication. But one could also imagine a connection between transmitter and receiver using a light beam from a laser. Lasers have the advantage of being a high frequency carrier and there capable of transmitting message with large bandwidths. Rajarshi Roy and his colleagues have shown the feasibility of such communication using chaotic lasers (Van Wiggern and Roy 1998) Yet having gone to the trouble of describing this communications method we must also note that encryption by chaotic coupled systems may not be completely secure. (See page 203 of Strogatz (2003).) Given enough time to analyze the transmission the message can be deciphered. However, for a short transmission, chaotically transmitted signals may be secure enough for many applications where long-term secrecy is not a requirement. We might imagine that such systems could be of practical use in providing secure recognition of military units on a changing battlefield or, more mundanely, keeping short "chaotic" cell phone conversations somewhat private.

7.3.3 Control of the chaotic pendulum

In this section we describe the essentials for a common method of controlling the chaotic pendulum, or any chaotic system. Here is the basic idea. Given a driven pendulum with parameter settings that would ordinarily cause it to act chaotically, the goal is to apply periodically a small perturbation that will change its motion from chaotic to periodic. Of course we could encourage periodic behavior by superposing a large periodic force on the pendulum. But our aim here is to achieve this same regular behavior with (a) a minimal disturbance of the system and (b) by using only a small input of energy. The secret to this method is feedback. The state of the pendulum is periodically sampled and information about that state is then fed back to the pendulum in order to counter the tendency to chaotic motion.[3]

[3] Similar techniques might apply to maintenance of stable flight for high performance aircraft. For example, it could be important to have an aircraft whose controls are very responsive to a pilot's touch. Yet this same aircraft could be so sensitive that the pilot might have trouble maintaining desired flight patterns. Therefore, some sort of feedback mechanism would then be necessary to provide sufficient stability.

How does control of a chaotic system relate to synchronization? Control involves feedback of information about the current state of the system—the pendulum—and synchronization also requires that information be exchanged between one system and another. In the case of synchronization neither system is stabilized to the other unless there is sufficient connection and information flow. And it is only when both chaotic systems are slightly out of synchronization that there is information to exchange. Similarly, being slightly out of control generates feedback and therefore information that, as in the case of synchronization, slightly modifies the dynamics of the system to push it toward the controlled state. The only distinction is that synchronization involves information exchange between two or more systems whereas with control information is part of a feedback loop within the system. Hence control and synchronization have much in common.

In 1990, Ott, Grebogi, and Yorke of the University of Maryland published a brief paper describing what was to become a popular method of controlling chaotic systems (Ott et al. 1990). Their method was quickly applied to a variety of systems including a parametric pendulum, both simulated and experimental (Starrett and Tagg 1995). In this section we demonstrate the method with our theoretical pendulum given by Eq. (6.3) that is rewritten here:

$$\frac{d^2\theta}{dt^2} + \frac{1}{Q}\frac{d\theta}{dt} + \sin\theta = A\cos\omega_D t. \tag{7.39}$$

We recall from the chapter on the chaotic pendulum that one way to represent the motion of the pendulum is by periodic sampling or strobing of the motion at intervals that equal the period of the sinusoidal forcing motion. This strobing process generates the points for the Poincaré section. Thus a pair of phase coordinates $(\theta_n, \omega_n = d\theta_n/dt)$ is mapped into a new set $(\theta_{n+1}, \omega_{n+1})$ at a time that is one forcing period later. We represent this mathematically as a mapping;

$$\begin{pmatrix} \theta_{n+1} \\ \omega_{n+1} \end{pmatrix} = M(\theta,\omega)\begin{pmatrix} \theta_n \\ \omega_n \end{pmatrix}, \tag{7.40}$$

where $M(\theta, \omega)$ is some unknown transformation acting on the vector (θ_n, ω_n) at each point in phase space. If the pendulum's motion is regular and has the same period as that of the forcing motion then the phase points are equal, the motion is referred to as period-one, and $M(\theta, \omega) = 1$. If the pendulum's motion is still regular but has a period that is twice that of the forcing motion then $(\theta_{n+2}, \omega_{n+2}) = (\theta_n, \omega_n)$, $M(\theta, \omega)^2 = 1$, and similarly for period-three and more complex, but still periodic, motions. When the motion is regular and periodic there are relatively few such orbits for the pendulum. However, when the motion is chaotic there are *infinitely* many periodic orbits but they are *unstable*. The goal in controlling the pendulum is to stabilize one of the unstable orbits so that the pendulum tracks with that specific periodic orbit. And the way to stabilize the orbit is by periodic adjustment of one of the pendulum's parameters. The damping parameter, Q, is a convenient choice. By studying the effect of changing Q on the pendulum dynamics we determine the appropriate periodic change that needs to be applied. In this section we develop a detailed model and a specific example of controlling a chaotic pendulum (Baker 1995).

Theoretically, it is possible to stabilize any other unstable periodic orbits. But some orbits are more stable than others and therefore we try to choose one such orbit by looking for partially stable *fixed points* in the Poincaré section. These are phase points (θ_n, ω_n), which recur approximately and successively a few times before the chaotic pendulum wanders off to another unstable orbit. Specifically, as the Poincaré section is generated, we look for phase points that almost repeat the previous value. Figure 7.21 shows a Poincaré section for the parameter set $A = 1.5, Q = 3.9$, and $\omega_D = 2/3$. The period -1 fixed point is depicted as a solid square at about $(1.5,\ -0.5)$. More precise values $(\theta_F = 1.523,\ \omega_F = -0.415)$ are found by averaging the coordinates of several returns.

Now changes in the parameter Q will cause the fixed point to change and for small changes we assume a linear approximation;

$$\begin{pmatrix} \theta'_F \\ \omega'_F \end{pmatrix} = \begin{pmatrix} \theta_F \\ \omega_F \end{pmatrix} + \delta Q \begin{pmatrix} \partial\theta_F/\partial Q \\ \partial\omega_F/\partial Q \end{pmatrix}. \tag{7.41}$$

By varying slightly Q in the region of the fixed point we determine the change in the fixed point coordinates and therefore the values of the derivatives near the fixed points. One finds that $(\partial\theta_F/\partial Q,\ \partial\omega_F/\partial Q) = (-0.41,\ -0.29)$. The control mechanism is not very sensitive to the particular numbers and variations of about 10% do not adversely affect control. Now we have some sense of the effect of changes in damping on the location of the fixed point. Let us then return to the Poincare mapping, $M(\theta,\ \omega)$. Close to the fixed point we assume that the mapping can be represented by linear changes. Remember that the fixed points are on orbits that tend to be more stable than most others. The linearized mapping is written as

$$\begin{pmatrix} \theta_{n+1} \\ \omega_{n+1} \end{pmatrix} \approx \begin{pmatrix} \theta_n \\ \omega_n \end{pmatrix} + M \begin{pmatrix} \theta_n - \theta_F \\ \omega_n - \omega_F \end{pmatrix} = \begin{pmatrix} \theta_n \\ \omega_n \end{pmatrix} + M \begin{pmatrix} \Delta\theta_n \\ \Delta\omega_n \end{pmatrix} \tag{7.42}$$

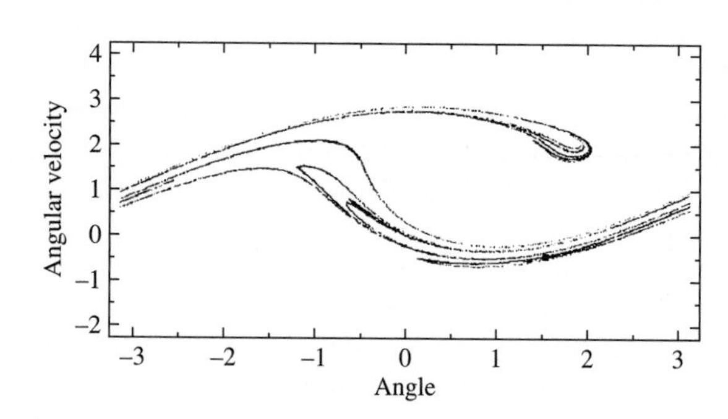

Fig. 7.21
A Poincaré section for the pendulum indicating—by a small square—an unstable fixed point (1.5, −0.5). (Reprinted with permission from Baker (1995, p. 834). © 1995 American Association of Physics Teachers.)

and therefore the mapping of the Poincare section *near the fixed point* becomes

$$\begin{pmatrix} \Delta\theta_{n+1} \\ \Delta\omega_{n+1} \end{pmatrix} = M \begin{pmatrix} \Delta\theta_n \\ \Delta\omega_n \end{pmatrix}. \tag{7.43}$$

Passing through the fixed point are two special curves. One is called the *stable manifold* and the other is called the *unstable manifold*. Phase points on the unstable manifold tend to successively leave the fixed point whereas points on the stable manifold tend to successively move toward the fixed point. (A fixed point with these properties is called *hyperbolic*.) Vectors that are located along the stable and unstable manifolds tend, to a first approximation, to keep their respective directions and lengths. Thus they are *eigenvectors* of the mapping M and solution of the corresponding eigenvalue problem leads to eigenvalues λ_S and λ_U and eigenvectors $\mathbf{e}_S$ and $\mathbf{e}_U$ respectively along the stable and unstable manifolds. This information can be used to determine the mapping, M. By observing the evolution of vectors in the region of the fixed point we may obtain the Poincaré mapping. Consider the four points that form the vertices of a square in the phase space shown in Fig. 7.22.

We determine the vectors by observing the places to which these points evolve during one forcing period. It is sufficient to consider a pair of mutually perpendicular vectors formed by two sides of the square. Since each of the original points has its own symbol it is clear where each point goes in one drive cycle. Placing the appropriate numbers in Eq. (7.43) leads to values for the matrix elements of the transformation,

$$M = \begin{pmatrix} -3.42 & -5.79 \\ -1.52 & -2.48 \end{pmatrix}. \tag{7.44}$$

The normalized eigenvectors and eigenvalues are then determined from this specific mapping and the results are

$$\begin{aligned} \lambda_U &= -5.85, \quad \mathbf{e}_U = (e_{U1}, e_{U2}) = (0.92, 0.40) \\ \lambda_S &= 0.050, \quad \mathbf{e}_S = (e_{S1}, e_{S2}) = (0.86, -0.52). \end{aligned} \tag{7.45}$$

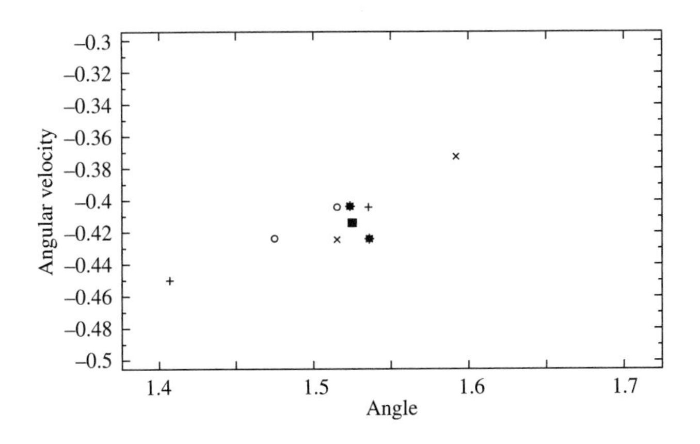

Fig. 7.22
Diagram showing the evolution of four points at the corners of a square through one forcing cycle. By treating pairs of points as vectors, the transformation matrix M can be determined. (Reprinted with permission from Baker (1995, p. 836). ©1995 American Association of Physics Teachers.)

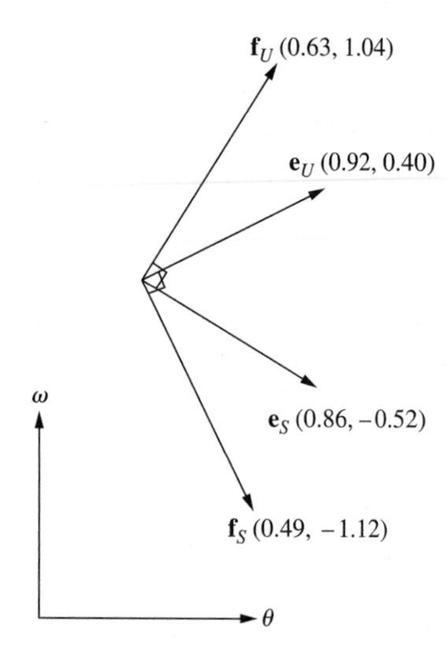

Fig. 7.23
Illustration of the calculated eigenvectors and the corresponding orthogonal vectors in the (θ, ω) phase plane. (Reprinted with permission from Baker (1995, p. 836). ©1995, American Association of Physics Teachers.)

The OGY method also requires calculation of vectors $\mathbf{f}_u$ and $\mathbf{f}_s$ that are perpendicular to the eigenvectors. These are determined according to the orthogonality and normalization relationships

$$\begin{aligned} \mathbf{f}_U \cdot \mathbf{e}_S = 0 \quad &\text{and} \quad \mathbf{f}_S \cdot \mathbf{e}_U = 0, \\ \mathbf{f}_U \cdot \mathbf{e}_U = 1 \quad &\text{and} \quad \mathbf{f}_S \cdot \mathbf{e}_S = 1, \end{aligned} \tag{7.46}$$

respectively. All the relevant vectors are shown in Fig. 7.23. Why do we determine all of these vectors? The general answer is that they will be used in the generation of the control algorithm. But first we must invert the process we just did—namely the finding of the vectors from the four numerical values of the transformation, M. That is, we now require a formula for M in terms of the vectors and eigenvalues.

The key step toward obtaining the elements of M is to determine the action of the transformation on a simple linear combination of the eigenvectors: thus

$$\begin{aligned} M\left[\begin{pmatrix} e_{U1} \\ e_{U2} \end{pmatrix} + \begin{pmatrix} e_{S1} \\ e_{S2} \end{pmatrix}\right] &= \lambda_U \begin{pmatrix} e_{U1} \\ e_{U2} \end{pmatrix} + \lambda_S \begin{pmatrix} e_{S1} \\ e_{S2} \end{pmatrix} \\ &= \lambda_U \begin{pmatrix} e_{U1} \\ e_{U2} \end{pmatrix} [\mathbf{f}_U \cdot \mathbf{e}_U + \mathbf{f}_U \cdot \mathbf{e}_S] \\ &\quad + \lambda_S \begin{pmatrix} e_{S1} \\ e_{S2} \end{pmatrix} [\mathbf{f}_S \cdot \mathbf{e}_S + \mathbf{f}_S \cdot \mathbf{e}_U], \end{aligned} \tag{7.47}$$

where we have used the orthogonality and normalization relations. Collection of terms leads to

$$M[\mathbf{e}_U + \mathbf{e}_S] = \left[\lambda_U \begin{pmatrix} e_{U1} \\ e_{U2} \end{pmatrix} \mathbf{f}_U + \lambda_S \begin{pmatrix} e_{S1} \\ e_{S2} \end{pmatrix} \mathbf{f}_S\right] \cdot [\mathbf{e}_U + \mathbf{e}_S] \tag{7.48}$$

and we finally arrive at an expression for the transformation,

$$M = \left[\lambda_U \begin{pmatrix} e_{U1} \\ e_{U2} \end{pmatrix} (f_{U1}\ f_{U2}) + \lambda_S \begin{pmatrix} e_{S1} \\ e_{S2} \end{pmatrix} (f_{S1}\ f_{S2})\right]. \tag{7.49}$$

With this expression in hand we now return to the business of obtaining the control algorithm.

The control mechanism attempts to force the pendulum's trajectory toward the fixed point in the phase space, (θ_F, ω_F). By adjustment of the friction parameter Q we push the intersection point of the trajectory with the Poincaré section onto the stable manifold near the fixed point. The natural attraction along the stable manifold will then pull the intersection point toward the fixed point. Figure 7.24 illustrates the Poincaré section near the fixed point.

As previously noted, the fixed point (θ_F, ω_F) is located at the intersection of the stable and unstable manifolds. If the damping parameter is adjusted, the fixed point moves to a new location (θ'_F, ω'_F), which is related to the old

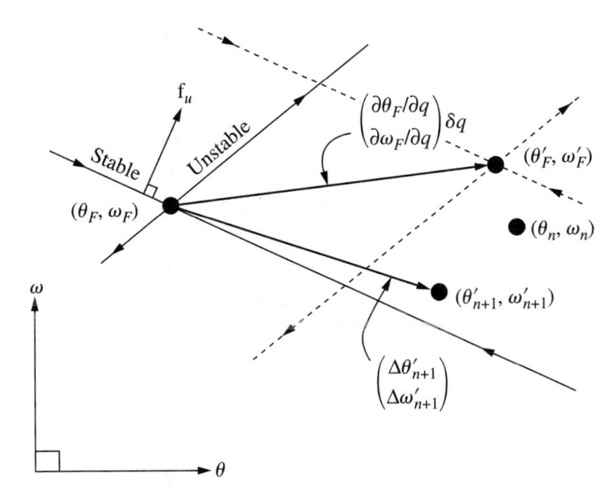

Fig. 7.24
A region of the (θ, ω) plane near the fixed point. the solid lines that intersect at the fixed point are the manifolds of the fixed point. Arrows pointing toward the fixed point mark the stable manifold and arrows pointing away from the fixed point mark the unstable manifold. The dashed lines that intersect at (θ'_F, ω'_F) represent the analogous configuration after a change of δQ in the friction parameter. The control mechanism forces the $(n+1)$ point toward the stable manifold at (θ_F, ω_F). (Reprinted with permission from Baker (1995, p. 836). ©1995, American Association of Physics Teachers.)

location by a linear transformation of Eq. (7.41) that is reproduced here,

$$\begin{pmatrix} \theta'_F \\ \omega'_F \end{pmatrix} = \begin{pmatrix} \theta_F \\ \omega_F \end{pmatrix} + \delta Q \begin{pmatrix} \partial\theta_F/\partial Q \\ \partial\omega_F/\partial Q \end{pmatrix}. \tag{7.50}$$

Once the fixed point is changed, the next point in the phase space becomes

$$\begin{pmatrix} \theta'_{n+1} \\ \omega'_{n+1} \end{pmatrix} = \begin{pmatrix} \theta'_F \\ \omega'_F \end{pmatrix} + M \begin{pmatrix} \theta_n - \theta'_F \\ \omega_n - \omega'_F \end{pmatrix} \tag{7.51}$$

for which we assume that the transformation M remains roughly in the region. The next point on the Poincare section is the result of both the dynamics of the pendulum and the movement of the fixed point to a new position (θ'_F, ω'_F). The vector from the new fixed point to the new position is then given by including both the damping change δQ and the transformation, M as follows:

$$\begin{aligned} \begin{pmatrix} \Delta\theta'_{n+1} \\ \Delta\omega'_{n+1} \end{pmatrix} &= \begin{pmatrix} \theta'_{n+1} \\ \omega'_{n+1} \end{pmatrix} - \begin{pmatrix} \theta'_F \\ \omega'_F \end{pmatrix} \\ &= \begin{pmatrix} \theta'_F \\ \omega'_F \end{pmatrix} - \begin{pmatrix} \theta_F \\ \omega_F \end{pmatrix} + M \begin{pmatrix} \theta_n - \theta'_F \\ \omega_n - \omega'_F \end{pmatrix} \\ &= \delta Q \begin{pmatrix} \partial\theta_F/\partial Q \\ \partial\omega_F/\partial Q \end{pmatrix} + M \left[\begin{pmatrix} \Delta\theta_n \\ \Delta\omega_n \end{pmatrix} - \delta Q \begin{pmatrix} \partial\theta_F/\partial Q \\ \partial\omega_F/\partial Q \end{pmatrix} \right]. \end{aligned} \tag{7.52}$$

Now we are ready to insert the complex form of the transformation as given by Eq. (7.49). leading to a new version of the previous equation;

$$\begin{pmatrix} \Delta\theta'_{n+1} \\ \Delta\omega'_{n+1} \end{pmatrix} = \delta Q \begin{pmatrix} \partial\theta_F/\partial Q \\ \partial\omega_F/\partial Q \end{pmatrix} + \left[\lambda_U \begin{pmatrix} e_{U1} \\ e_{U2} \end{pmatrix} (f_{U1}\ f_{U2}) + \lambda_S \begin{pmatrix} e_{S1} \\ e_{S2} \end{pmatrix} (f_{S1}\ f_{S2}) \right] \times \left[\begin{pmatrix} \Delta\theta_n \\ \Delta\omega_n \end{pmatrix} - \delta Q \begin{pmatrix} \partial\theta_F/\partial Q \\ \partial\omega_F/\partial Q \end{pmatrix} \right]. \tag{7.53}$$

For control we want the new point $(\theta'_{n+1}, \omega'_{n+1})$ to move toward the stable manifold and therefore the vector $(\Delta\theta'_{n+1}, \Delta\omega'_{n+1})$ should move toward alignment with the stable manifold. Therefore the change in the damping δQ should be such as to make the dot product of the vector $(\Delta\theta'_{n+1}, \Delta\omega'_{n+1})$ and $\mathbf{f}_U$ equal to zero;

$$(f_{U1}\ f_{U2}) \begin{pmatrix} \Delta\theta'_{n+1} \\ \Delta\omega'_{n+1} \end{pmatrix} = 0. \tag{7.54}$$

Inserting this result into the previous equation and solving for the appropriate change in damping leads to

$$\delta Q = \frac{\lambda_U}{\lambda_U - 1} \frac{(f_{U1}\ f_{U2}) \begin{pmatrix} \Delta\theta_n \\ \Delta\omega_n \end{pmatrix}}{(f_{U1}\ f_{U2}) \begin{pmatrix} \partial\theta_F/\partial Q \\ \partial\omega_F/\partial Q \end{pmatrix}}. \tag{7.55}$$

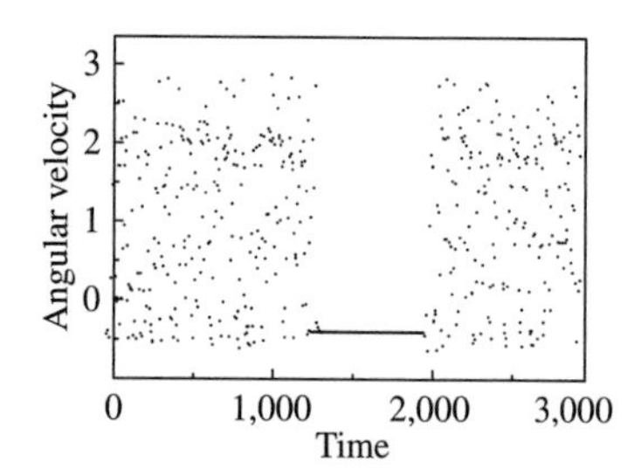

Fig. 7.25
A graph of the angular velocity at the beginning of each forcing cycle. Control is initiated at $t = 1{,}000$, and then eventually turned off

This expression gives the required change in the damping needed at each forcing cycle. Only the top right vector $\begin{pmatrix} \Delta\theta_n \\ \Delta\omega_n \end{pmatrix}$ changes with each cycle—all other quantities remaining constant near the fixed point. The feedback δQ must, as expected, be recalculated for each cycle. This algorithm only works in the region of the fixed point in question, and therefore once the algorithm is activated it takes a little time for the chaotic motion to be rendered periodic. There are algorithms that can reduce this delay but these are not part of our subject here. Figure 7.25 shows the initiation of control and the time delay in establishing control. In this figure, the pendulum's angular velocity is displayed as in a Poincaré section. The control algorithm is turned on at $t = 1000$ and a small time delay occurs. Figure 7.26 shows the full (not sampled at the Poincaré rate) time series of the angular velocity on an expanded time scale and again control is delayed for a few cycles.

We have focused on controlling the chaotic pendulum. But it is quite evident that control by feedback is a broadly applicable technique, and it has long been used in many fields, including aviation, manufacturing, electronics and so forth. The application to the stabilization of chaotic systems is perhaps less well known but this application also would appear to be of significant use to society. In many technical areas we seek systems that are fast and responsive. Such flexibility often comes at the price of having some instability, and possibly chaos. In the last decade a variety of

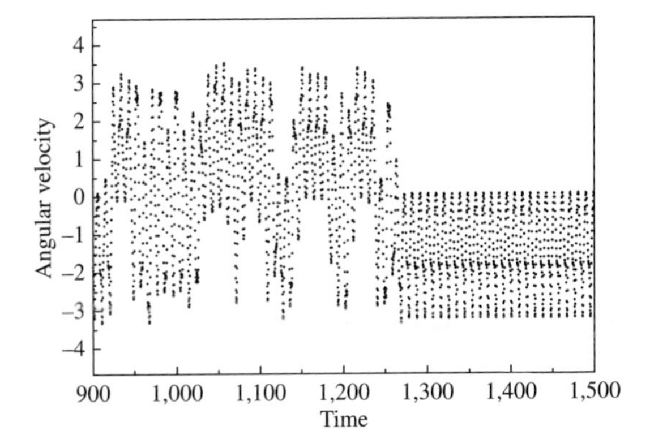

Fig. 7.26
Time series of the pendulum's angular velocity. Control is initiated at $t = 1{,}000$ and becomes apparent after several cycles.

control applications have been found. These include control of chaotic lasers by Rajarshi Roy and coworkers at Georgia Tech (Roy et al. 1994), control of chaotic chemical reactions by Roger Rollins and coworkers at Ohio University, and some interesting work involving heart arrhythmias by Mark Spano and coworkers at the Naval Surface Warfare Center in Silver Springs, Maryland. Research on control of heart arrhythmias might lead to better treatment of cardiac fibrillation. Some further possibilities include the use of control of an oscillator to carry information in a military context, somewhat in the spirit of the secure communications schema described above. The review article by Ott and Spano (1995) contains references to this and other work. It is quite likely that the various control schemes for chaotic systems will be increasingly important in diverse technological applications.

7.3.4 A final weirdness

We began this tale of synchronized pendulums with the phenomenon of a large number of pairs of hands clapping to create noise. What about the synchronization of a large number of pendulums? Are there special effects when a large number of pendulums are coupled together? Whether a large number of coupled pendulums will synchronize depends on several factors: the strength of the coupling, the inertia of each pendulum, and the amount of forcing given to each pendulum. Weakly driven pendulums with strong coupling are likely to synchronize. Strongly driven chaotic pendulums with weak coupling are not likely to synchronize. These conclusions seem consistent with our discussion of two coupled pendulums.

But there can be surprises when many pendulums are involved. Let us consider a system of pendulums that are coupled by some sort of torsion springs. Following the work of Braiman et al. (1995) we examine the simple model equations of motion given by

$$ml^2\frac{d^2\theta_n}{dt^2} + \gamma\frac{d\theta_n}{dt} + mgl\sin\theta_n = F\sin\omega t + \kappa(\theta_{n+1} - \theta_n) + \kappa(\theta_{n-1} - \theta_n), \tag{7.56}$$

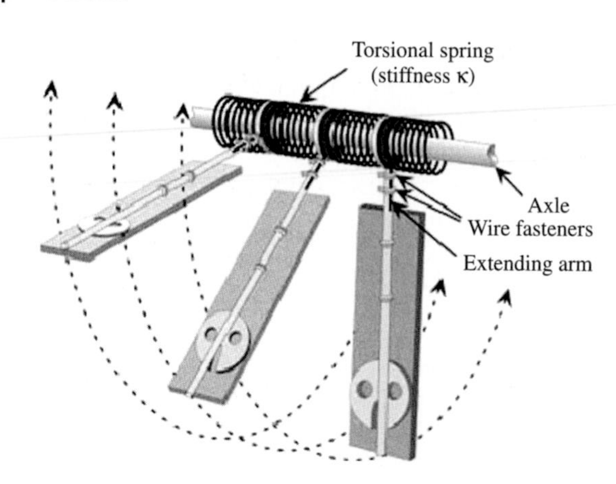

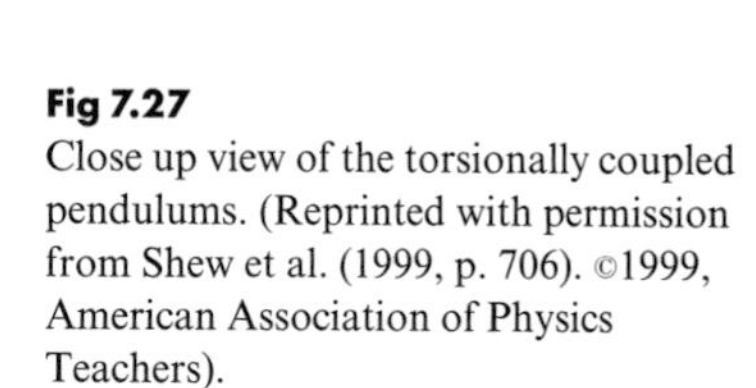

Fig 7.27
Close up view of the torsionally coupled pendulums. (Reprinted with permission from Shew et al. (1999, p. 706). ©1999, American Association of Physics Teachers).

where m is the mass, l is the length, γ is the friction coefficient, F is the forcing amplitude and κ is the torsion constant of the springs connecting the pendulums. A relatively close physical representation of this equation is shown in Fig. 7.27 (Shew et al. 1999) in which the torsional coupling is via springs connecting the pendulums near the pivot points. Let us suppose that the pendulums are all identical and that their parameters are adjusted so that, individually, the pendulums are chaotic, then it turns out that the array of pendulums will be chaotic. On the other hand, if some degree of disorder is introduced into the system by, for example, randomly shortening or lengthening the pendulums, then the pendulums will synchronize *and* the motion will become periodic. That is, the motion is both synchronized and controlled in one stroke. Figure 7.28 illustrates the situation. The parameters of the pendulums are set so that each one is chaotic. In (a) all quantities are equal whereas in (b) the lengths of the pendulums are varied randomly. It is clear that the introduction of the random lengths or disorder into the pendulum array has caused the chaos to cease. This is another example of the control of chaos. This type of chaos control is therefore referred to as *disorder taming chaos*. The introduction of disorder causes spatial symmetry breaking in the array of pendulums.

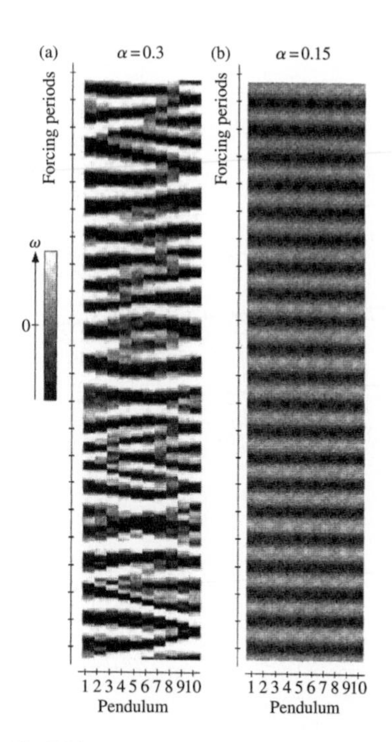

Fig 7.28
Experimental spatiotemporal angular velocity plots, for (a) homogeneous array and (b) heterogeneous array. (Reprinted with permission from Shew et al. (1999, p. 707). ©1999, American Association of Physics Teachers.)

The idea that the introduction of spatial disorder can cause regular behavior in otherwise coupled chaotic systems may seem strange. If anything we might expect the motion of such a variegated collective system to become even more disordered. But let us consider more carefully. By adding a lot of identical pendulums (or any chaotic subsystems) we increase the number of possible periodic motions and harmonics or subharmonics of frequencies that are characteristic of the pendulums. We also know that one route to chaos is through a series of period doublings or increasingly longer subharmonics. Thus the system of coupled identical chaotic pendulums is also going to be chaotic. On the other hand the motion of a group of nonidentical pendulums, as in the disordered array, is less likely to be chaotic because the parameter differences between one pendulum and the next tend to damp out the unstable motions. In the disordered system it is unlikely that a recurring sequence of subharmonics will be generated as the constituent pendulums will have different fundamental frequencies. Similar control of chaos is sometimes observed through the addition of noise to an otherwise deterministic system. Noise prevents the buildup of the necessary chain of subharmonics leading to chaos. Although the phenomenon is not the same, we might consider a certain correspondence with the process of tempering steel. Iron is heated and cooled in order to create defects (spatial disorder) in the lattice of the iron and give the lattice more rigidity. Similarly, Carbon is added to strengthen steel. Again the carbon causes spatial disordering of the lattice thereby inhibiting long range motion of the material.

In this chapter we have seen an example of how the whole is sometimes greater than the sum of the parts. By coupling individual pendulums together we create more degrees of freedom and therefore more possibilities for variety of motion. In the next chapter we turn to a very simple pendulum, but one where very small size brings a new richness to the story of the pendulum.

7.4 Exercises

1. The linearized coupled pendulums described by Eq. (7.1) connect the variables x_1 and x_2 together in these equations of motion. By a suitable change of coordinates the equations may be uncoupled and therefore solved separately as simple harmonic oscillators. Let $q_1 = x_1 - x_2$ and $q_2 = x_1 + x_2$. (a) Show that the equations of motion for x_1 and x_2 may be written as

$$\frac{d^2 q_1}{dt^2} + m\omega_1^2 q_1 = 0$$
$$\frac{d^2 q_2}{dt^2} + m\omega_2^2 q_2 = 0,$$

where $\omega_2 = \omega_0$ and $\omega_1 = \omega_0\sqrt{1 + (2k/m\omega_0^2)}$. The coordinates q_1 and q_2 are called the *normal coordinates* and the motions described by these coordinates are called the *normal modes* of vibration since the coordinates so-defined oscillate harmonically. Normal mode analysis is important in complex systems of coupled harmonic oscillators and is used in, for example, the study of molecular vibrations to sort out the various harmonic motions of the constituent atoms. (b) Solve the normal mode equations of motion subject to the initial conditions

$$q_1(0) = Q_1, \quad \dot{q}_1(0) = 0$$
$$q_2(0) = Q_2, \quad \dot{q}_2(0) = 0.$$

2. Continuing the discussion of Problem 1. (a) Write the original physical coordinates, x_1 and x_2, in terms of the q_i, and substitute the solutions arrived at in (b) in order to obtain the time dependence of x_1 and x_2. (b) Consider only the case where $Q_1 = Q_2 = Q$, and show that with these initial conditions, each pendulum oscillates with higher frequency equal to $\omega_1 + \omega_2$ but with an amplitude that is modulated at a lower frequency, $\omega_2 - \omega_1$. In this way an arbitrary displacement of the coupled pendulums leads to a periodic transfer of energy between the two normal modes of oscillation. However, specific mechanical configurations may favor one mode over another. Thus, Huygen's coupled pendulum clocks , always tended toward the out-of-phase motion.

3. Here is another approach to modeling the probability that a system is synchronized using the two state model. Consider the coupled system over a long period of time T, which may be subdivided into a large number N of short time intervals ΔT. Suppose that during any given time interval ΔT the (unknown) probability of the system being synchronized is h_S and the (unknown) probability of being unsynchronized is h_U. (Note that $h_S + h_U = 1$.) The number of time intervals in synchronization is denoted by N_S and the number of time intervals when the system is unsynchronized is N_U, such that $N_S + N_U = N$. If the events from one time interval to the next are taken as independent then the process is a Bernoulli process and obeys the binomial distribution. (a) Write the binomial distribution for the probability $P_r(N_S, h_S, N)$ of N_S intervals of synchronization. (b) Using formulas for the mean and standard deviation of the binomial distribution, write the mean and standard deviation for N_S in terms of h_S and N. Note that the mean divided by N is really just the fraction of time that the system is synchronized.
4. In statistical mechanics the entropy of a system is often maximized in order to arrive at the equilibrium distribution for a system. Entropy, like missing information, is a measure of the probability for a system to be in a particular state. Systems are most likely to be in or will spend most of the time in states with maximum probability. (See, for example, Levine and Tribus (1979).) Using the maximum entropy formalism we can calculate the probabilities h_S and h_U and therefore compute the mean and standard deviations using the answers from question 3. First, define the entropy as $S = k \ln P_r(N_S, h_S, N)$ where the distribution is the binomial distribution found in the previous problem. (a) Using Stirlings approximation for large numbers: $\ln x! = x \ln x - x$, write an expression for the entropy. (b) Differentiate the entropy with respect to N_S and thereby find optimal values of h_S and h_U. (c) Using the results of (b) and the expressions for the mean and standard deviation, show that $\bar{N}_S = N_S$ and $\sigma_{N_S} = \sqrt{N_S(N - N_S)/N}$.
5. In the text Eq. (7.11), the differential equation for the difference angle θ_-, was derived from the original equation of motion for the magnetically coupled pendulums, Eqs. (7.8) and (7.9). (a) Using these same equations, find an equation of motion involving primarily the θ_+ coordinate. (b) Now assume that the system is strongly synchronized. How does this change the equation, and how does this new system relate to the equation of one of the original pendulums? What change occurs in the resulting system if the pendulums are not synchronized?
6. Using a computer graphics package or otherwise, plot the two functions $P_1(t) = e^{-t/\tau_1}$ and $P_2(t) = t^{-3/2}e^{-t/\tau_2}$ on the same graph. (Try letting $\tau_1 = 0.7$ and $\tau_2 = 1$.) Remember that these functions are the probability distributions of the synchronization times from the two models described in the text. Note any differences. Can you see why these models give similar results for the longer times? Try different combinations of τ_1 and τ_2 and achieve a closer fit of the two models for longer times. Remember that $\tau_1 = 1/\alpha$, the inverse of the (phenomenological) birth rate for synchronization, whereas in the random walk/diffusion model, $\tau_2 = D/(\lambda_\perp)^2$, thus connecting to a (phenomenological) diffusion constant and the transverse Lyapunov exponent.
7. In the latter part of this chapter we described a system of nonlinear pendulums coupled together by torsional springs. A linear variant on that system allows us to develop some idea of the notion of *normal modes*. In this exercise we work through some of the steps leading to normal modes for a linearized system. Let us simplify Eq. (7.56) considerably by assuming that the gravitational restoring force may be linearized, that there is no dissipation, and that there is no forcing. We have then a set of coupled linearized pendulums that, if disturbed, should execute some ongoing motion. Therefore Eq. (7.56) may be considerably simplified to

$$\theta_p'' + \omega_0^2\theta_p = \omega_{00}^2\left[(\theta_{p+1} - \theta_p) + (\theta_{p-1} - \theta_p)\right], \tag{7.57}$$

where there are now two basic frequencies, $\omega_0 = \sqrt{g/l}$ and $\omega_{00} = \sqrt{\kappa/(ml^2)}$, and p is the position number for a given pendulum. The normal mode concept rests upon the idea that the motions of a complex structure are composed of a set of oscillatory modes each having the same frequency and a fixed phase relationship of all particles within a given mode. Let us therefore assume that the pth pendulum oscillates in the nth mode according to the relation

$$\theta_{p,n} = A_{p,n} \cos \omega_n t.$$

(a) Substitute this solution into the equation of motion and thereby show that this solution leads to the relation

$$\frac{A_{p+1,n} + A_{p-1,n}}{A_{p,n}} = \frac{-\omega_n^2 + \left(\omega_0^2 + 2\omega_{00}^2\right)}{\omega_{00}^2}. \tag{7.58}$$

We will assume boundary conditions such that the right most and left most torsion springs are fixed rather than being connected to another pendulum. This means that $A_{0,n} = A_{N+1,n} = 0$. Also, the right side of the above relation must be a constant for any value of n and therefore independent of p, the particle position. One substitution that works for these constraints is letting $A_{p,n} = C_n \sin p\beta$. (b) Substitute this equation into the above relation and show that the left side equals 2 cos β. Now let us use the boundary conditions. It is obvious that the expression for $A_{p,n}$ is zero when $p = 0$. For $A_{N,n}$ to equal zero $\sin(N+1)\beta = 0$. (c) Show that this constraint leads to $\beta = (n\pi)/(N+1)$. We now have the various parts of $A_{p,n}$. What about ω_n? Substitute your final expression for $A_{p,n}$ into Eq. (7.58) and, through some clever trigonometry, show that the frequency of the nth normal mode for *all* pendulums is

$$\omega_n = \omega_0 \sqrt{1 + 4\left(\frac{\omega_{00}}{\omega_0}\right)^2 \sin^2\left(\frac{n\pi}{2(N+1)}\right)}.$$

If the coupling between the pendulums vanishes, then there is only the single frequency, ω_0. (d) Put all the pieces together, add a phase factor δ_n for each mode and show that the angular displacement of the pth pendulum is given by

$$\theta_p = \sum_{n=1}^{N} C_n \sin\left(\frac{pn\pi}{N+1}\right) \cos\left[\omega_0 \sqrt{1 + 4\left(\frac{\omega_{00}}{\omega_0}\right)^2 \sin^2\left(\frac{n\pi}{2(N+1)}\right)}\, t + \delta_n\right].$$

Therefore the motion of any given pendulum is simply a sum over all possible normal modes (oscillations).

8. In some cases, difference coordinate, e_i, may be used to demonstrate the tendency of coupled dynamical systems to synchronize. Consider the unidirectional coupling defined by Eqs. (7.37) and (7.38). Assume a strong coupling whereby $\theta_T = \theta_R$, or the difference coordinate $e_\theta = \theta_T - \theta_R = 0$ identically. Furthermore the system is driven by a common forcing term so that $\phi_T = \phi_R$. (a) Using these constraints and the difference coordinate $e_\omega = \omega_T - \omega_R$ show that coupled pendulums of Eqs. (7.37) and (7.38) obey the equations

$$\dot{e}_\theta = 0$$

$$\dot{e}_\omega = -\frac{1}{q} e_\omega$$

$$\dot{e}_\phi = 0.$$

(b) Under these same conditions show that ω_T approaches ω_R exponentially.

9. Follow the procedures of Problem 8 with the coupled pendulums described by Eqs. (7.8) and (7.9). (a) First convert these second order differential equations to a set of six first order equations. (b) Then, as before, assume strong coupling such that $\theta_1 = \theta_2$. Recall that the common drive requires that $\phi_1 = \phi_2$, as before. Show that the differential equation for the difference coordinate is $\dot{e}_\omega = -(b + 2c)e_\omega$ and solve this equation to show exponential decay toward equilibrium. Note that the factor $(b + 2c)$ is the dissipation factor for the synchronization machine. Note also that the result may also be obtained from solving Eq. (7.11) in the case when $\theta_- = 0$. These exercises demonstrate that when one variable is synchronized, all variables tend to synchronization.

The quantum pendulum

8

The *classical* pendulum has a long and rich history in physics. On the other hand, the history of the *quantum* pendulum can only stretch back as far as the appearance of quantum mechanics itself in 1925–1926. Yet within three years of that exciting time, a paper on the quantum pendulum appeared in *Physical Review* authored by one of the pioneers of atomic physics, Edward Condon (Condon 1928). Condon's paper laid out the basic problem and the means to its solution. Contributions to the study of the quantum pendulum continue to be made even in recent times. See, for example, (Pradhan and Khare 1973; Aldrovandi and Ferreira 1980; Cook and Zaidins 1986; Baker et al. 2002). In this tale we describe some of the novel and surprising features of the pendulum when it is analyzed according to the rules of quantum physics.

8.1 A little knowledge might be better than none

Quantum mechanics was not stated in the preface as one of the prerequisites for this book and therefore we begin our discussion with a minimal introduction to the basic notions of quantum physics. This introduction is necessarily limited and many quantum concepts are non-intuitive. Thus, certain ideas will just have to be taken as working postulates. For those seeking a deeper understanding, there are a host of available quantum mechanics texts at various levels (e.g. (Anderson 1971; Saxon 1968; Liboff 1980).)

The early part of the twentieth century was a remarkably fruitful time for the development of modern physics. The first three decades witnessed the birth of Einstein's special and general theories of relativity as well as the even more far reaching revolution encompassed by the various realizations of quantum physics. Probing ever more deeply into nature, late nineteenth and early twentieth century laboratory experimentation increasingly yielded phenomena for which classical physics provided no adequate explanation. Some of the most difficult puzzles included the photoelectric effect, the unphysical ultraviolet light distribution predicted by classical physics for the electromagnetic spectrum of a heated object, various experiments where electrons and other small objects behaved like waves rather than particles, the discrete structure of absorption and emission spectra in atomic radiation, and so forth. The development of a comprehensive new theory was gradual, and spanned the first quarter of the twentieth century, culminating in the years 1925–1926 with the publication of Erwin Schrodinger's (1887–1961) wave mechanics and the matrix mechanics of Max Born (1882–1970), Werner Heisenberg (1901–1976) and Pascual Jordan (1902–1980). These two approaches are different, but equivalent, mathematical formulations of the new paradigm of quantum mechanics.

Quantum mechanics is a truly scientific theory in the sense that it predicts and deals with observations but does not attempt to describe the underlying reality in terms of primitive notions or metaphysical constructs. In fact, the philosophy of logical positivism had tangible influence on the early development of the interpretation of quantum mechanics. Even today, there is no completely satisfactory idea as to how the microscopic processes impact those macroscopic instruments that routinely measure the outcome of these underlying processes. Nevertheless, for the working physicist, quantum mechanics is enormously successful and, in the last eight decades, no successful alternatives have been found to replace it. It does predict the events of the microcosm with astonishing accuracy. Thus it is that when we enter the world of molecular, atomic, and nuclear dimensions, we must abandon classical mechanics and turn to the methodology of quantum mechanics.

Classical physics still provides the imagery. We have now become used to viewing reality at the micro level as sometimes acting like particles and sometimes acting like waves. We do not know what the microworld objects really are until we measure them. And even then, we only develop a set of characteristics with which to categorize the material of our experiments. These include energy levels, values of angular momentum, spin, charge, and so on. For example, the hydrogen atom is characterized by a set of discrete energy levels, angular momentum states, and electron spin states. Evidence to support this description comes from various spectra produced with atomic hydrogen as the medium. (Further confirmation comes from spectra beamed from the stars which, because such spectra fit well with earthbound hydrogen spectra, lead us to conclude that stars are mostly hydrogen.) We do our calculations as if the hydrogen atom were made up of particles—an electron and a proton—but the rationale for the particular equations used depends on the electron (at least) having both particle and wave like properties. Thus we speak of the *wave-particle duality* of matter.

As noted above, the rationale for the wave-particle duality in quantum physics rests upon the results of various experiments performed in the first two decades of the twentieth century. More recent refinements of these early experiments as well as new experiments have confirmed the same picture. Under the right circumstances, electrons (particles) can display dynamics that exactly mirror interference effects from wave optics. Similarly radiation often interacts with matter as if the radiation consisted of little bundles (particles) of energy. In his 1923 Ph.D. thesis, a French graduate student, Prince Louis deBroglie, suggested his now-famous relationship between a particle's momentum p, and its corresponding wavelength λ:

$$p = \frac{h}{\lambda}, \tag{8.1}$$

where $h = 2\pi\hbar = 6.6260755 \times 10^{-34}$ J s is Planck's constant, named after the German physicist Max Planck (1858–1947) who first proposed the quantization of an oscillator's energy. The deBroglie relationship provides a tangible connection between particle and wave pictures.

Physical quantities such as energy are called *observables* and their prediction by equations requires a new formalism. For example, the energy characterization of a micro system is, like that of the solution of a linear differential equation with boundary conditions, by means of eigenvalues and eigenfunctions. One formulates an eigenvalue problem. In order to develop the correct eigenvalue problem corresponding to a given physical variable, quantum mechanics requires that there be a special mathematical *operator* corresponding to each physical variable. Let us designate the operator that corresponds to the classical momentum variable p as p_{op}. Similarly the operator that corresponds to potential energy V we designate as V_{op}. Sometimes operators take on a familiar look identical to their classical counterparts. In other cases the operators seem quite non-intuitive. Consider, for example, the one-dimensional harmonic oscillator. In terms of spatial coordinates, its quantum mechanical operators are

$$p_{\text{op}} = \frac{\hbar}{i}\frac{d}{dx} \quad \text{and} \quad V_{\text{op}} = \frac{m\omega^2}{2}x^2. \tag{8.2}$$

The total energy operator is the sum

$$E_{\text{op}} = \frac{p_{\text{op}}^2}{2m} + V_{\text{op}} = \frac{-\hbar^2}{2m}\left(\frac{d}{dx}\right)^2 + \frac{m\omega^2}{2}x^2. \tag{8.3}$$

The potential energy operator has the same form as its classical counterpart, but the momentum operator is a surprise. However, we will just suspend our disbelief and accept this expression as part of the arcane knowledge of quantum physics. In a physical equation, the operators do not stand on their own but operate or act upon the as-yet unknown eigenfunction, often denoted by ψ. Therefore the energy eigenvalue problem is expressed as

$$E_{\text{op}}\psi = E\psi,$$

where E, in the time independent case, is a numerical eigenvalue of the final equation. There are typically several eigenvalues and, in this case, the various eigenvalues correspond to allowed energy values. Combining these bits of formulae we find that the eigenvalue problem for the energy of an oscillator becomes a linear differential equation—an example of the famous (time independent) *Schrodinger* equation:

$$\frac{-\hbar^2}{2m}\frac{d^2\psi}{dx^2} + \frac{m\omega^2 x^2}{2}\psi = E\psi. \tag{8.4}$$

The same development can be applied to create the Schrodinger equation for the pendulum. Since the independent variable is the angle θ of the pendulum rather than a position coordinate, there is a slight change in the momentum operator. (Actually we use an angular momentum operator.) Therefore the appropriate quantum operators are

$$p_{\theta\text{op}} = \frac{\hbar}{i}\frac{d}{d\theta} \quad \text{and} \quad V_{\text{op}} = V_0(1 - \cos\theta). \tag{8.5}$$

$V_0 = mgL$ for the pendulum in a gravitational field but V_0 could, as we shall see, take a different form if some other force were the dominant interaction. Therefore the energy operator becomes

$$E_{\text{op}} = -\frac{\hbar^2}{2mL^2}\left(\frac{d}{d\theta}\right)^2 + V_0(1 - \cos\theta) \tag{8.6}$$

and the Schrodinger equation for the pendulum is then

$$-\frac{\hbar^2}{2mL^2}\frac{d^2\psi}{d\theta^2} + V_0(1 - \cos\theta)\psi = E\psi. \tag{8.7}$$

For this limited introduction, we just flatly state that the eigenfunction or wavefunction ψ does not, remarkably, correspond to any *physical variable*. However the *product* of the wavefunction with its complex conjugate ψ^* does have correspondence with physical reality. The product $\psi\psi^* = |\psi|^2$ is a probability density function. In our example, $\psi\psi^*$ is the probability density for the location of the bob of the pendulum. By itself, ψ is called a probability amplitude. (The wave-particle duality provides a context for the name "probability amplitude." The *intensity* of radiation (waves) corresponds to the probability density for particles, whereas the *amplitude* of the radiation—which is proportional to the square root of the intensity—corresponds to the probability amplitude or wave function. But unlike the amplitude of radiation which is really the amplitude of an electric field, the wave function (probability amplitude) has no physical meaning. Thus the nomenclature arises only by analogy with wave concepts.)

When solved, our version of the Schrodinger equation will tell us both the energy levels E (or E_n) of the pendulum and the probability for the bob's location in any small region of space. Quantum physics is therefore deterministic about some things and probabilistic about others. There is also a somewhat more complex time-dependent version of the Schrodinger equation that is deterministic in the sense that it predicts exactly how the probabilities will change in time. But we will not consider this extra complication. Finally, we will see that the notion of probability will lead to uncertainty in the measurement of some physical variables.

8.2 The linearized quantum pendulum

In our initial treatment of the pendulum, we described the linearized pendulum, the pendulum of small amplitude motion. We noted that its motion is equivalent to the harmonic oscillator or to any situation where the force acting on the system can be modeled by the linear term of its power series expansion. The Schrodinger equation for the *quantum* oscillator (and therefore the linearized pendulum) may be solved analytically. But for large amplitudes, the quantum pendulum is, like its classical counterpart, much more problematic. Therefore we first discuss the linearized pendulum. Equation (8.7) may be linearized, as in Chapter 2, first by using a series expansion for the cosine term and then by dropping all but the first nonvanishing term for the potential energy. This procedure leads to another Schrodinger equation

$$-\frac{\hbar^2}{2mL^2}\frac{d^2\psi}{d\theta^2}+V_0\frac{\theta^2}{2}\psi=E\psi. \tag{8.8}$$

Simplification is achieved by using dimensionless quantities

$$U_0=V_0\frac{2mL^2}{\hbar^2} \quad \text{and} \quad \epsilon=E\frac{2mL^2}{\hbar^2}, \tag{8.9}$$

which, in turn, leads to a dimensionless form of Schrodinger's equation,

$$-\frac{d^2\psi}{d\theta^2}+U_0\frac{\theta^2}{2}\psi=\varepsilon\psi. \tag{8.10}$$

Except for notational differences, the solution of the linearized pendulum follows exactly the solution to the harmonic oscillator problem. Let us begin our solution of Eq. (8.10) with the trial solution, the Gaussian curve

$$\psi=e^{-\alpha\theta^2}, \tag{8.11}$$

where α is a parameter to be determined. This trial solution has several virtues: (a) it works, and (b) both ψ and $d\psi/d\theta$ vanish when θ get large. Remember that the potential energy well is just a parabola and there should be a high probability that the motion stays inside the potential well. In classical physics the bob of an oscillator is completely confined inside the potential well—we cannot imagine anything else. But in quantum physics, there may be a small but finite probability that the bob is somehow outside the well. (Note that the while the solution $\psi=e^{+\alpha\theta^2}$ will also satisfy Eq. (8.10), it diverges for large values of θ. This form of the wavefunction would therefore lead to high probabilities outside the potential well—an obviously nonphysical solution.)

Substitution of the trial wavefunction into the Schrodinger equation gives a value to the parameter α and provides a fixed energy for the system. Thus

$$\alpha=\sqrt{\frac{U_0}{8}} \text{ and therefore } \varepsilon=2\alpha=\sqrt{\frac{U_0}{2}}. \tag{8.12}$$

(We recall that while ε and U_0 both look like energies, they are actually dimensionless parameters.)

Our trial solution predicts a single energy for the pendulum. We can find other possibilities by looking for a more complete solution of the differential equation. Let us suppose, as is customary with the oscillator, that the wavefunction is a product of our original gaussian function and a power series,

$$H(\theta) = \sum_{n=0}^{\infty} a_n \theta^n, \tag{8.13}$$

where the notation is in honor of the French mathematician Charles Hermite (1822–1901). We substitute the solution $\psi(\theta) = e^{-\alpha\theta^2} H(\theta)$ into Eq. (8.10) and obtain a new differential equation for $H(\theta)$:

$$H'' - 4\alpha\theta H' + (\epsilon - 2\alpha)H = 0, \tag{8.14}$$

where the dash indicates differentiation with respect to θ. Then, substituting the solution, Eq. (8.13), into Eq. (8.14) and following the usual methods for a power series solution, we obtain the expansion

$$H(\theta) = a_0\left(1 + \frac{a_2}{a_0}\theta^2 + \frac{a_4}{a_2}\frac{a_2}{a_0}\theta^4 \cdots\right) + a_1\left(\theta + \frac{a_3}{a_1}\theta + \frac{a_5}{a_3}\frac{a_3}{a_1}\theta^5 \cdots\right), \tag{8.15}$$

where a_0 and a_1 are constants. The various ratios of the coefficients are found by forcing the power series solution to fit the differential equation. This constraint produces an equation for the ratios called the indicial equation (the subscript on the power series term is an "index") of the form

$$\frac{a_{n+2}}{a_n} = \frac{2\alpha + 4\alpha n - \epsilon}{(n+1)\,(n+2)}, \quad n \geq 0. \tag{8.16}$$

It seems straightforward to use the indicial equation with the power series solution to develop the wavefunction ψ. But there is a surprising problem with this series when it is combined with the Gaussian solution. This problem is crucial to forcing the final energy spectrum. Despite the fact that the Gaussian vanishes for large θ, the combination of the gaussian function and the power series together gives a function that does *not* vanish for large values of θ and therefore provides an unphysical wavefunction. This dilemma may be avoided by excluding all solutions for which the series is infinite and retaining only those solutions for which all the coefficient ratios vanish beyond a certain index. For example, a_2 is zero when $\epsilon = 2\alpha$, a_3 is zero when $\epsilon = 6\alpha$, a_4 is zero when $\epsilon = 10\alpha$, and so forth. Furthermore, if any given even-indexed coefficient is zero than all subsequent even-indexed coefficients are zero also and the series now becomes a polynomial of finite degree, n with even powers. Furthermore if an even-indexed coefficient is made equal to zero, then the constant a_1 is also set to zero and that final polynomial only has even powers of θ. The case for odd indexed coefficients is the same. These conditions result in the Hermite polynomials and energy levels given in Table 8.1.

Table 8.1 Hermite polynomials with their corresponding dimensionless energy levels

n	$H_n(\theta)$	ϵ_n
0	1	2α
1	θ	6α
2	$1 - 4\alpha\theta^2$	10α
3	$\theta - \frac{4\alpha}{3}\theta^3$	14α
...	...	...
n		$4\alpha(n + \frac{1}{2})$

One of the fascinating things to emerge from these calculations is the fact that the conditions for physically meaningful states result in *discrete* energy levels. The system is not allowed to exist at energies that are intermediate between the prescribed values. And this restriction has been confirmed in countless laboratories. Molecular physics is a major area for applications of the quantum oscillator. It is as if the atoms in the molecules are connected by little springs (or linearized pendulums). Figure 8.1 shows the energy levels of a few states against a background of the parabolic potential well of the linearized pendulum. The figure also displays the modulus squared of each wavefunction ψ_n (the probability density function) superposed on the corresponding energy level. As the energies increases the number of peaks increase. Note also the probabilities are nonvanishing outside the potential well, behavior that is counter-intuitive and peculiar to quantum physics. This phenomenon is called tunneling and will be discussed more fully in the next chapter.

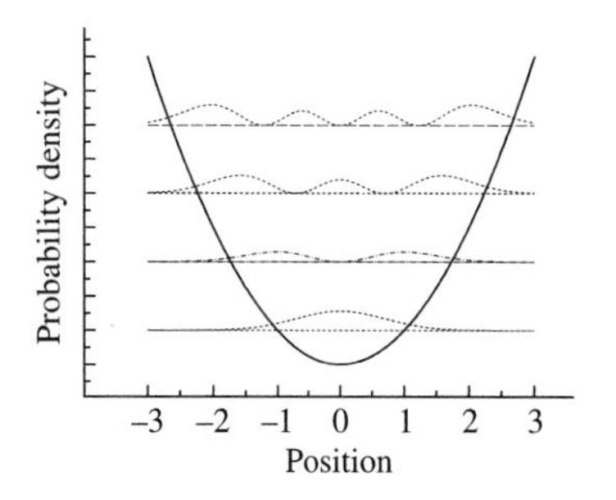

Fig. 8.1
Probability density functions for the lowest four energy states located on their corresponding energy levels.

The use of dimensionless parameters and variables makes the exercise seem a little abstract. Let us convert the energy parameter back to a physical energy. Equation (8.12) gives $\alpha = \sqrt{U_0/8}$ and in our definition of the dimensionless parameters we set $U_0 = V_0(2mL^2)/\hbar^2$ and $\epsilon = E(2mL^2)/\hbar^2$. Therefore, the ground state energy dimensionless energy ϵ_0 becomes the ground state energy

$$E_0 = \frac{\hbar}{2mL}\sqrt{mV_0}, \tag{8.17}$$

which is perhaps not very helpful. But by imposing a particular interaction on the pendulum so that V_0 takes a concrete form we may achieve a more helpful result. First let us suppose that the pendulum is subject to a gravitational restoring force just like a typical macroscopic pendulum. Then $V_0 = mgL$, and we also note that the angular frequency for a gravity pendulum is $\omega = \sqrt{g/L}$. These two expressions lead to a ground state of

$$E_0 = \frac{\hbar\omega}{2}. \tag{8.18}$$

Given the fact that $\hbar = 1.05457266 \times 10^{-34}$ Js, and ω may be of the order of 10 rad/s for a macroscopic pendulum, the lowest energy is about 10^{-33} Joules, a *very* small number. Furthermore, from the table it can be shown that the general expression for energy, for this interaction is

$$E_n = \left(n + \frac{1}{2}\right)\hbar\omega. \tag{8.19}$$

Therefore the energy spacing between levels of $\Delta E = \hbar\omega$, is again a very small number. We may safely conclude that for a macroscopic pendulum, the discrete nature of the energy levels would be completely unobservable. Rather, the ordinary observer would see a continuum of possible energy states. This particular example illustrates what is known as Neils Bohr's (1885–1962) correspondence principle. In this context, it may be stated as, "Quantum mechanics must give the same result as classical mechanics in the appropriate classical limit." (Anderson 1971, p. 135).

The torsion pendulum is another example of a linearized pendulum. In this case $V_0 = \tau$, the torsional rigidity described in Chapter 5, and the angular frequency is $\omega = \sqrt{\tau/I}$, where I is the moment of inertia of the torsion pendulum. For the simple pendulum, acting as a torsion pendulum, the axis of rotation is through the pivot point and therefore $I = mL^2$.

Surprisingly, the expression for the ground state energy again becomes $E_0 = \hbar\omega/2$. Perhaps *not* surprisingly, the energy levels for the linearized pendulum are, in this form, identical with those of the harmonic oscillator. As with the linearized gravitational pendulum, if the system is macroscopic, the frequency will be relatively low and the quantized nature of the energy spectrum will be unobservable. On the other hand, at the molecular and smaller dimensions, there are oscillations that are many orders of magnitude faster, in part because the oscillators are much smaller and in part because the interactions are much stronger. For our discussion, the question is whether there are any *microscopic* objects that correspond to pendulums at this deep level of nature. But before addressing that question, let us consider some further characteristics of the generic quantum linearized pendulum.

8.3 Where is the pendulum?—uncertainty

The probability density functions shown in Fig. 8.1 highlight the fact that even in principle, one cannot predict exactly where the pendulum will be found. For the lowest energy state, the pendulum is most likely to be near the equilibrium position at the center of the well. Yet for the second energy level, the pendulum is most likely to be located in one of the two symmetrically spaced areas on either side of the equilibrium position. As the energy increases so too does the number of favored locations. It seems reasonable that the pendulum is more likely to be further from the equilibrium position as energy increases. Is this also true for a classical pendulum?

The classical probability density function can be readily calculated. Suppose that a classical linearized pendulum's motion is given by

$$\theta = \theta_0 \cos \omega t, \tag{8.20}$$

where θ_0 is the amplitude of the motion and therefore related to the energy. The probability density $P(\theta)$ is proportional to the ratio $dt/d\theta$, which gives the relative amount of time dt spent in each small interval of angular displacement, $d\theta$. With some manipulation it can be shown that the probability for the pendulum to be at an angle between θ and $\theta + d\theta$ is given by

$$P(\theta)d\theta = \frac{d\theta}{\pi\sqrt{\theta_0^2 - \theta^2}}. \tag{8.21}$$

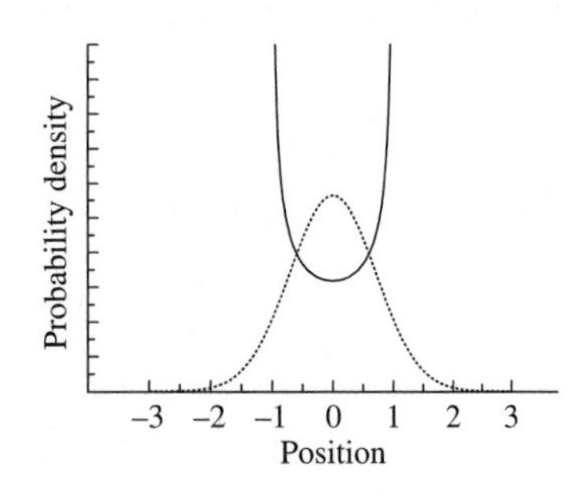

Fig. 8.2
Classical (upper) and quantum mechanical (lower) probability density functions for the ground state of the linearized pendulum.

The factor of π results from the need to normalize the probability to unity. The form of $P(\theta)$ shows that, classically, the pendulum has to be somewhere in the potential well. To actually plot this function let us return to the previously defined dimensionless energies in Table 8.1. The classical energy expression involving amplitude is $\epsilon = \frac{1}{2}U_0\theta_0^2$ with $U_0 = 2$ in our example. Therefore the angular amplitude θ_0 is numerically equal to $\sqrt{\epsilon}$. Figure 8.2 shows the classical and quantum probability densities for an energy equal to the quantum ground state:

$$\text{Classical } P(\theta)_{Cl} = \frac{1}{\pi\sqrt{1-\theta^2}}, \quad \text{Quantum } P(\theta)_{Qm} = \frac{e^{-\theta^2}}{\sqrt{\pi}}. \tag{8.22}$$

The two density functions do not match at all. The classical density is fairly level until it nears either end of its range at ± 1 where it increases rapidly. Furthermore, the classical density is not defined outside of this range, thus ensuring that the pendulum stays inside the well as expected. On the other hand, the quantum density peaks at the middle of the well and is nonzero outside the classical range and therefore outside the well—again that unique quantum tunnelling effect.

In Fig 8.3 the classical and quantum probability densities are shown for the fourth energy state. The peaks of the quantum probability now somewhat follow the classical density. This trend becomes more pronounced for higher energies as more and more peaks appear in the quantum density. The increasingly better approximation of the quantum density to the classical density is a further confirmation of Bohr's correspondence principle.

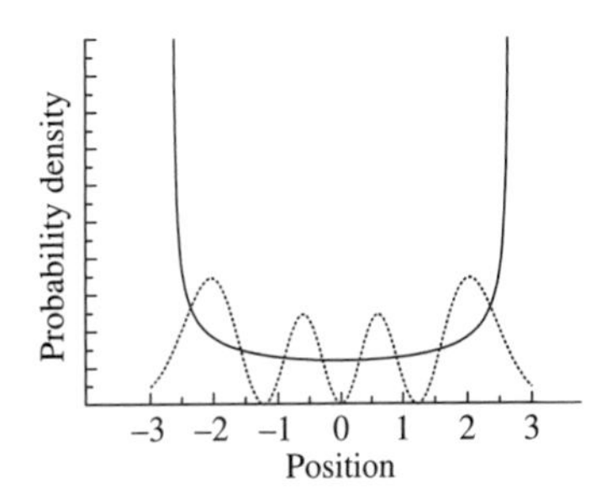

Fig. 8.3
Classical (upper) and quantum probability (lower) density functions for the third excited state.

Once again the classical distribution is confined within the well, $(-\sqrt{7}, \sqrt{7})$, whereas the quantum distribution allows the pendulum bob to exist outside the well.

There are a variety of characteristic statistical calculations that can be made once a probability distribution is known including mean and standard deviation. For some generic physical variable x, probability theory prescribes that these quantities are defined as

$$\text{Mean} = \overline{x} = \int_R xP(x)\, dx \tag{8.23}$$

$$\text{Standard deviation} = \text{Uncertainty in } \Delta x = \sqrt{\overline{x^2} - \overline{x}^2}, \tag{8.24}$$

where

$$\overline{x^2} = \int_R x^2 P(x)\, dx \tag{8.25}$$

is sometimes called the "second moment" of the distribution because of the power of two under the bar. Let us calculate the mean and the uncertainty in angular position for the two states illustrated above in Figs. 8.2 and 8.3, $n=0$ and $n=3$, respectively. We use the previously obtained density functions to do the calculation with the modification that the integral over the total domain of each function must be normalized to unity. Tables 8.2 and 8.3 give the results.

Because of the reflection symmetry (even parity) of the quadratic potential well, the mean angular position is always zero. In fact, all odd moments of the distribution will be zero. (This result simplifies the calculation of the standard deviation.) The standard deviation $\Delta\theta$, which measures the uncertainty in the angular position, increases with increasing quantum state. Remarkably, the classical and quantum density functions both give the same uncertainties for a given state. (However, the higher order classical and quantum even moments diverge with increase in the order of the moment.) What are the implications of this exercise? While it is interesting to do the classical calculation of

Table 8.2 Mean angular displacement for two states calculated using classical and quantum density functions

n	Type of mechanics	$\overline{\theta}$
0	Classical	$\int_{-1}^{1} \frac{\theta}{\pi\sqrt{1-\theta^2}}\, d\theta = 0$
0	Quantum	$\int_{-\infty}^{\infty} \frac{1}{\sqrt{\pi}} \theta e^{-\theta^2}\, d\theta = 0$
3	Classical	$\int_{-\sqrt{7}}^{\sqrt{7}} \frac{\theta}{\pi\sqrt{7-\theta^2}}\, d\theta = 0$
3	Quantum	$\int_{-\infty}^{\infty} 1.69260\theta(\theta - \frac{2\theta^3}{3})^2 \times e^{-\theta^2}\, d\theta = 0$

Table 8.3 Standard deviation (uncertainty) calculations for two states using classical and quantum denisty functions

n	Type of mechanics	$\Delta\theta$
0	Classical	$\sqrt{\int_{-1}^{1} \frac{\theta^2}{\pi\sqrt{1-\theta^2}}\, d\theta} = \frac{1}{2}\sqrt{2}$
0	Quantum	$\sqrt{\int_{-\infty}^{\infty} \frac{1}{\sqrt{\pi}}\theta^2 e^{-\theta^2}\, d\theta} = \frac{1}{2}\sqrt{2}$
3	Classical	$\sqrt{\int_{-\sqrt{7}}^{\sqrt{7}} \frac{\theta^2}{\pi\sqrt{7-\theta^2}}\, d\theta} = 1.8708$
3	Quantum	$\sqrt{\int_{-\infty}^{\infty} 1.69266\theta^2 \left(\theta - \frac{2\theta^3}{3}\right)^2 e^{-\theta^2}\, d\theta} = 1.8708$

uncertainty, there is little use for the result. Classical physics is premised on being able to measure the pendulum's position to an arbitrary degree of precision. But the situation is very different in quantum physics. The maximum information is contained in the wavefunction and therefore the probability distribution. Hence, prediction of the pendulum's position is uncertain, *in principle*, and uncertainty becomes an important experimental reality.

Can we calculate statistical quantities for other physical variables? The answer is yes provided we make some adjustments in our definition of these quantities as given by Eq. (8.23). We recall that the potential energy operator contained the pendulum's angle of deviation, and therefore one might expect that the operator θ_{op} is just the angle itself. This is indeed the case, and this fact allowed us to calculate $\Delta\theta$ from Eq. (8.23). However the general case, needed for the angular momentum p_θ, is more involved. Rather than just using the variable itself in the calculation, we must use the operator. Thus we have the following revision of Eq. (8.23):

$$\text{Mean } \overline{x} = \frac{\int \psi^* x_{\text{op}} \psi \, dx}{\int \psi^* \psi \, dx} \quad \text{and Uncertainty } \Delta x = \sqrt{\overline{x^2} - \overline{x}^2}, \tag{8.26}$$

where

$$\overline{x^2} = \frac{\int \psi^* x_{\text{op}}^2 \psi \, dx}{\int \psi^* \psi \, dx}. \tag{8.27}$$

At first glance the integrals do not look much different from the previous expressions. They all contain ψ and ψ^*, the ingredients of the probability density functions. But there are some new features. The integrals that contain the operators must be evaluated exactly as shown. The operators must first act on ψ and then that result is multiplied by ψ^* for the integration. The integral in the denominator ensures that even if the probability density functions $\psi^*\psi$ are not properly normalized to unity, the result is correct.

In 1927 Heisenberg published his famous uncertainty principle; if two complementary variables, such as angle and angular momentum p_θ, were simultaneously measured, then the product of their uncertainties could never be less than a number of the order of Planck's constant, $\hbar$. Let us test this relationship for the ground state of the linearized pendulum. All variables must be expressed in terms of real physical quantities rather than as dimensionless quantities. Let us substitute our previous trial wave function $\psi = e^{-\alpha\theta^2}$ into the original Schrodinger equation, Eq. (8.8). We find that

$$\alpha = \frac{mL^2\omega}{2\hbar}, \tag{8.28}$$

where $\omega = 2\pi/T$ is the angular frequency of the pendulum. For the ground state both $\overline{\theta}$ and $\overline{p_\theta}$ are zero because of the symmetry of the potential well. Therefore we find that $\Delta\theta = \sqrt{\overline{\theta^2}}$ and $\Delta p_\theta = \sqrt{\overline{p_\theta^2}}$. These quantities become

$$\begin{aligned} \Delta\theta = \sqrt{\overline{\theta^2}} &= \sqrt{\frac{\int_{-\infty}^{\infty} \theta^2 e^{-2\alpha\theta^2}\,d\theta}{\int_{-\infty}^{\infty} e^{-2\alpha\theta^2}\,d\theta}} \\ &= \frac{1}{2\alpha}\frac{\int_{-\infty}^{\infty} x^2 e^{-x^2}\,dx}{\int_{-\infty}^{\infty} e^{-x^2}\,dx} = \sqrt{\frac{\hbar}{2mL^2\omega}} \end{aligned} \tag{8.29}$$

and[1]

$$\begin{aligned} \Delta p_\theta = \sqrt{\overline{p_\theta^2}} &= \sqrt{\frac{\int_{-\infty}^{\infty} e^{-\alpha\theta^2}(\hbar d/id\theta)^2 e^{-\alpha\theta^2}\,d\theta}{\int_{-\infty}^{\infty} e^{-2\alpha\theta^2}\,d\theta}} \\ &= \sqrt{\frac{\int_{-\infty}^{\infty} -\hbar^2[4\alpha^2\theta^2 - 2\alpha]e^{-2\alpha\theta^2}\,d\theta}{\int_{-\infty}^{\infty} e^{-2\alpha\theta^2}\,d\theta}} \\ &= \sqrt{-4\hbar^2\alpha^2\frac{\int_{-\infty}^{\infty} \theta^2 e^{-2\alpha\theta^2}\,d\theta}{\int_{-\infty}^{\infty} e^{-2\alpha\theta^2}\,d\theta} + 2\alpha\hbar^2\frac{\int_{-\infty}^{\infty} e^{-2\alpha\theta^2}\,d\theta}{\int_{-\infty}^{\infty} e^{-2\alpha\theta^2}\,d\theta}} \\ &= \sqrt{\frac{mL^2\omega\hbar}{2}}. \end{aligned} \tag{8.30}$$

These two outcomes lead to the remarkably simple result,

$$\Delta p_\theta \Delta\theta = \hbar/2. \tag{8.31}$$

In general this relationship is expressed as $\Delta A \cdot \Delta B \geqslant \hbar/2$ where A and B are a pair of complementary variables. This inequality holds for all quantum systems. For the pendulum, the complementary pair consisted of angle and angular momentum. An even more common pair is position x, and linear momentum p. In a subtle way, energy E and time t also form a complementary pair.

[1] Pascual Jordan (1927) pointed out that, unlike the case of position and linear momentum, there are problems with the meaning of the uncertainty principle as applied to angle and angular momentum. Unlike a spatial coordinate, angle is limited to the interval $[-\pi, \pi]$ and therefore its uncertainty is limited. Thus for small uncertainty in angular momentum the uncertainty in angle would be impossibly large. However, for small angle oscillations, using the linearized pendulum, the uncertainty in angle is small for a macroscopic pendulum. Thus the calculation of uncertainties for a large pendulum in the ground state is still valid.

The most astonishing implication of this relationship occurs if it is desired to measure one variable of the pair. We calculate numerical values for the uncertainties of a *macroscopic* pendulum in the ground state. Let $m = 1$ kg, $L = 1$ m, and $\omega = \sqrt{g/L} = 0.32$ rad/s. Then Eqs. (8.29) and (8.30) yield

$$\Delta\theta \simeq 10^{-17} \text{ rad and } \Delta p_\theta = 10^{-17} \text{ J s.} \tag{8.32}$$

Numerically, the uncertainties in each quantity are about equal. On the other hand, the same sort of calculation for *microscopic* objects—mass of a few protons, length equal to a typical molecular bond length, and frequency equal to a typical molecular frequency—leads to a much larger uncertainty in angle of $\Delta\theta \approx 1$ rad. We can also calculate the value of α and therefore the sharpness of the ground state wave function in each case. For the macro pendulum, the wave function is very sharp ($\psi \simeq e^{-10^{34}\theta^2}$), whereas for the micropendulum the wave function is relatively broad ($\psi \simeq e^{-10^{-1}\theta^2}$). Given that there is some probability that the pendulum will have an angle of up to a radian or so, it may be that the linearization approximation is not especially appropriate for the microscopic pendulum. With these issues in mind, we turn to a discussion of the *nonlinear* quantum pendulum.

8.4 The nonlinear quantum pendulum

The Schrodinger equation for the nonlinear pendulum, first partially solved by Condon (1928) in 1927 and later by other authors, is given by Eq. (8.7)

$$-\frac{\hbar^2}{2mL^2}\frac{d^2\psi}{d\theta^2} + V_0(1 - \cos\theta)\psi = E\psi. \tag{8.33}$$

The cosine function is a serious complication. If θ is small, then the linear approximation is sufficient, as it is for the classical pendulum. For large angles, the approximation breaks down as is evident in the comparison of the two potential wells in Fig. 8.4.

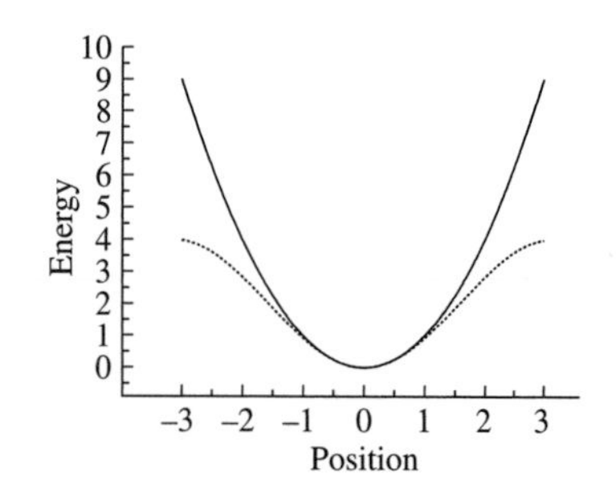

Fig. 8.4
The potential wells for the linearized (upper) and nonlinear pendulum (lower). The nonlinear well "softens" as it nears the top and, while not shown in the figure, is periodic, and will extend infinitely, with the same pattern, on both sides of the linearized potential.

Let us again convert Schrodinger's equation to dimensionless form

$$-\frac{d^2\psi}{d\theta^2} + U_0(1 - \cos\theta)\psi = \varepsilon\psi. \tag{8.34}$$

This equation is essentially equivalent to a Mathieu equation, first proposed by Emile Mathieu (1835–1890) in 1868 for the study of vibration of elliptically shaped membranes (Mathieu 1868). The Mathieu equation and its solutions, Mathieu functions, have a long history of development with contributions from a variety of mathematicians (McLachlan 1947). Derivation of these functions is rather complex and the interested reader is referred to McLachlan's book or more recently, the handbook of mathematical functions (Abramowitz and Stegun 1972). However, we give a little of the flavor of Mathieu functions and then, for actual solution of the pendulum problem, resort to numerical methods.

8.5 Mathieu equation

The Mathieu equation lends itself to many applications (Ruby 1996). These include the original application to the vibrations of an elliptic drum, the motion of an inverted pendulum, the motion of an ion in a radio frequency quadrupole field used for focusing low-energy ion beams, and many other examples. In the present case, we apply it to the quantum pendulum. The Mathieu equation has the form

$$\frac{d^2y}{dx^2} + [a - 2q\cos 2x]y = 0. \tag{8.35}$$

The substitution $\theta = 2x$ in Eq. (8.34) leads exactly to the Mathieu equation with the parameters a and q expressed in terms of the dimensionless energies:

$$a = 4(\varepsilon - U_0) \text{ and } q = 2U_0. \tag{8.36}$$

The parameter q is fixed and depends only on the dimensionless potential, U_0. The parameter a depends upon both U_0 and ε, and is therefore undetermined. However, like many differential equations in physics, physically realizable solutions exist for the Mathieu equation only for certain values of ε and, consequently, a. Thus the Mathieu equation is, like the Schrodinger equation, an eigenvalue problem. A set of solutions, eigenfunctions, and a set of eigenvalues may be generated. In the usual notation, a represents a set of eigenvalues $\{a_i, b_i\}$. We note that in the original dimensionless Schrodinger equation Eq. (8.34) the potential is invariant under the coordinate transformation $\theta = \theta + 2\pi$ and it turns out that, for this circumstance, only the even numbered eigenvalues of the corresponding Mathieu equation are retained; a_0, a_2, b_2, a_4, b_4, and so forth.

Solutions to the Mathieu equation are complex and we make only a few remarks to give some sense of them. (The actual solution of the pendulum Schrodinger equation will be done numerically.) First we observe that if $q = 0$, the pendulum acts like a free rotor and the Mathieu equation reduces to

$$\frac{d^2y}{dx^2} + ay = 0. \tag{8.37}$$

This equation is readily solved to give $y = \cos mx$, $y = \sin mx$, or the complex version $y = e^{imx}$, where $a = m^2$ prescribes the appropriate eigenvalue. Therefore when the pendulum energy is well above the height of the potential well $a \gg q$, then the solution is rather simple. However, when q becomes important the eigenvalue is determined using an appropriate number of terms from an infinite series. Each term of the series contains some power of q:

$$a = m^2 + \alpha_1 q + \alpha_2 q^2 + \alpha_3 q^3 + \cdots \tag{8.38}$$

and there are similar expansions for the eigenfunctions. For example the "cosine" solutions might look like

$$y = \cos mx + qc_1(x) + q^2c_2(x) + q^3c_3(x) + \cdots, \tag{8.39}$$

where the c_i are combinations of cosine functions. For the lowest order eigenfunction and eigenvalues, corresponding to $m=0$, the solution is given by infinite series whose first few terms are

$$\begin{aligned} ce_0(x,q) = 1 - \frac{1}{2}q\cos 2x + \frac{1}{32}q^2\cos 4x \\ - \frac{1}{128}q^3\left(\frac{1}{9}\cos 6x - 7\cos 2x\right) + \cdots \end{aligned} \tag{8.40}$$

and the eigenvalue is

$$a_0 = -\frac{1}{2}q^2 + \frac{7}{128}q^4 - \frac{29}{2304}q^6 + \cdots, \tag{8.41}$$

respectively. The functions whose series begin with a cos mx term are labeled $ce_i(x,q)$, whereas functions that begin with the sin mx are labeled $se_i(x,q)$. The expressions for Mathieu eigenfunctions and eigenvalues become increasingly complex for the higher indices, and truncation of a given series usually gives a good approximation only if q is relatively small. However, we will describe some quantum pendulums for which q is relatively large and therefore we use commercial numerical software to provide the eigenvalues which, in turn, give the energy levels of the pendulum. Figure 8.5 shows the Mathieu eigenvalues appropriate to the pendulum as a function of q and are generated using the IMSL Fortran subroutine package[2].

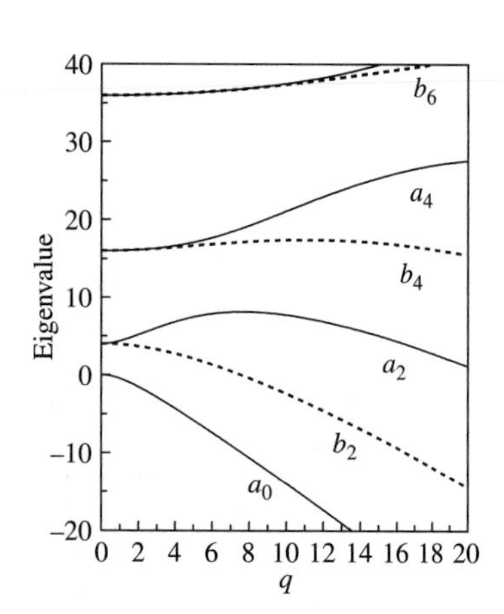

Fig. 8.5
Plot of the eigenvalues of the Mathieu equation as a function of the parameter q. (Reprinted with permission from Baker et al. (2002, p. 526). ©2002, American Association of Physics Teachers.)

There are two limiting cases illustrated in Fig. 8.5; (a) when the pendulum's energy is much less than the height of the barrier, and (b) when the pendulum's energy is much greater than the height of the barrier. Classically these cases correspond to (a) the pendulum oscillating with very small angular amplitude, and (b) the pendulum rotating rapidly and almost freely such that it slows only slightly going over the top of its motion. Case (a), in the limit of small oscillation becomes the linearized pendulum or harmonic oscillator. Case (b), in the limit of unbounded energy, becomes a free rotor. We have already discussed (a), the low energy case, and found that the energy levels increase according to the sequence

$$\frac{1}{2}\hbar\omega,\ \frac{3}{2}\hbar\omega,\ \frac{5}{2}\hbar\omega,\ \cdots, \tag{8.42}$$

with a uniform spacing between the levels. Case (b), the free rotor, is new to our discussion. Let us deal with it.

In classical physics the rotor is simply a rigid object with a certain moment of inertia that rotates without friction about some axis. Classically, it may have an arbitrary energy—all values are possible. But the quantum version must obey the Schrodinger equation. This

[2] Product of Visual Numerics, Inc, Houston, Texas.

equation takes an especially simple form because there is no potential energy term;

$$-\frac{\hbar^2}{2ML^2}\frac{d^2\psi}{d\theta^2} = E\psi, \quad (8.43)$$

or, written in dimensionless form,

$$-\frac{d^2\psi}{d\theta^2} = \varepsilon\psi, \quad (8.44)$$

or, as a much simplified Mathieu equation,

$$\frac{d^2y}{dx^2} + ay = 0. \quad (8.45)$$

It is easy to verify that, as noted above, the solutions are of the form $e^{\pm imx}$, where $m = \sqrt{a}$ or, in the dimensionless Schrodinger equation, $e^{\pm im\theta}$, where $m = \sqrt{\varepsilon}$, or, in physical terms, $m = (L\sqrt{2mE})/\hbar$. At this point we are not surprised that this solution forces the energy levels to take certain values. Here is the rationale. Consider a solution to the dimensionless Schrodinger equation, $e^{im\theta}$. This solution is periodic; substitution of $\theta + 2\pi$ for θ must leave the function—and therefore the dynamics—unchanged. That is, a rotation of 2π by the rotor leaves its orientation unchanged. Therefore the parameter m is confined to

$$m = 0, \pm 1, \pm 2, \pm 3, \cdots \quad (8.46)$$

The energy of the free rotor is quantized and, unlike the harmonic oscillator, the difference between energy levels is not constant, but increases with increasing energy:

$$\Delta E = E_{m+1} - E_m = \frac{\hbar^2}{2ML^2}\left[(m+1)^2 - m^2\right] = \frac{\hbar^2}{2ML^2}(2m+1). \quad (8.47)$$

How do these two limiting cases relate to the eigenvalues, a_i, b_i and ultimately the energy levels? In Fig. 8.5 the left side of the graph where $q = 0$ corresponds to the free rotor limit and the right side of the graph roughly corresponds to the harmonic oscillator limit. Using the relationship $\varepsilon = \frac{a}{4} + \frac{q}{2}$ and the data used to generate Fig. 8.5, a graph of energy ε versus q is shown in Fig. 8.6. Again the left side is the free rotor limit and the right side approximates the harmonic oscillator limit. In this figure it is now evident that (a) the spacing between oscillator energy levels is approximately equal, (b) the rotor energy spacing increases linearly with m, and (c) the rotor energy levels are doubly degenerate as expected from Eq. (8.46). Intermediate between the two limiting cases is the nonlinear quantum pendulum. With these ideas and the image of Fig. 8.6 in hand, we now have some understanding of the energy levels of the quantum pendulum.

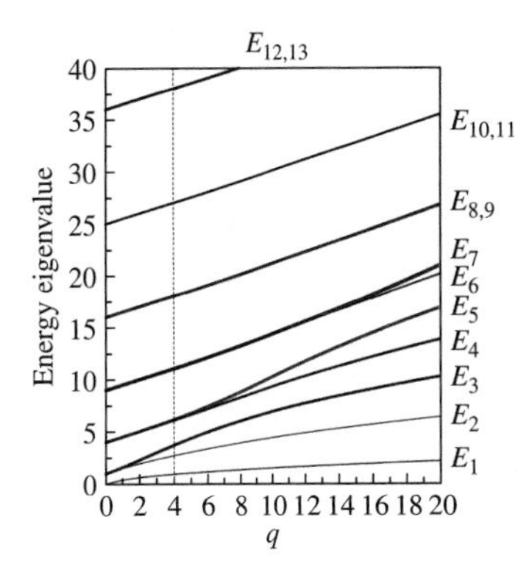

Fig. 8.6
Plot of energy eigenvalues versus q. The left side, where $q = 0$ corresponds to free rotor behavior, whereas the right side tends to represent harmonic oscillator (or linearized pendulum) behavior.

8.6 Microscopic pendulums

Are there microscopic pendulums at the atomic and molecular levels of nature? Or did we do this mathematics just as an exercise? There may not be microscopic pendulums per se, but there are molecular units that act like pendulums. That is, their potential energies have mathematical forms, $V(\theta) = V_0(1 - \cos n\theta)$, similar to the potential of the pendulum. (n is an integer that depends on the symmetry of the molecule.) Molecules act as if their constituent atoms or subcomplexes are connected by little springs—harmonic oscillators. Sometimes a subcomplex of atoms within a molecule will behave like a torsion pendulum. Again the potential is that of an harmonic oscillator. However, there are occasions where the torsional behavior of such a complex can have a rather large amplitude and, unlike the perfect torsion pendulum, occasionally flip to a new state of oscillation about a new equilibrium angle. In such cases the *form* of the potential energy is *not* linear like a perfect torsion pendulum, but is nonlinear like an ordinary gravity pendulum. But, unlike the ordinary pendulum, the source of the potential is the electromagnetic interaction between the constituent atoms rather than gravity. If the motion of the complex is to execute a series of such "flips," then the motion is said to be a *hindered rotation*. Typically, the complex rotates, but not all the time and not completely or freely.

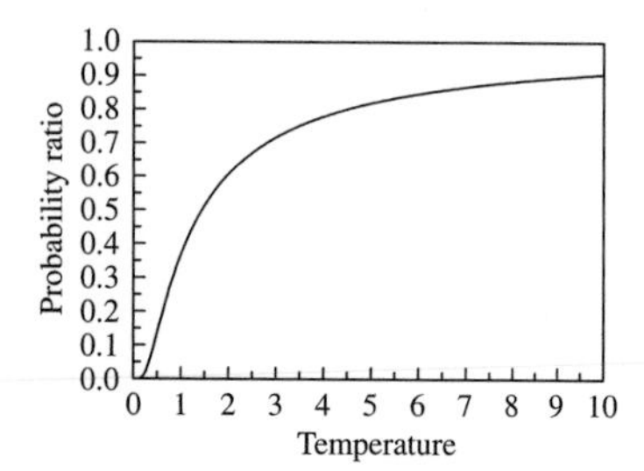

Fig. 8.7
Plot of the Boltzmann relative probability distribution of particles in two energy levels as a function of temperature.

Whether a complex executes hindered rotations frequently or only occasionally depends on two factors. One factor is how stiffly the complex is held within the molecule. In turn the stiffness depends both on the strength of the intramolecular interactions between the complex and the rest of the molecule, and also on the moment of inertia of the complex. The other factor is the amount of thermal energy in which the complex is situated. If the ambient temperature is high, the ensemble of molecules will have a lot of energy and the ability, on a statistical basis, to occupy higher energy levels. Thus, heat provides energy and encourages rotation rather than just oscillation. A molecular complex in this condition occupies energy levels that tend to be away from the harmonic oscillator limit and more toward the free rotor limit.

Thermal energy acts statistically and there is a formula that gives the relative probability of, for example, a molecule being in a certain energy state. Suppose there are two states of energies E_b and E_a. We denote the probability of the molecule occupying state a as P_a and the probability of the molecule being in state b as P_b. Then statistical physics tells us that the ratio of these two probabilities is

$$\frac{P_a}{P_b} = e^{-(E_a - E_b)/kT}, \tag{8.48}$$

where $k = 1.38 \times 10^{-38}$ J/K is called Boltzmann's constant, and T is the ambient temperature. Figure 8.7 shows a sketch of the ratio as a function of temperature when $E_a > E_b$.

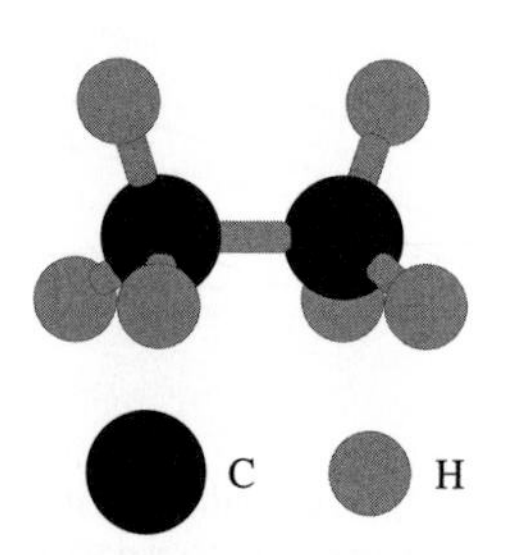

Fig. 8.8
The molecular structure of Ethane in the "eclipsed" form where the opposing hydrogen atoms align. (Reprinted with permission from Baker et al. (2002, p. 528). ©2002, American Association of Physics Teachers.)

Increasing temperature rapidly increases the probability of occupation of the more energetic state. We now consider two examples of hindered rotary complexes with quite different behaviors.

8.6.1 Ethane—almost free

The chemical ethane (C_2H_6) has the molecular structure shown in the following two figures. In Fig. 8.8 the structure is shown in the aligned or "eclipsed" form where the hydrogen atoms match each other in facing pairs, whereas the structure in Fig. 8.9 shows opposing hydrogen atoms nonaligned or "staggered". The eclipsed structure requires about 2.9 kcal/mol (9.73×10^{-21} J/molecule) compared to the "staggered" conformation. This interaction is due to the slight extra repulsion of the electron clouds of carbon–hydrogen bonds as they pass at closer quarters in the eclipsed conformation. The potential function takes account of the fact that there are three identical positions of minimum energy and is expressed as

$$V = V_0(1 - \cos 3\theta), \tag{8.49}$$

an expression that is very similar to the quantum pendulum. The angle θ, between the orientations of the two CH_3 complexes, is zero when the molecule is in the staggered formation. The appropriate conversion to our usual dimensionless units requires that

$$U_0 = \frac{V_0 I}{\hbar^2} \quad \text{and} \quad \varepsilon = \frac{EI}{\hbar^2}, \tag{8.50}$$

where I is the moment of inertia due to three hydrogen atoms with an H–C bond length of 1.11 Angstrom units. The H–C bond is at an angle of 111° with the C–C bond, with the resultant moment of inertia, $I = 5.39 \times 10^{-47}$ kg m^2. The dimensionless Schrodinger equation is now

$$-\frac{d^2\psi}{d\theta^2} + U_0(1 - \cos 3\theta)\psi = \varepsilon\psi, \tag{8.51}$$

where $U_0 = 47.2$. This equation can then be converted to the corresponding Mathieu equation where now

$$a = \frac{9}{4}(\varepsilon - U_0) \quad \text{and} \quad q = \frac{2}{9}U_0 \tag{8.52}$$

are the appropriate eigenvalue and parameter expressions. Some of the (dimensionless) energy levels are given in Table 8.4.

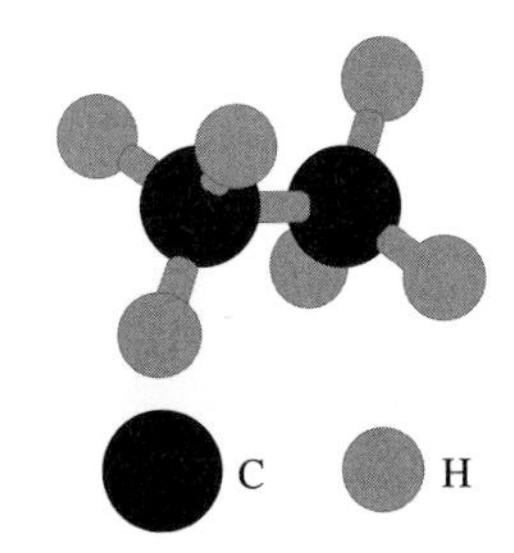

Fig. 8.9
The molecular structure of Ethane in "staggered" form where opposing hydrogen atoms are out of alignment. (Reprinted with permission from Baker et al. (2002, p. 528). ©2002, American Association of Physics Teachers.)

Table 8.4 Some dimensionless energy levels for ethane and comparison with oscillator and rotor-like states (Only even numbered indexed states are used for the rotor-like calculation.)

State, m, $(m+1)$	$a_i, b_i, U_0 = 47.25$	Pendulum energy $\varepsilon = \frac{9a_i}{4} + U_0$	Oscillator energy $\varepsilon_1 = 3\sqrt{\frac{U_0}{2}}$	Rotor–like energy $\varepsilon = \frac{9m^2}{4} + U_0$
1	$a_0 = -14.780$	13.95	14.6	
2	$b_2 = -2.9118$	40.65	43.8	
3	$a_2 = 7.5425$	64.17	73.0	
4	$b_4 = 17.403$	86.36		
5	$a_4 = 21.579$	95.75		
6	$b_6 = 37.559$	131.7		
7	$a_6 = 37.709$	132.0		
8(9)	$a_8, b_8 = 64.884$	193.2		191
10(11)	$a_{10}, b_{10} = 100.56$	273.5		272

For the low energy states, the approximation to a harmonic oscillator is fair, whereas for the higher energies the rotor-like states are an increasingly good approximation. Furthermore, the energy of the comparison rotor-like states is not just quadratic in m, but is augmented by the U_0 which, in turn, is the midpoint of the potential energy barrier height. U_0 provides a baseline for the rotor energies.

There are some other things we can learn from this model. For example, using the known value of V_0, the harmonic oscillator approximation will yield an estimate of the oscillatory frequency of about 4.5×10^{12} vibrations per second (Hertz). Using Eq. (8.48) and some basic ideas about probability we can estimate how much time the ethane molecule spends in rotor-like states. One might crudely allow states above the height of the potential barrier $2V_0$ to be rotor type states. It can be shown that the fraction F of time that the molecule spends in rotary type states is given by

$$F = \frac{\sum_{E_i > 2V_0} e^{-E_i/kT}}{\sum_{All\ E_i} e^{-E_i/kT}}. \tag{8.53}$$

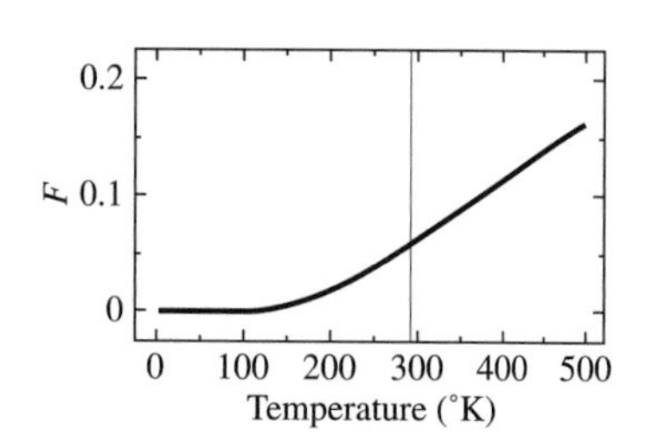

Fig. 8.10
The fraction of ethane molecules above the potential barrier. The vertical line highlights the behavior at room temperature. (Reprinted with permission from Baker et al. (2002, p. 528). ©2002, American Association of Physics Teachers.)

If F is of the order of unity, then the molecule spends most of its time in rotary type states. But if F is much less than one, then the CH_3 complexes execute primarily harmonic motion. Figure 8.10 shows the amount of rotary behavior for ethane. As expected the fraction increases with increasing temperature. At room temperature ethane spends part of its time in rotary motion. (Theoretically, the sums in the fraction F require an infinite number of energy states. However, for ethane, the value of F converges for about 2000 energy states. Some of these states come from numerically solving the Mathieu equation but most of the states may be calculated using the rotor-like approximation.) The hindered rotation behavior in ethane is substantiated by experimental work.[3]

In contrast to the behavior of ethane we now discuss a molecular complex that is hardly ever free to execute hindered rotations.[4]

8.6.2 Potassium hexachloroplatinate—almost never free

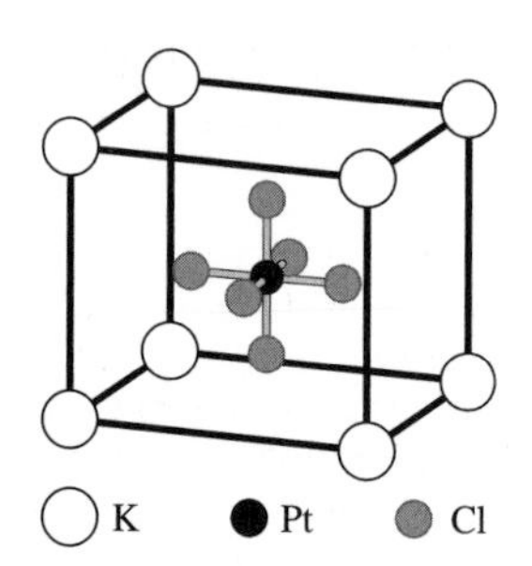

Fig. 8.11
The $PtCl_6$ complex surrounded by a cubic cage of potassium ions. Only alternate cages of potassiums contain the complex ion. (Reprinted with permission from Baker et al. (2002, p. 529). ©2002, American Association of Physics Teachers.)

Let us now analyze a microscopic "pendulum" that is predominantly oscillatory. Our example is a crystalline lattice of K_2PtCl_6 molecules. The $PtCl_6$ complex is situated in surrounding cubic cage of K ions (see Fig. 8.11) The coupling of the complex to the potassium ions is relatively loose. If we think of a torsion or spring constant it is rather weak and the motion associated with this interaction is of relatively low frequency. But inside the tightly connected $PtCl_6$ complex there are internal motions that represent stronger interactions and therefore have associated motions that are relatively high in frequency. However, our interest is in the weak interaction of the complex as a whole with its surroundings and its corresponding motion. We are interested in the torsional motion of the

[3] See Ouellete and Rawn (1996, p. 168).

[4] The motions of the ethane molecule are somewhat more complex than our simple model suggests, although the hindered rotation motion is real enough.

whole complex inside the cubic cage and, from Fig. 8.11, note that it has fourfold symmetry about any one of its three straight Cl–Pt–Cl axes. Therefore the potential may be represented as

$$V = V_0(1 - \cos 4\theta). \tag{8.54}$$

Unlike the ethane, we do not know the value for the height of the barrier, $2V_0$, and therefore must find another way. Fortunately, as we will see, the $PtCl_6$ complex executes relatively small amplitude oscillations and therefore the linearized Schrodinger equation, Eq. (8.8), may be used,

$$-\frac{\hbar^2}{2I}\frac{d^2\psi}{d\theta^2} + V_0\frac{\theta^2}{2}\psi = E\psi, \tag{8.55}$$

which leads to a harmonic oscillator solution. Furthermore the potential barrier height can be simply expressed in terms of the oscillator frequency and moment of inertia:

$$V_0 = I\omega^2, \tag{8.56}$$

using earlier equations in the chapter. The moment of inertia is known to be 1.28×10^{-44} kg m^2. The angular frequency is less straightforward. It has been calculated, from fundamental considerations of electromagnetic force fields between the complex and the surrounding potassium ions, to be about 1.89×10^{12} Hz. On the other hand a different approach, using a technique called *Nuclear Quadrupole Resonance*, suggests that the number of vibrations per second is 1.75×10^{12}. We will compromise and let $\omega = 2\pi \times 1.8 \times 10^{12}$ rad/s. Compared to ethane, the frequency of oscillation for K_2PtCl_6 is less by a factor of 3, whereas the moment of inertia is higher by a factor of almost 300. Therefore it is much more difficult for the $PtCl_6$ complex to get over its potential barrier and rotate by 90° to a new equilibrium position about which it may oscillate. We then expect that the complex will spend the vast majority of its time in energy states *below* the potential barrier. The fraction of time spent above the barrier may be calculated using Eq. (8.53) and the fraction is plotted as a function of temperature in Fig. 8.12. A comparison of Fig. 8.10 for ethane with Fig. 8.12 is striking and bears out our intuition that, for room temperatures, the $PtCl_6$ complex acts as a small amplitude torsional oscillator rather than as a hindered rotor.

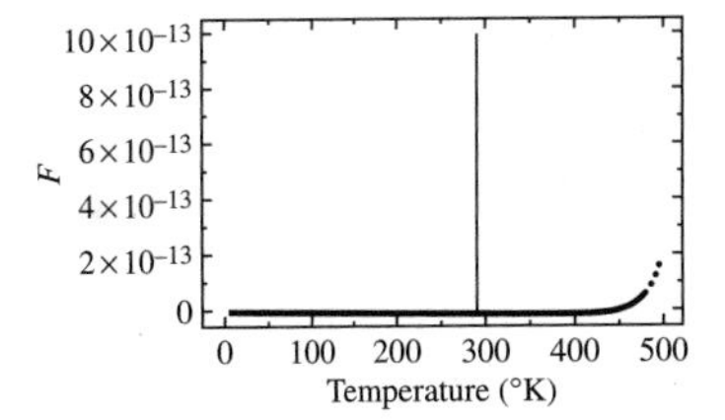

Fig. 8.12
The fraction of complex ions in states above the potential barrier. The complexes are strongly confined to oscillatory behavior. (Reprinted with permission from Baker et al. (2002, p. 529). ©2002, American Association of Physics Teachers.)

The use of numerical software is essential in the case of the chloroplatinate complex. Values for the physical parameters lead to a large barrier height; that is, the dimensionless potential becomes $U = 235{,}820(1 - \cos 4\theta)$. The dimensionless Schrodinger equation is

$$-\frac{d^2\psi}{d\theta^2} + U_0(1 - \cos 4\theta)\psi = \varepsilon\psi, \tag{8.57}$$

and conversion to the Mathieu equation leads to the relations

$$a = \frac{4}{16}(\varepsilon - U_0) \quad \text{and} \quad q = \frac{2}{16}U_0. \tag{8.58}$$

Table 8.5 Some dimensionless energy levels for potassium hexachloroplatinate and comparison with oscillator and rotor-like states. (Only even number indexed states are used for the rotor-like calculations.)

State, m, $(m+1)$	$a_i, b_i, U_0 = 235820$	Pendulum energy $\varepsilon = \frac{16a_i}{4} + U_0$	Harmonic oscillator energy $\varepsilon_1 = 4\sqrt{\frac{U_0}{2}}$	Rotor–like energy $\varepsilon = \frac{16m^2}{4} + U_0$
1	$a_0 = -58,613$	1,368	1,373	
2	$b_2 = -57,927$	4,112	4,120	
3	$a_2 = -57,242$	6,850	6,850	
4	$b_4 = -56,559$	9,586	9,611	
Near barrier top				
218	$b_{218} = 58,743$	470,754	597,255	425,916
219	$a_{218} = 58,889$	471,375	600,001	
Well above barrier				
996(997)	$b_{996}, a_{996} = 992,452$	4,205,636		4,203,884
998(999)	$b_{998}, a_{998} = 996,440$	4,221,580		4,219,836

Some of the dimensionless energy levels are given in Table 8.5 along with comparisons to the harmonic oscillator and the rigid rotor.

If we allow entities that have potential energy functions of the form $(1 - \cos n\theta)$ to be called pendulums, then the model of the quantum pendulum has application on the microscopic level. In our illustrations of the quantum pendulum we have found one example, ethane, where the pendulum commonly goes "over the top" and another example, K_2PtCl_6, where the pendulum is a tightly bound torsional pendulum and confines its motion to the lower parts of the potential well.

Let us now look at some further implications of applying the quantum pendulum model to a large scale gravity pendulum.

8.7 The macroscopic quantum pendulum and phase space

In a previous section we explored the idea of a *linearized* macroscopic pendulum in a gravitational field. We found expressions for the ground state energy, the energy levels in general and some idea of the magnitude of these energies. That is,

$$E_0 = \frac{\hbar\sqrt{mV_0}}{2mL} = \frac{\hbar\sqrt{m^2gL}}{2mL} = \frac{\hbar}{2}\sqrt{\frac{g}{L}} = \frac{\hbar\omega}{2} \text{ and } E_n = \hbar\omega\left(n + \frac{1}{2}\right). \tag{8.59}$$

Furthermore, we also found expressions for the quantum uncertainty in angle and angular momentum that, for the gravity pendulum, lead to

$$\Delta\theta = \sqrt{\frac{\hbar}{2mgL}} \text{ and } \Delta p_\theta = \sqrt{\frac{mgL\hbar}{2}}. \tag{8.60}$$

For a large pendulum of mass $m = 1$ kg and length $L = 1$ m, these formulas yield very small uncertainties, $\Delta\theta \approx 10^{-17}$ and $\Delta p_\theta \approx 10^{-17}$. In earlier chapters we had thought of trajectories in phase space as continuous well defined curves. But these uncertainties suggest that a better representation of the trajectory would be a slightly fuzzy curve. Alternatively, one could think of phase space as consisting of a grid of small cells, each having a length $\Delta\theta$ and a width Δp_θ. The area of a each cell approximates Planck's constant. According to the uncertainty principle, a pendulum's phase coordinates would only be known up to a resolution equal to the size of a phase cell.

While cells of such small dimensions are not measurable with physical instrumentation, these uncertainties do have consequences for the study of classical dynamics and especially for classical chaotic dynamics. For nonchaotic systems we could content ourselves with having a miniscule degree of fuzziness in a simulated trajectory. Such systems are characterized by regular or periodic orbits. Uncertainty only leads to some broadening of the phase space trajectory. But for unstable chaotic systems the situation may be quite different. The computer is the preeminent tool for exploring the mathematical terrain of a chaotic system, and it is not unusual, in mapping out strange attractors in phase space, to simulate very long trajectories. Therefore, when the trajectory is very sensitive to its previous history, the use of long trajectories calculated with double precision may lead to incorrect conclusions as the limitations of quantum uncertainty come into play. Trajectories may be completely erroneous through taking a series of incorrect twists caused by quantum uncertainty. In our previous discussion of classical chaos, we learned that if the system has a positive Lyapunov exponent, prediction of its position quickly becomes impossible. The situation is even less predictable when quantum uncertainty is included. It is sometimes said that when nature is examined in detail, quantum mechanics washes out chaos.[5]

8.8 Exercises

1. For the wavefunction $\psi = Ae^{-\beta\theta^2}$ determine A by normalizing the wave function on the interval $(-\infty, \infty)$ and determine β by substitution into Eq. (8.8).
2. Calculate $\overline{p_\theta^2}$ for the quantum linearized gravity pendulum in the ground state. Use $p_{\theta_{op}} = -\hbar/i\ (\partial/\partial\theta)$ and $\psi = Ae^{-\beta\theta^2}$, where A and β are as determined in question 1.
3. Using the results of exercise 1 show that for the ground state of the linearized pendulum that $\overline{p} = 0$. Hint: You may need to break up the interval of integration into one integration over negative values of θ, and the other integration over positive values.
4. Consider a very small gravity pendulum of length $L = 10^{-9}$ m. (a) What is the upper limit on its mass in atomic mass units if the maximum uncertainty, in the ground state, in angle is $\Delta\theta = 10°$? (1 amu $= 1.66 \times 10^{-27}$ kg) (b) Find the corresponding value of Δp_θ and use it as an estimate of the pendulum's momentum. With this estimate, calculate the maximum velocity of the bob of the pendulum. (c) Now calculate the temperature that such a velocity would correspond to. Hint: Use the relation $1/2(mv^2) \approx 1/2(kT)$ and the linear pendulum approximation.

[5] Lest there be any confusion, we note that the quantum pendulum as we have described it is not a chaotic system. Yet the effects of quantum uncertainty are universal and therefore our quantum pendulum provides a legitimate illustration of such effects.

5. Suppose that the uncertainty in position for a macroscopic chaotic pendulum is $\Delta\theta \approx 10^{-17}$ radians and that the positive Lyapunov exponent is $\lambda_+ = 0.01\, s^{-1}$. Use the prediction time formula from Chapter 6 to determine how long before the memory of the pendulum bob's original position is lost.
6. For the linearized gravitational quantum pendulum with angular frequency ω (a) calculate the relative probabilities of occupation of the nth state and the ground state, p_n/p_0, using Eq. (8.48) and the end result given by Eq. (8.59). (b) Using the normalization relation, $\sum_{n=0}^{\infty} p_n = 1$, show that $p_n = e^{-n\hbar\omega/kT}/(1 - e^{-\hbar\omega/kT})$.
7. Using the probabilities calculated in exercise 6 and the following expression for the average energy of N particles

$$\overline{E} = N\sum_{n=0}^{\infty} p_n E_n,$$

   (a) calculate the average energy for N linearized pendulums at temperature T,
   (b) What does $\overline{E}$ approach as $T \to \infty$?
   (c) What does $\overline{E}$ approach as $T \to 0$?
8. (a) At room temperature (293 K) what is the average thermal energy of a macroscopic gravity pendulum for which $L = 1$ m and $m = 1$ kg?
   (b) If $\overline{E} \approx \overline{n}\hbar\omega$, where $\overline{n}$ is an "average" quantum number, how big is $\overline{n}$ at room temperature?
   (c) Does it seem likely that thermal energy will cause the macroscopic pendulum to go "over the top"?
9. Following the general argument in the text derive the formula for U_0 in terms of known physical parameters I and ω in the general case where the potential is given by $V = V_0(1 - \cos n\theta)$. Use the linearized pendulum.

9 Superconductivity and the pendulum

There is a quite unexpected connection between the classical pendulum—chaotic or otherwise—and quantum mechanics when it functions on a macroscopic scale, as happens in superconductors. More specifically, the connection arises through something known as the Josephson effect. It turns out that there is an exact correspondence between the dynamics of Josephson devices and the dynamics of the classical pendulum. To uncover this curious relationship, we first review some essential ideas about superconductors and superconducting devices.

9.1 Superconductivity

The Dutch physicist H. Kamerlingh Onnes , noted in an earlier chapter for his Ph.D. dissertation on the Foucault pendulum, made his mark in the world of science for great discoveries in low temperature physics. Three years after receiving his degree in 1879, Onnes ascended to the chair in experimental physics at the University of Leiden in Holland. His research interest was low temperature physics and, like others in England and on the continent, he worked to achieve ever lower temperatures through the liquefaction of gases. Just two years earlier, on December 16, 1877, a French scientist Louis-Paul Cailletet demonstrated the liquefaction of oxygen at 93 K. Cailletet performed the liquefaction publicly at the *École Normale* in order to boost his candidacy as a corresponding member of the *Académie des Sciences.* Two decades later in 1898, the British physicist James Dewar, inventor of the Dewar flask (popularly known as the "thermos"), liquefied hydrogen gas at 23 K. In 1908 Onnes became the first to liquefy helium at the extremely low temperature of 4.216 K above absolute zero. This was the lowest temperature to be achieved through liquefaction; future lower temperatures would require different approaches (Schachtman 1999). (However, we note that some lower temperatures can be readily achieved. Just as the temperature of boiling water falls as the atmospheric pressure is reduced, so also can the boiling point of liquid helium be reduced to less than 1 K by simply evacuating the gaseous helium from the boiling liquid surface. Kamerlingh Onnes achieved a temperature of 0.9 K and, at that moment, justified the notion that Leiden was the coldest spot on earth.)

The liquefaction of helium made possible a host of new and exciting experiments. By 1908, it was well established that the chilling of a metal caused changes in its ability to conduct electricity. Prior to the twentieth century, scientists found that some metals such as copper, iron, and aluminum, that were moderate electrical conductors at room temperature, became, at low temperatures, better conductors than some very good room temperature conductors such as silver, zinc, and gold. Hence there was already a vision of lowering the temperature to improve the conductivity of materials. Higher conductivities meant that, in applications, less electrical energy would be lost as heat and more electrical energy would be transferred to do useful work for humankind. Other types of experiments were also being done at low temperatures. Good insulators were found to become better insulators at low temperatures. Magnetic properties of matter also changed at low temperatures, but the results were varied. Magnetization increased in some cases, whereas iron, for example, was found to be much harder to magnetize after being subjected to cold. Rigidity of some metals increased at low temperatures. The behavior of chemical reactions also changed significantly at low temperatures. Some reactions would not even proceed at low temperatures. Optical properties of materials are also modified at low temperatures. Sometimes color would fade or materials might even change color.

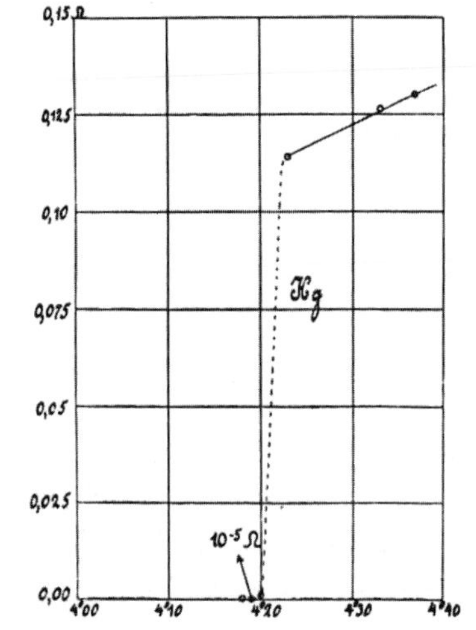

Fig. 9.1
Original data from Kamerlingh Onnes' experiment on the temperature dependence of the resistance of a sample of mercury. From Kamerlingh Onnes (1911*c*, p. 23).

After their triumph in liquefying helium, Kamerlingh Onnes and his assistants at the laboratory in Leiden experimented with the conductivity of metals. Accurate data on metals requires that the metal be free of impurities. Unlike solid metals, liquid mercury could be readily purified. In Kamerlingh Onnes' now-famous 1911 experiment[1], purified liquid mercury was placed in a low temperature bath. Electrodes extended from the liquid through which an electric current could flow. As the temperature dropped the resistance decreased. At around 4.2 K the resistance dropped precipitously and over a small fraction of a degree, as shown in Fig. 9.1, at 4.154 K, the resistance went to zero. The mercury was now *superconducting*. Onnes was a careful experimenter and did not at first identify the state as superconducting. He reported the results in a paper entitled "On the sudden rate at which the resistance of mercury disappears," (Kamerlingh Onnes 1911*a*, *b*, *c*) and only later used the term "superconductivity" (Schachtman 1999). We now are used to the fact that many materials exhibit superconductivity and that there are a host of interesting effects due to, and many applications are made possible by, superconductivity. For his discovery of superconductivity, the liquefaction of helium, and his general contributions to the low temperature physics, Kamerlingh Onnes was awarded the 1913 Nobel prize in physics.

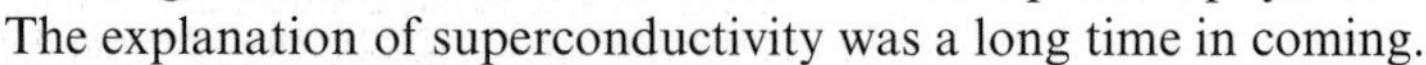

The explanation of superconductivity was a long time in coming.

The origin and nature of superconductivity lie in the realm of quantum mechanics, a theory that was not invented until more than a decade after Onnes' experiments. Even quantum mechanics did not at first hint at such an

[1] The actual experiment was performed by Gilles Holst. See (Matricon and Waysand, 2000) p. 25

astonishing phenomenon. The distinguishing feature of a metal is that for each atom, an outer shell electron, called a *conduction electron*, is able to absorb energy in even tiny amounts, become mobile, and move with other conduction electrons throughout the solid lattice. The ensemble of conduction electrons forms a gas of charged particles. Of course, all of the other electrons belonging to each atom are bound to the nucleus and do not contribute to conductivity. When a voltage is applied to the metal, the conduction electrons drift through the metal in a direction defined by the imposed electric field. The conventional early twentieth century model of resistivity attributed the energy losses to the scattering of charged carriers (electrons) by lattice ions and/or by impurities.[2] Lowering the temperature of a sample would be expected to reduce lattice scattering because lattice vibrations would be diminished as the material was cooled. However, it was also thought that the scattering contribution from impurities and lattice defects would not be temperature dependent. Therefore, it was anticipated that the resistivity would flatten out at very low temperatures, approaching a finite limit known as the *residual resistivity*. This behavior was encapsulated in what is called "Matthiessen's rule," (Matthiessen 1862) and it was reflected in experimental data of many materials. But it certainly was not correct for superconductors whose resistance was seen to disappear absolutely.

In 1935 the theoretical physicist Fritz London (London and London 1935; London 1961) suggested that superconductivity is a state of long range order for the conduction electrons—that is, superconductivity is quantum mechanics acting, quite uncharacteristically, on a *macroscopic* scale. This was a revolutionary notion. The idea was that at a low enough temperature, somehow the conduction electrons are able to condense into a new state in which they act entirely in concert rather than as individual particles. In this correlated ensemble, individual electrons could not be scattered—it is all or nothing and below the transition temperature there would not be enough thermal energy to scatter the entire group. Ergo: no resistance.

The idea of *long range* correlations and therefore a sort of macroscopic quantum mechanical wave function was completely novel in 1935 and was a radical departure from the usual expectation that quantum mechanics would be limited to the atomic scale.

Two decades after London's publication, the team of Bardeen, Cooper, and Schrieffer (BCS) (Bardeen et al. 1957) developed a complete, microscopic quantum theory of superconductivity that fully explained the nature of the phenomenon. For this work Bardeen shared the 1972 Nobel prize (his second) with Cooper and Schreifer. The BCS theory is very complex, but a central role is played by what are called "Cooper pairs."

Normally the conduction electrons repel each other because of their like negative electrical charges. However, a conduction electron also will locally distort the fixed lattice of positive ions. The distortion takes the form of a very slight squeezing-in of ions around the conduction electron. To another electron wandering in this neighborhood, the compression of the lattice appears as a region of slightly enhanced positive charge which will

[2] See Kittel (1996, p. 161).

preferentially attract the second electron. In other words, one electron can generate a small attractive force on a second electron, mediated by the lattice distortion. But this attractive force is very weak and is easily disrupted by thermal agitation. Therefore it is only at low temperatures that such an attractive force can possibly overcome thermal motion of the lattice. Similarly, the paradox that good conductors tend not to easily become superconductors is also explained. The conduction electrons in good conductors are *not* as closely coupled to the lattice (less resistance) and therefore their ability to cause distortions and thereby create the required attractive force is smaller than in poorer conductors.

When conditions are such that there is a net attraction, then Cooper pairs are formed. In a Cooper pair, the electrons have equal and opposite momentum and spin. It is remarkable that individual members of a pair can be thousands of atoms apart, about one millionth of a meter. Because the *pair* is an entity with zero spin, a so-called Boson, the Pauli exclusion principle no longer forbids the occupation of a common ground state. The condensed sea of Cooper pairs defining the superconducting phase becomes perfectly correlated via the pairing and no longer scatters in the conventional resistive manner. Thus there is a common multiparticle wave function: in the superconducting state, quantum mechanics goes macroscopic.

As described in Chapter 8 the behavior of a particle is fully described by a complex quantity ψ called the wavefunction and as enunciated by Max Born, the real number $\psi^*\psi$ is a probability density. The unique long range order that lies at the heart of superconductivity suggests that in such materials the coherent state of all the Cooper pairs can be represented by a single macroscopic wavefunction of the form $\psi = \sqrt{\rho}e^{i\varphi}$, a complex quantity whose phase is φ. The real quantity $\psi^*\psi$ is now just the particle density ρ. This phenomenological picture can be applied to some simple configurations and, as we shall see, it predicts some startling results.

9.2 The flux quantum

From a wavefunction ψ, the corresponding electric current density $\vec{J}$ associated with a flow of charged particles may be written

$$\vec{J} = \frac{ie\hbar}{2m}(\psi\nabla\psi^* - \psi^*\nabla\psi), \tag{9.1}$$

where m and e are the particle mass and charge, respectively. This is a standard result in quantum theory (Liboff 1980). It turns out that when a magnetic vector potential $\vec{A} = \nabla \times \vec{B}$ is present, the gradient operator ∇ must be replaced by its more general form $\nabla - i\frac{e}{\hbar}\vec{A}$.[3] Then, with our macroscopic wavefunction $\psi = \sqrt{\rho}e^{i\varphi}$,

$$\vec{J} = \left(\frac{e\hbar\rho}{m}\right)\nabla\varphi - \left(\frac{e^2\rho}{m}\right)\vec{A}. \tag{9.2}$$

[3] see Section 21.1 in Feyman's Lectures (Feynman 1965). This is equivalent to generalizing the quantum mechanical momentum operator from $p_{\rm op} = -i\hbar\nabla$ to $p_{\rm op} = -i\hbar\nabla - e\vec{A}$ for situations where there is a magnetic vector potential.

Now consider a superconducting ring, as illustrated in Fig. 9.2. No current will flow within the interior of the superconductor, so, for example, along the dashed path

$$\nabla\varphi = \frac{e}{\hbar}\vec{A}. \tag{9.3}$$

Therefore, considering a line integral of $\nabla\varphi$ taken completely around the dashed circuit,

$$\oint \nabla\varphi \cdot d\vec{s} = \frac{e}{\hbar}\oint \vec{A} \cdot d\vec{s} = \frac{e}{\hbar}\Phi, \tag{9.4}$$

using Stokes theorem. Φ is the total magnetic flux contained in the non-superconducting center hole. The integral $\oint \nabla\varphi \cdot d\vec{s}$ around the ring gives the change in phase φ of the wave function around the loop. However, the single valuedness of the wavefunction ψ constrains the net change in the phase around a complete circuit to equal a multiple of 2π. Consequently,

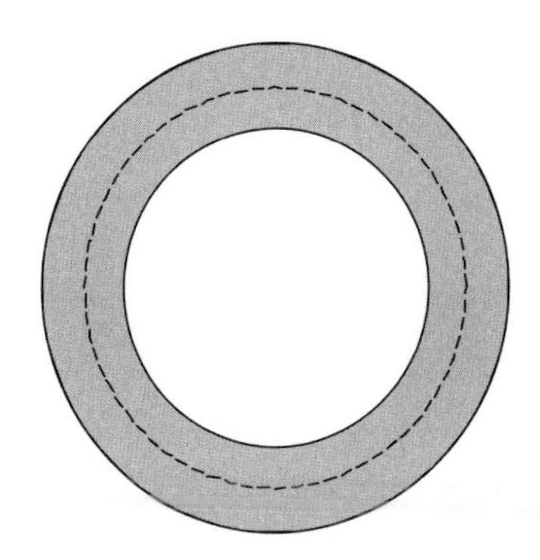

Fig. 9.2
A superconducting ring. The dashed path lies entirely within the superconducting material.

$$\Phi = n\left(\frac{h}{e}\right) \qquad n = 1, 2, \ldots \tag{9.5}$$

This equation predicts that the hole in the superconductor can contain magnetic flux only in integer multiples of the amount h/e. When this hypothesis was tested experimentally in the early 1960s, quantization was indeed confirmed, but the relationship was found instead to be

$$\Phi = n\left(\frac{h}{2e}\right) = n\Phi_0. \tag{9.6}$$

The extra factor of 2 was caused by the fact that the "particles" were not actually single electrons, but rather were the Cooper pairs of the microscopic BCS theory. The *flux quantum* $\Phi_0 = h/2e$ depends only on the ratio of a pair of fundamental constants, the charge and mass of the electron, and has the value $2.06783461 \times 10^{-15}$ Wb. This is a very small quantity. For comparison, the amount of flux from the earth's relatively weak magnetic field over an area of just 1 mm^2 is about 0.5×10^{-10} Wb, approximately 10,000 flux quanta.

9.3 Tunneling

In the previous chapter, we saw that the wavefunction of the quantum oscillator could have a nonzero value in regions where the potential energy was too high to allow for classical motion. That is, there was a finite probability that the oscillator would be found in a state of motion that was forbidden by classical physics. This effect results in some interesting behavior. Let us consider the situation illustrated in Fig. 9.3 where a particle or beam of particles is incident on a potential barrier that is higher than the particle energy. According to classical physics such a particle cannot reach the other side of the barrier. But according to quantum physics, if the barrier is not too high or thick, then the wave function can leak through, allowing a finite amplitude traveling wave solution to exist

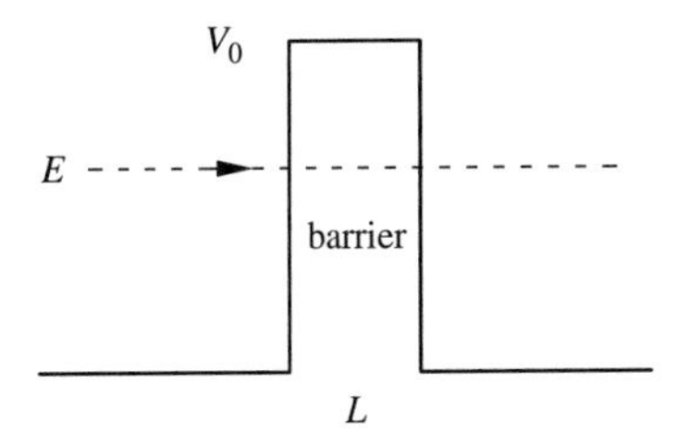

Fig. 9.3
A particle of energy E incident on a potential barrier of height $V_0 > E$. Tunneling is the quantum mechanical process by which the particle may actually transfer to the other side, having gone through the barrier rather than over it.

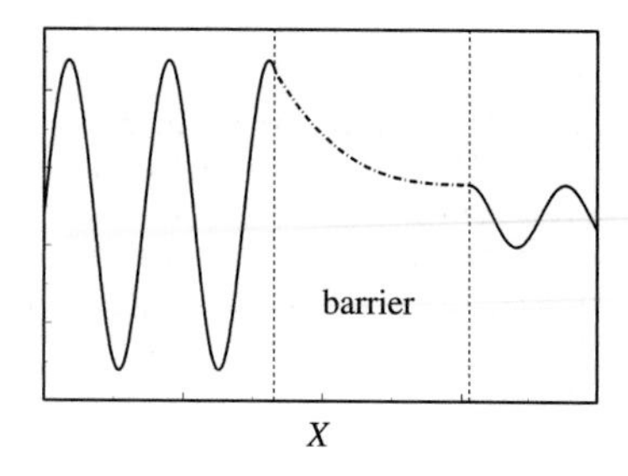

Fig. 9.4
Qualitative appearance of the wave function. The solutions must match smoothly at the edges of the tunneling barrier.

on the other side, as suggested in Fig. 9.4. This surprising phenomenon is known as *tunneling* and is the physical basis of, for example, alpha particle decay from a nucleus, and electron leakage from a metal surface in the presence of an electric field. The probability that a particle might tunnel to the other side of the barrier is derived in most standard texts on quantum mechanics,[4] the result is

$$T = \frac{1}{1 + (V_0^2/4E(V_0 - E))\sinh^2(\alpha L)} \tag{9.7}$$

with

$$\alpha = \sqrt{\frac{2m(V_0 - E)}{\hbar^2}}.$$

Classically, of course, a particle with energy less than the height of the barrier would simply be reflected and the transmission probability would equal zero. At a macroscopic scale, this remains essentially true, but at the atomic scale the situation can change. For example, consider electrons of energy 7 eV directed at a barrier of height 10 eV. If the barrier thickness is 50 Å, then $T = 0.96 \times 10^{-38}$, whereas with $L = 10$ Å, $T = 0.66 \times 10^{-7}$. These numbers emphasize the extreme sensitivity of tunneling probabilities to barrier thickness. They also reveal that appreciable tunneling currents can occur, with sufficiently thin barriers, as occur, for example, with thin oxides formed at metal or semiconductor surfaces.

There would also be a bit of a tunneling effect in the examples of hindered rotations that we considered earlier in Chapter 8.[5]

9.4 The Josephson effect

In 1961, Brian Josephson, then a graduate student at Cambridge University, concluded that Cooper pairs might be able to tunnel from one superconductor to another, provided the barrier was thin enough. Figure 9.5 shows a typical configuration for a Josephson junction made from two superconducting metal films separated by a thin oxide layer.

Based on theoretical arguments, Josephson predicted that, by a special manifestation of quantum mechanical tunneling, Cooper pairs could transfer across a sufficiently weak barrier separating a pair of superconductors, thus creating a *supercurrent*. He showed that his supercurrent was governed by two unusual relationships involving the *difference* in the phases of the wavefunctions of the two bulk regions, namely

$$I = I_c \sin \phi \tag{9.8}$$

[4] See, for example, (Serway et al. 1989).

[5] Most of the hindered rotation to new positions of equilbrium was due to thermal agitation of the quantum "pendulum" to energy states above the potential well. But, although we did not take this into account in our earlier discussion, there is a bit of the hindered rotation due to tunneling of the pendulum from energy states that are near, but not quite at, the top of the potential barrier.

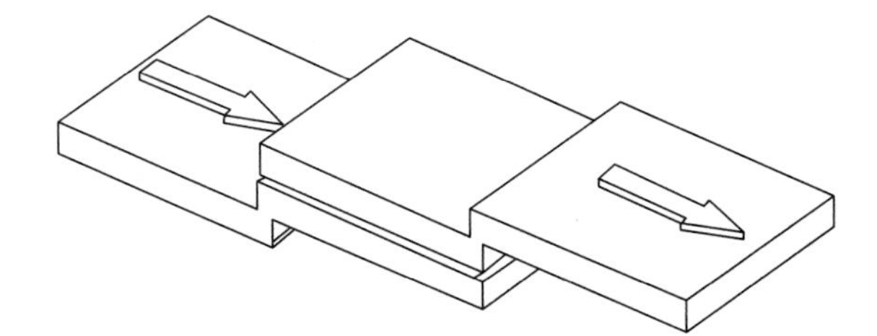

Fig. 9.5
A Josephson junction consisting of upper and lower superconducting films separated by a very thin insulator. As indicated by the arrows, current flows into the junction area from the lower electrode and exits through the upper electrode.

and

$$\frac{d\phi}{dt} = \frac{2e}{\hbar} V. \qquad (9.9)$$

The constant I_c was expected to depend on the size of the junction, the types of superconductors, and temperature. These expressions have some rather startling implications.

First let us consider the arrangement illustrated in the upper portion of Fig. 9.6. The box is imagined to contain a Josephson junction (sufficiently cooled). An external current source is set to feed bias current into the box. Classically, of course, the insulating barrier would break the circuit and no current could actually pass through the box. However, Josephson's first equation says that a supercurrent can indeed flow through the junction from one superconductor to the other, via the thin barrier. More surprisingly, this supercurrent should exist without any associated voltage appearing across the junction, so the voltmeter in the figure would read zero. To the observer, it is as if the box contained simply a superconducting wire. The key to the process was the automatic self adjustment in the system of the barrier phase difference ϕ so as to satisfy Eq. (9.8). Of course, $\sin\phi$ can never exceed unity, so the supercurrent has a maximum value of I_c.

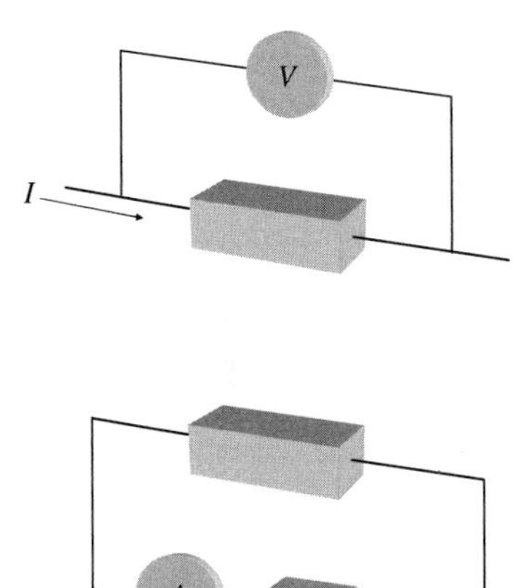

Fig. 9.6
Depiction of two arrangements for a box containing a Josephson junction. The element V denotes a voltmeter and A is an ammeter. The upper illustration shows a fixed current fed through the box; the lower arrangement has a fixed voltage applied across the terminals of the box.

Next consider the second arrangement depicted in Fig. 9.6. In this case, a fixed voltage is applied across the terminals of the box. From Eq. (9.9), it is clear that the phase must have a time dependence of the form

$$\phi = \left(\frac{2e}{\hbar} V\right) t. \qquad (9.10)$$

Now the ammeter A will record a supercurrent $I_c \sin\phi$ oscillating sinusoidally at a frequency

$$f = \frac{2e}{h} V. \qquad (9.11)$$

In other words, a steady bias voltage has led to a time-varying current. Note the significant fact that the Josephson frequency depends only on the value of the DC applied voltage together with the ratio of two fundamental constants of nature: the charge of the electron and Planck's constant. This relationship has the precise numerical value $4.8359767 \times 10^{14}\,\text{Hz}\,\text{V}^{-1}$.

These two results are known as the DC Josephson Effect (small constant current carried without the appearance of any junction voltage), and the

AC Josephson Effect (alternating current when a constant voltage is applied across the junction).

As will be apparent later, in the event that a total bias current I greater than I_c is forced through a junction, there must be an additional normal current component (i.e. a current associated with conventional dissipation); that is, $I = I_c \sin \phi + I_{\text{normal}}$. In such a case, a voltage V will appear across the junction. This voltage is associated with a time dependence in the junction phase ϕ, according to the second of Josephson's fundamental equations.

The English physicist and Josephson's advisor, Brian Pippard (whom we met earlier in connection with the Foucault pendulum) suggested that the effect was likely unobservable because the probability of pair correlation tunneling was quite small. While the probability of a single electron tunneling was quite low, say 10^{-10}, the probability of tunneling by a pair would be extremely small since, according to Pippard, it should be proportional to the *square* of that small number; that is 10^{-20}. Fortunately for Josephson, it later turned out that this conjecture was incorrect (McDonald 2001).

Richard Feynman, winner of the 1965 Nobel prize for his work in quantum electrodynamics, developed a simple derivation of the equations for the two Josephson effects. His derivation connects the quantum mechanical picture with the macroscopic Josephson effects (Feynman 1965) (vol III, chapter 21, pp. 14–16). Feynman supposed that for the superconducting material on either side of the insulating barrier (Fig. 9.7) there was a single wave function ψ_1 on the left, say, and another wavefunction ψ_2 on the right, each of which characterizes all the electrons on each side of the barrier. Furthermore, Feynman represented the coupling across the insulating gap or junction by an energy term K such that the time dependent Schrödinger equation for each side is given by

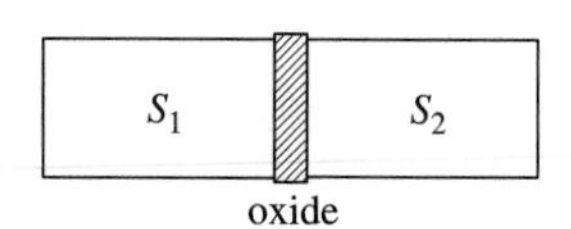

Fig. 9.7
Depiction of two superconductors separated by a thin insulating barrier.

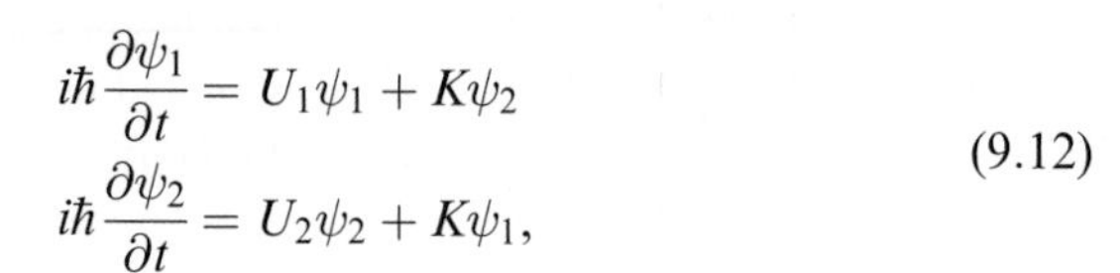

$$i\hbar \frac{\partial \psi_1}{\partial t} = U_1 \psi_1 + K \psi_2$$
$$i\hbar \frac{\partial \psi_2}{\partial t} = U_2 \psi_2 + K \psi_1, \qquad (9.12)$$

where $U_{1,2}$ are the respective ground state energies of each side of the gap. If both sides were identical then U_1 and U_2 would be identical. We are mostly interested in differences in energy that develop when a voltage is applied. Therefore, let us connect a battery, with voltage V across the whole system. Then, for a charge $q = 2e$ (of a Cooper pair) the energy difference is now $U_1 - U_2 = qV$. We can take the zero of energy as the midway energy point between the $U_{1,2}$ such that the previous set of Schrodinger equations can be rewritten as

$$i\hbar \frac{\partial \psi_1}{\partial t} = \frac{qV}{2} \psi_1 + K \psi_2$$
$$i\hbar \frac{\partial \psi_2}{\partial t} = \frac{-qV}{2} \psi_2 + K \psi_1. \qquad (9.13)$$

We learned earlier that square of the magnitude of the wavefunction, $|\psi|^2$ is the probability density for the electron, ρ, and is proportional to the probability of finding an electron in a particular region of space. Thus the real part of the wave function $|\psi|$ is the square root of the density function, $\rho^{1/2}$. The wave function can then be written in terms of the density function and the phase of the wavefunction in the form

$$\psi = \rho^{1/2} e^{i\varphi}, \tag{9.14}$$

where φ is the phase that we had introduced earlier. Putting this relation into Eqs. (9.13) and remembering that the phase shift across the insulating barrier is $\phi = \varphi_2 - \varphi_1$, we find the relations

$$\begin{aligned}
\frac{d\rho_1}{dt} &= +\frac{2K}{\hbar}\sqrt{\rho_1\rho_2}\sin\phi \\
\frac{d\rho_2}{dt} &= -\frac{2K}{\hbar}\sqrt{\rho_1\rho_2}\sin\phi \\
\frac{d\varphi_1}{dt} &= \frac{K}{\hbar}\sqrt{\frac{\rho_1}{\rho_2}}\cos\phi - \frac{qV}{2\hbar} \\
\frac{d\varphi_2}{dt} &= \frac{K}{\hbar}\sqrt{\frac{\rho_1}{\rho_2}}\cos\phi + \frac{qV}{2\hbar}
\end{aligned} \tag{9.15}$$

by equating real and imaginary parts. It looks as if the rates of change in pair density (first two equations) on either side of the junction are equal in magnitude and opposite in sign; that is, as pair density increases on one side of the junction, it must decrease on the other side. But this cannot be the case since a charge imbalance would result. The reality is that the rest of the circuit—with voltage V—that is connected to the Josephson junction replaces the electrons to maintain charge balance. Thus there is a current flow of the general form $I = I_c \sin\phi$—the *dc Josephson effect*—as was stated earlier in Eq. (9.8). What are the consequences of the third and fourth equations? The parameter K and therefore I_{max} depend on the properties of the gap. By subtracting the third equation from the fourth equation we obtain the *ac Josephson effect*:

$$\frac{d\phi}{dt} = \frac{qV}{\hbar} = \frac{2eV}{\hbar} \tag{9.16}$$

as was stated in Eq. (9.9).

For an historical perspective, let us return to developments of the early 1960s.

The elements of the story as recounted in McDonald (2001) are as follows. Prior to any experimental confirmation of the Josephson effects, there was a now famous debate between Josephson, the graduate student, and John Bardeen. Bardeen was the leading theorist in solid state physics who had stated, in a note added in proof to a paper published in Physical Review Letters (Bardeen 1962), that "pairing does not extend into the barrier, so that there can be no such superfluid flow." He therefore dismissed Josephson's prediction as completely erroneous. However,

Bardeen's view was not universally held and thus the stage was set for a showdown. The debate took place in September, 1962 at Queen Mary College in London. At this point Josephson's paper had already appeared (Josephson 1962). The occasion was the Eighth International Conference on Low Temperature Physics. Bardeen delivered the opening address known as the Fritz London Memorial lecture. Both men gave talks during the session on tunneling. Josephson spoke first followed by Bardeen. Josephson interrupted Bardeen several times, defending his new ideas. The exchange was quite civil and ended somewhat inconclusively. Within a few months the issue was settled by experimental observation; the predicted Josephson effects were confirmed by Rowell and Anderson at Bell labs in New Jersey, and other researchers around the world. In 1973, Josephson shared the Nobel prize in physics for his predictions.[6]

A typical experimental arrangement for the Josephson junction is shown in Fig. 9.5. Two small crossed strips of superconducting material such as tin are separated by a thin layer of oxide, a few angstrom units thick, deposited on the lower piece of tin. The upper piece of tin is placed on top of the oxide layer.

Josephson junctions have appeared in a variety of practical applications and have also been used to further knowledge of fundamental physics. Applications include a superconducting device formed from two Josephson junctions. This superconducting quantum interference device or SQUID is used in the sensitive measurement of small magnetic fields. In this application the junctions measure very small fields and electrical currents in the heart and the brain. These measurements help determine the origins of, for example, epilepsy. Josephson devices have been proposed for use as high speed switches in computer applications. Finally, the frequency measurements of the ac Josephson effect have allowed new precision in the determination of the ratio e/h, two important fundamental physical constants (Clarke 1970).

9.5 Josephson junctions and pendulums

These accounts of the development of the concepts of quantization of magnetic flux and of the Josephson junction, while interesting in their own right, have not yet revealed any connection with the pendulum. The relationships between pendulums and superconducting quantum devices will now be discussed.

9.5.1 Single Junction: RSJC Model

An ideal Josephson junction is characterized by the expression $I_J = I_c \sin \phi$. However, in a real device, the finite size of the superconducting films and

[6] A fascinating review of the early days of this great achievement is provided by A. B. Pippard in chapter 1, *The Historical Context of the Josephson Discovery*, of (Schwartz and Foner 1977). Josephson's Nobel Prize address was published as "The Discovery of Tunneling Supercurrents" in *Science* **184**, 527–530 (May 1974).

the unavoidable leakage resistance of the oxide barrier imply that a more complete model should include both shunt capacitance and resistance.[7] This combination is illustrated in the equivalent circuit shown in Fig. 9.8. The total current flowing through this device consists of, first, the supercurrent of paired electrons which is proportional to sin ϕ, second, a dissipative normal current through the resistance of the junction, and third, a displacement current through the capacitor. Taking these in turn, and using Eq. (9.16)

$$I_J = I_c \sin\phi, \tag{9.17}$$

and

$$I_R = \frac{V}{R} = \frac{\hbar}{2eR}\frac{d\phi}{dt}, \tag{9.18}$$

where Eq. (9.9) has been used at the second step. Finally

$$I_C = C\frac{dV}{dt} = \frac{\hbar C}{2e}\frac{d^2\phi}{dt^2}. \tag{9.19}$$

where again Eq. (9.9) has been applied to the last step.

If a Josephson junction is biased from an external current source, as in Fig. 9.9

then

$$\frac{\hbar C}{2e}\frac{d^2\phi}{dt^2} + \frac{\hbar}{2eR}\frac{d\phi}{dt} + I_c \sin\phi = I_{dc}, \tag{9.20}$$

where I_{dc} is the bias current.

Now we recall the equation governing the dynamics of a torque driven pendulum.

$$I\frac{d^2\theta}{dt^2} + b\frac{d\theta}{dt} + mgr\sin\theta = \Gamma, \tag{9.21}$$

where θ is the angular coordinate of the pendulum bob whose mass is m, r is its distance from the axis of rotation, I is the total moment of inertia of all rotating components of the pendulum, and b is a parameter specifying the strength of a frictional damping that depends on the angular velocity of the pendulum. The term mgr is just the critical torque necessary to raise the pendulum through an angle of 90°, and Γ is the net applied torque. The natural frequency of undamped small amplitude oscillations is

$$\omega_0 = \sqrt{\frac{mgr}{I}}. \tag{9.22}$$

A term by term comparison of the electrical and mechanical equations provides the results shown in Table 9.1.

For a mechanical pendulum ω_0 is typically of the order of a few radians per second, but for the Josephson junction it is of the order of 10^{11} rad/s. In effect the pendulum is a slow motion mechanical analog of the Josephson Junction.

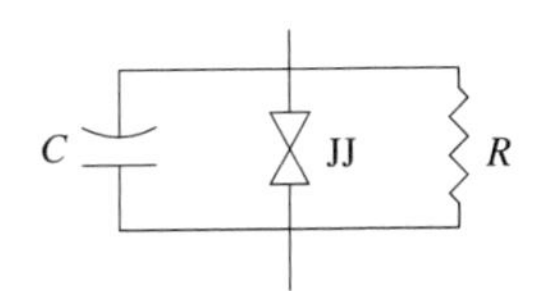

Fig. 9.8
RSJC equivalent circuit of a Josephson junction; the components are a shunt capacitance C, leakage resistance R, and the phase dependent element JJ represented by the bowtie symbol.

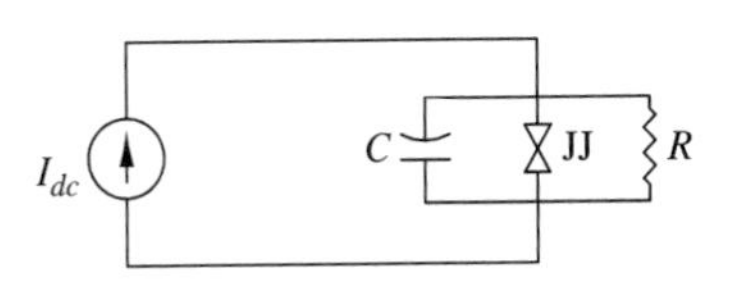

Fig. 9.9
Equivalent circuit of a Josephson junction with an applied dc bias current.

[7] RSJC is the acronym for Resistively Shunted Junction with Capacitance.

Table 9.1 Comparison of pendulum and Josephson junction quantities

	Pendulum	Josephson junction
Coordinate	Pendulum angle θ	Junction phase ϕ
Natural frequency (ω_0)	$\sqrt{mgr/I}$	$\sqrt{2eI_C/\hbar C}$
Damping	b/I	$1/RC$
Forcing	Applied torque Γ	Bias current I_{dc}
Critical forcing	$\Gamma_c = mgr$	I_c
Time dependence	$\omega = d\theta/dt$	$V = \frac{\hbar}{2e}\frac{d\phi}{dt}$

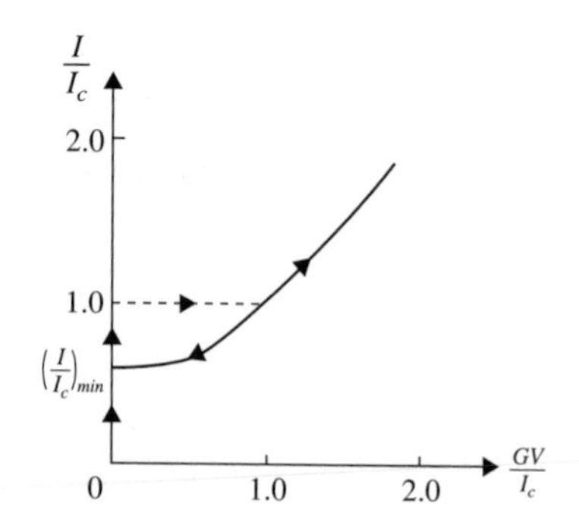

Fig. 9.10
Current–voltage characteristic for a Josephson junction with only dc bias current (from Van Duzer and Turner (1981, p. 173), ©1999. Reprinted by permission of Pearson Educations, Inc., Upper Saddle River, NJ.)

Consider the situation when a Josephson junction is biased with a dc source that is *slowly* increased, starting from a value of zero. Considering Fig. 9.9, it is apparent that as long as the external bias is smaller than the junction critical value I_c, the phase ϕ can adjust to a value such that all of the bias will be carried entirely by the Josephson element in the form of a dc supercurrent and the voltage across the junction will be zero. The shunt resistance and capacitance are, effectively, bypassed. When eventually the bias just reaches I_c, then $\sin \phi$ reaches its maximum value of unity. Any slight increase of the bias beyond this point must cause the junction to begin oscillating. The resulting dynamic state activates the other two branches in the equivalent circuit (shunt capacitance and resistance) and a finite *time-averaged* voltage will abruptly appear across the device. This sequence of events is observed in the current–voltage characteristics of Josephson junctions, as illustrated in Fig. 9.10. Further increases in the dc bias take the device up the indicated characteristic. When the process is slowly reversed, the $I-V$ characteristic is retraced until a lower bias threshold is reached, at this point, the junction will abruptly snap back to the zero voltage state. The difference between the upper switching point at $I = I_c$ and the lower snap back at $I = I_{\min}$ is known as *hysteresis*.

Figure 9.11 is a photograph of an experimental apparatus for simulating a dc biased Josephson junction by means of a driven pendulum.

Fig. 9.11
Apparatus used by Sullivan and Zimmerman to model a dc biased Josephson junction as a driven pendulum analog. A motor spins two small magnets in proximity to the large aluminum disc, inducing a constant torque. (Reprinted with permission from Sullivan and Zimmerman (1971, p. 1509). ©1971, American Association of Physics Teachers.)

The aluminum disc projects between the pole pieces of a large horseshoe magnet. As the pendulum moves, so does the disc, and magnetically induced eddy currents produce damping. The small electrical motor is used to spin small permanent magnets near the edge of the aluminum disc, thus inducing an approximately constant torque on the pendulum. A simplified drawing of this same apparatus is shown in Fig. 9.12.

The analog of a Josephson junction current–voltage characteristic is, for the driven pendulum, a plot of time-averaged angular velocity versus applied dc torque. Results obtained from this apparatus are shown in Fig. 9.13.

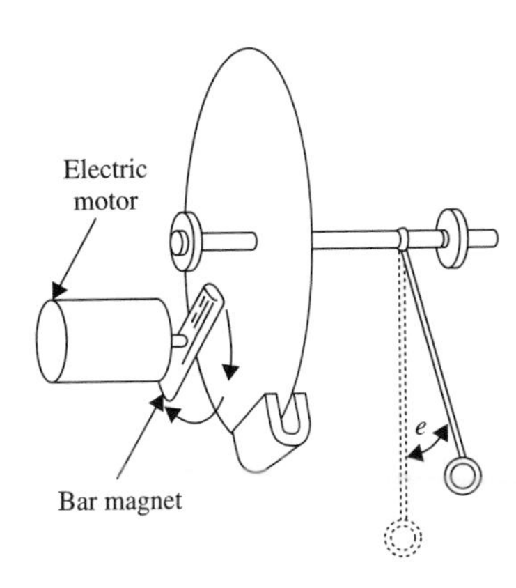

Fig. 9.12
Conceptual drawing of the apparatus shown in the preceding photograph. The viewing direction is reversed from that in the photo. (from Barone and Paterno (1982, p. 126), ©1982. This material is used by permission of John Wiley & Sons, Inc.)

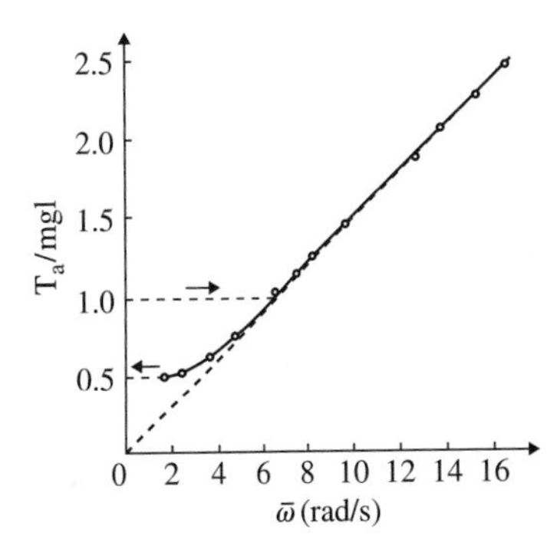

Fig. 9.13
Plot of time-averaged angular velocity as a function of applied dc torque for a damped pendulum (from Van Duzer and Turner (1981, p. 179), ©1999. Reprinted by permission of Pearson Education, Inc. Upper Saddle River, NJ).

The behavior can be understood as follows. For steady torques smaller than the critical value mgr, the pendulum angle will adjust to a value that satisfies $mgr \sin \theta = \Gamma$. For a torque just slightly larger than the critical value mgr, the angle θ will exceed 90° and the pendulum will begin to spin in a modulated fast–slow–fast fashion. The average angular velocity will be some finite number. Increasing the applied dc torque will increase the spin and its time average. Reversing the process by slowly decreasing the torque will retrace the characteristic, but the system will not switch back to a static state when the critical value mgr is again reached. This is because the spinning pendulum has momentum which causes the spinning state to persist until some lower torque value is reached, as indicated in the plot. As before, the system exhibits hysteresis.

It is sometimes useful to rewrite Eq. (9.21) in the form

$$I\frac{d^2\theta}{dt^2} + b\frac{d\theta}{dt} = -mgr\frac{d}{d\theta}\left[-\left(\frac{\Gamma}{mgr}\right)\theta - \cos\theta\right]. \tag{9.23}$$

Recalling that the negative gradient of a potential is the resulting force, we are led to identify the associated dimensionless part of the potential here as

$$U(\theta) = -\left[\left(\frac{\Gamma}{\Gamma_c}\right)\theta + \cos\theta\right]. \tag{9.24}$$

This function consists of a linear term with negative slope, together with a cosine modulation—a so-called "washboard" potential as shown in Fig. 9.14. With zero applied torque, the potential is a horizontal cosine shape and the system acts like a "particle" trapped in the minimum at $\theta = 0$. As the applied torque is gradually turned up, the potential tips and the minimum of the well moves slightly towards small values of θ. Eventually, when the critical torque is reached, the well becomes flattened at an angle of 90° and any infinitesimal perturbation will cause the pendulum to start spinning. Because of the additional viscous damping, the downward motion of the "particle" eventually reaches a terminal condition in its bouncing motion, so a definite value for the time average of the angular velocity of the pendulum is achieved. Hysteresis is seen as the tendency of the downward bouncing to continue even when the applied torque has been slowly reversed to the critical value. Indeed, the applied torque must then

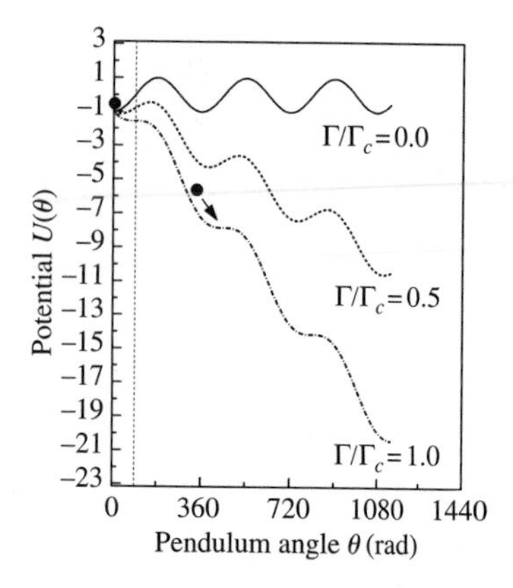

Fig. 9.14
Plot of the washboard potential for three different applied torques. As the constant torque increases in strength, the potential progressivly tips until at the critical torque, the potential well at the origin flattens and the "particle" bounces down the surface.

be further reduced to a value such that wells are created that are deep enough to finally stop the bouncing "particle".

The similarity between Fig. 9.13 and Fig. 9.10 confirms the close correspondence between these two seemingly very different systems, one classically mechanical and the other quantum mechanical.

9.5.2 Single junction in a superconducting loop

It was demonstrated earlier that a superconducting ring with a hole can contain only integer multiples of the flux quantum. What happens if such a ring is broken by a Josephson junction? This situation is depicted in Fig. 9.15. Again, the dashed line traces a path that lies entirely within the superconductor. When Eq. (9.3) is corrected for the fact that the charge is really $2e$ to account for the fact that the carriers are Cooper pairs, then within the interior of a bulk superconductor

$$\nabla\varphi = \left(\frac{2\pi}{\Phi_0}\right)\vec{A}. \tag{9.25}$$

The total phase change going completely around the dashed path in Fig. 9.15, in a counterclockwise sense, will then be governed by

$$\oint \nabla\varphi \cdot d\vec{s} = -2\pi\frac{\Phi_\downarrow}{\Phi_0}, \tag{9.26}$$

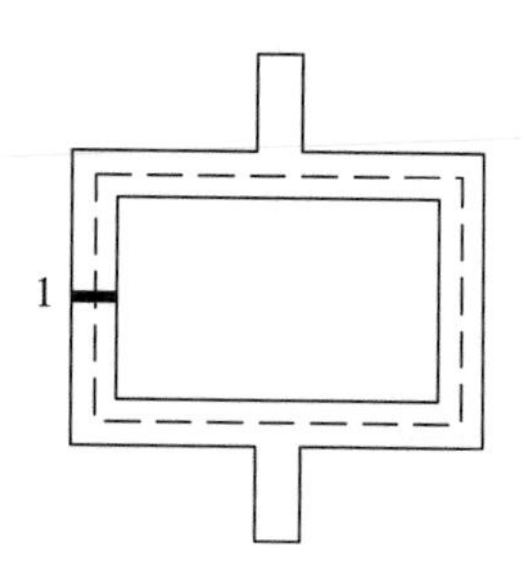

Fig. 9.15
Depiction of a single Josephson junction inserted into a superconducting loop.

where $\Phi_\downarrow$ indicates that we have adopted the convention that positive flux points down into the loop. But this does not include the phase jump ϕ across the Josephson junction. Adding this extra phase contribution and recalling that the wavefunction must be single valued, we see that

$$\phi - 2\pi\frac{\Phi_\downarrow}{\Phi_0} = 2n\pi. \tag{9.27}$$

Clearly the loop does not obey perfect flux quantization as before, because the junction "weakens" the ring.

The equivalent circuit of this arrangement is shown in Fig. 9.16. Suppose a bias current I_e is injected into the ring at the upper electrode and extracted from the bottom contact. To make the situation general, let us assume that the contacts are positioned asymmetrically such that the loop inductance L is divided into unequal portions between the left and right halves of the ring. If the total current flowing through the junction is I, then the current through the right hand branch must be $I_e - I$. The induced downward flux in the loop is $\Phi_\downarrow = (I_e - I)\alpha L - I(1-\alpha)L$, or

$$\Phi_\downarrow = (\alpha I_e - I)L. \tag{9.28}$$

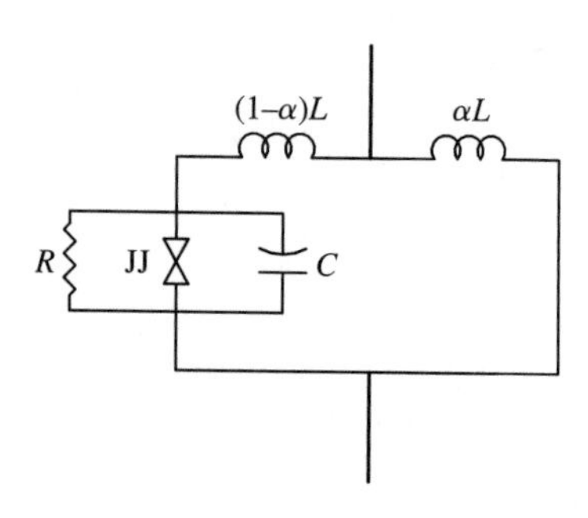

Fig. 9.16
RSJC equivalent circuit of a single Josephson junction embedded in a superconducting ring of inductance L.

Combining Eqs. (9.28) and (9.20), we obtain the dynamical equation governing this system,

$$\Phi_\downarrow = \left[\alpha I_e - \frac{\hbar C}{2e}\frac{d^2\phi}{dt^2} - \frac{\hbar}{2eR}\frac{d\phi}{dt} - I_c \sin\phi\right]L. \tag{9.29}$$

However, Eq. (9.27) relates junction phase and loop flux, so

$$\frac{\hbar C}{2e}\frac{d^2\phi}{dt^2}+\frac{\hbar}{2eR}\frac{d\phi}{dt}+I_c\sin\phi=\alpha I_e-\frac{\Phi_0}{2\pi L}\phi. \tag{9.30}$$

This equation of motion for junction phase has its own pendulum analogue—to be described below. But first we demonstrate that slightly different physics leads to an equation for the magnetic flux through the loop that is identical in form to Eq. (9.30).

If, instead of an external bias current, an external bias magnetic field is applied to the loop, then the total flux in the loop is composed of the applied value Φ_e (positive downward) plus a contribution from the circulating current flowing through the loop self inductance. Hence $\Phi=\Phi_e-Li$. In this case, the meaningful variable for an equation of motion is the net loop flux;[8] thus

$$LC\frac{d^2\Phi^*}{dt^2}+\frac{L}{R}\frac{d\Phi^*}{dt}+\frac{LI_c}{\Phi_0}\sin(2\pi\Phi^*)=\Phi_e^*-\Phi^*, \tag{9.31}$$

where Φ^* signifies the normalized quantity $\Phi^*=\Phi_\downarrow/\Phi_0$.

Now consider the pendulum illustrated in Fig. 9.17. For this mechanical system, the equation of motion is nearly the same as Eq. (9.21), but an additional restoring torque is generated by the differential twist of the torsion spring.

$$I\frac{d^2\theta}{dt^2}+b\frac{d\theta}{dt}+mgr\sin\theta=K(\theta_x-\theta) \tag{9.32}$$

which quite obviously is a direct analog of a single Josephson junction in a superconducting loop (Eq. (9.30)) with the pendulum angle taking the role of junction phase. The torsion spring contributes an equivalent of inductance to the mechanical system. Similarly, Eq. (9.31), that describes the loop flux, is also isomorphic with Eq. (9.32).

Here, the torsion spring is given an initial offset θ_x, so the torque due to its net twist will be proportional to $(\theta_x-\theta)$. Simply altering θ_x is equivalent to changing the bias current I_e or applied flux Φ_e in the case of the superconducting loop.

In correspondence with our earlier discussion of the driven simple pendulum, we note that this new equation of motion implies a potential function of the form

$$U(\theta)=-\left[\frac{K}{\Gamma_c}\left(\theta_x\theta-\frac{\theta^2}{2}\right)+\cos\theta\right] \tag{9.33}$$

or

$$U(\theta)=\frac{1}{2}\frac{K}{\Gamma_c}\left[(\theta-\theta_x)^2-\theta_x^2\right]-\cos\theta. \tag{9.34}$$

This function is plotted in Fig. 9.18. Imagine that the pendulum is initially hanging straight down, that the torsion spring is untwisted with

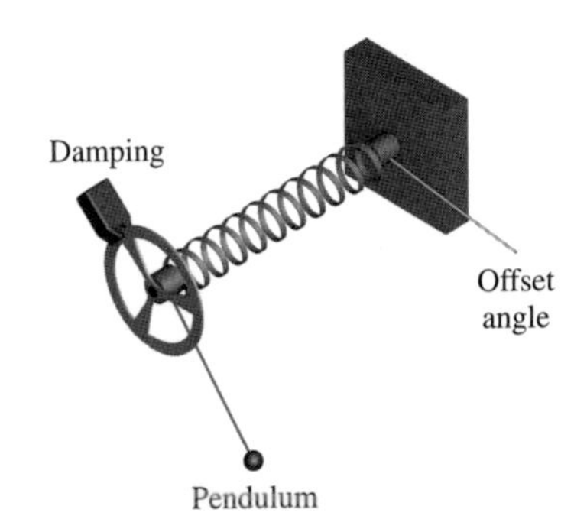

Fig. 9.17
A pendulum with a torsion spring that provides a restoring torque which is proportional to the pendulum angle measured relative to a settable offset: $\theta-\theta_x$. Velocity dependent damping is provided by eddy currents induced in the rotating metal disc as it moves between the poles of a permanent magnet mounted at its rim.

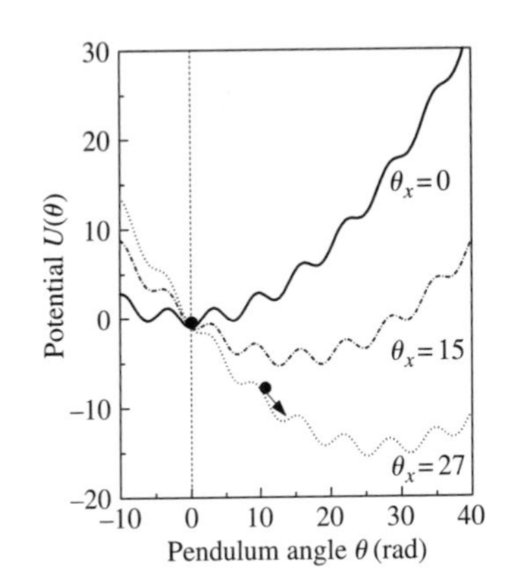

Fig. 9.18
Potential for the pendulum analog of a Josephson junction in a superconducting ring. For this example, the function is $U(\theta)=0.02\left[(\theta-\theta_x)^2-\theta_x^2\right]-\cos\theta$.

[8] See Barone and Paterno (1982, equation 12.2.15, p. 365), and Smith and Blackburn (1975).

the lever arm at the far end of the torsion spring (see Fig. 9.17) also pointing down. If the lever arm is now slowly rotated, then the pendulum mass will rise, but because of the ability of the torsion spring to twist, the pendulum angle θ will lag the offset angle θ_x. If the torsion spring is quite "soft," then it might happen that the lever will have to be turned through several full turns before the pendulum mass rises to a horizontal orientation. When $\theta = 90°$, the pendulum will quickly flip over the top and possibly start spinning. This sequence of events is illustrated in the plot of Fig. 9.18. As θ_x increases, the modulated parabolic potential changes as shown until at a critical value of $\theta_{xc} \approx 27$, the small well at $\theta = 0°$ flattens out and the "particle" begins bouncing down the wavy surface. This can be recognized as a variation of the "washboard" potential discussed earlier. What happens subsequently depends on the amount of friction in the system. For heavy damping, the particles will just drop into the next lower well. For lighter damping, the particle will bounce down to the bottom, passing n wells on its way. Very light damping could result in motion down to the bottom, up the other side, followed by a reversal. These scenarios would be observed as (a) one flip of the pendulum, (b) multiple flips of the pendulum, (c) multiple flips winding in one direction followed by flips unwinding in the reverse direction.

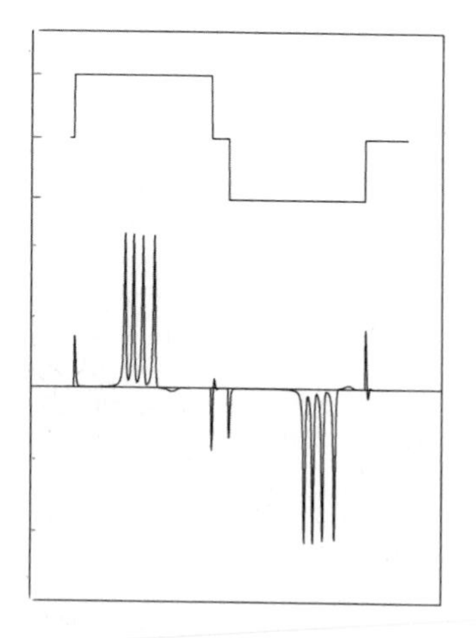

Fig. 9.19
Flux quanta entering then leaving a superconducting loop which is interrupted by a single Josephson junction. The upper plot is the bias signal; the lower plot is the time-dependent voltage across the junction. These are the results of a computer solution of Eq. (9.30).

The critical value of the forcing angle can be estimated from the condition

$$K\left(\theta_{xc} - \frac{\pi}{2}\right) = mgr \tag{9.35}$$

from which

$$\theta_{xc} = \frac{mgr}{K} + \frac{\pi}{2}. \tag{9.36}$$

For the example chosen in Fig. 9.18, $mgr/K = 25$, and so θ_{xc} should be 26.6.

Recalling the correspondence between this pendulum and magnetic flux in a superconducting ring that contains a Josephson junction, the pendulum flips are seen to be equivalent to flux quanta entering or leaving the loop. The results of a numerical simulation of the flux dynamics governed by Eq. (9.30), shown in Fig. 9.19. In this example, a bundle of four flux quanta is alternately forced in and out of the loop.

9.5.3 Two junctions in a superconducting loop

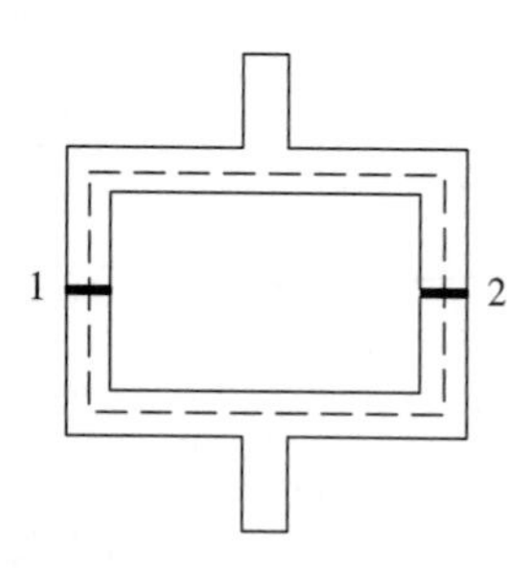

Fig. 9.20
Conceptual drawing of a double Josephson junction interferometer. The two junctions are labeled 1 and 2. The dotted path follows the interior of the superconducting branches and is used to determine the total phase change around the loop.

The previous section dealt with the case of a single Josephson junction inserted into a superconducting ring. Now we look at what happens when a ring is interrupted by two junctions, as illustrated in Fig. 9.20. Such configurations are analogous to variations on the coupled pendulums discussed in Chapter 7.

Assuming as before that the current feed is placed asymmetrically in the ring, then the equivalent circuit is given as in Fig. 9.21.

Let the Josephson currents be I_{J1} and I_{J2}. Of course, the bias current just divides between the left and right branches, so

$$I_e = I_{J1} + I_{J2}. \tag{9.37}$$

The net induced flux directed down into the loop is

$$\Phi_{\downarrow} = \alpha L I_{J2} - (1 - \alpha) L I_{J1}. \tag{9.38}$$

Again using the expression for the phase change around the loop

$$\oint \nabla\varphi \cdot d\vec{s} = -2\pi \frac{\Phi_{\downarrow}}{\Phi_0}, \tag{9.39}$$

the phase equation, Eq. (9.27), is now modified to account for the phase jumps at the two junctions,

$$(\phi_1 - \phi_2) - 2\pi \frac{\Phi_{\downarrow}}{\Phi_0} = 2n\pi. \tag{9.40}$$

Equations (9.37), (9.38), and (9.40) yield

$$\begin{aligned} \frac{\hbar C}{2e}\frac{d^2\phi_1}{dt^2} + \frac{\hbar}{2eR}\frac{d\phi_1}{dt} + I_c \sin\phi_1 &= -\frac{\Phi_0}{2\pi L}(\phi_1 - \phi_2) + \alpha I_e \\ \frac{\hbar C}{2e}\frac{d^2\phi_2}{dt^2} + \frac{\hbar}{2eR}\frac{d\phi_2}{dt} + I_c \sin\phi_2 &= \frac{\Phi_0}{2\pi L}(\phi_1 - \phi_2) + (1 - \alpha) I_e. \end{aligned} \tag{9.41}$$

This pair of coupled differential equations is isomorphic to the pair of equations describing two pendulums linked by a torsion spring that generates an action–reaction torque proportional to the angular *difference* $(\theta_1 - \theta_2)$. The constant bias terms αI_e and $(1 - \alpha) I_e$ in Eq. (9.41) are represented in the pendulum system by a constant total torque Γx applied in the proportions α: $(1 - \alpha)$ to pendulums 1 and 2, respectively. Hence, for the pendulums,

$$\begin{aligned} I\frac{d^2\theta_1}{dt^2} + b\frac{d\theta_1}{dt} + mgr\sin\theta_1 &= -K(\theta_1 - \theta_2) + \alpha\Gamma_x \\ I\frac{d^2\theta_2}{dt^2} + b\frac{d\theta_2}{dt} + mgr\sin\theta_2 &= K(\theta_1 - \theta_2) + (1 - \alpha)\Gamma_x. \end{aligned} \tag{9.42}$$

The term by term correspondence between the junction equations (Eq. (9.41)) and the pendulum equations (Eq. (9.42)) is readily apparent. As with the case of a single junction in a superconducting ring, the role of loop inductance is played by the torsion spring.

In a real system, the damping, as before, might be provided by magnetically induced eddy current losses in corotating discs (one for each pendulum). The required torque generation could be accomplished, as in the experimental design of Sullivan and Zimmerman (1971), by a rapidly spinning permanent magnet positioned close to the edge of the metal damping disk. An alternative, employed in the Daedalon Chaotic Pendulum (Daedalon 2000), makes use of a brushless slotless linear motor that has the necessary property of producing an output torque that is directly proportional to an applied voltage (Blackburn et al. 1989*a*).

A diagram of such a double pendulum system is shown in Fig. 9.22. The results of an electronic simulation of a double Josephson junction loop (Blackburn et al. 1988) are shown in Fig. 9.23. In the particular mode illustrated here, the parameters are such that a flux bundle first enters the loop through one junction, then the same bundle exits the

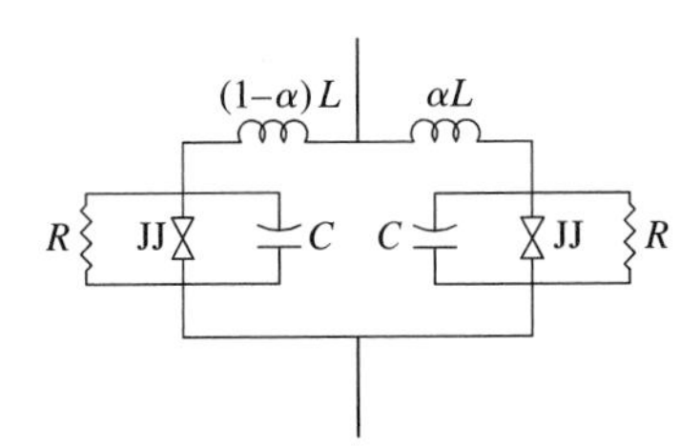

Fig. 9.21
Equivalent circuit of a double Josephson junction interferometer. Bias current is injected asymmetrically into the loop whose total inductance is L.

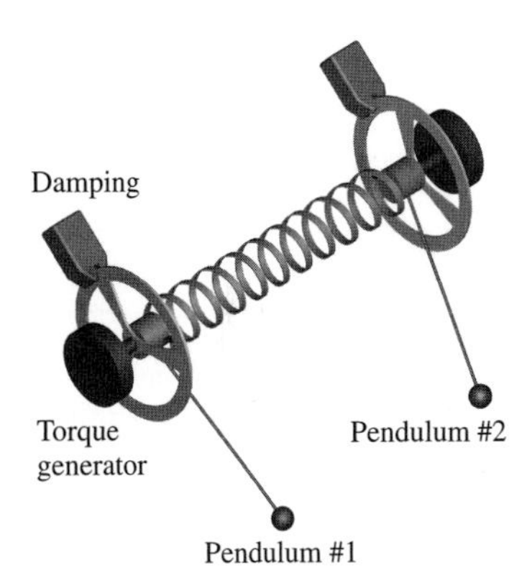

Fig. 9.22
Two pendulums coupled by a torsion spring. Torques are applied to the individual pendulums in a prescribed ratio as explained in the text.

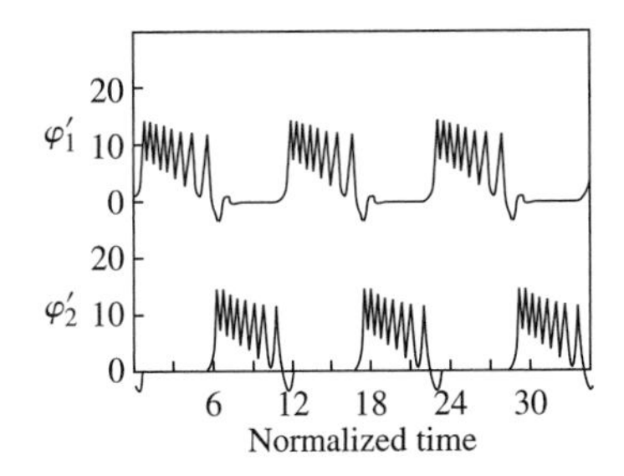

Fig. 9.23
Voltage (phase velocity) on each of the Josephson junctions in a superconducting interferometer as bundles of flux quanta alternately enter and exit the loop.

loop through the other junction. For the equivalent coupled pendulum system, this would be represented by one pendulum flipping n times, after which the second pendulum catches up by flipping the same number of times. Other dynamical modes can arise for other parameter choices.

9.5.4 Coupled Josephson junctions

Over the years, there has been a sustained interest in the dynamics that can occur when two or more Josephson junctions are coupled in some fashion and thus can interact. The general motivation for such interest lies in the potential of multijunction configurations in superconducting electronics. Applications include soliton oscillators, logic switches, and SQUID arrays for ultrasensitive magnetometers.[9] Many arrangements have been discussed in the research literature. We present here one particular and straightforward configuration, which will illustrate the way in which coupled Josephson junctions may be modeled by coupled pendula.

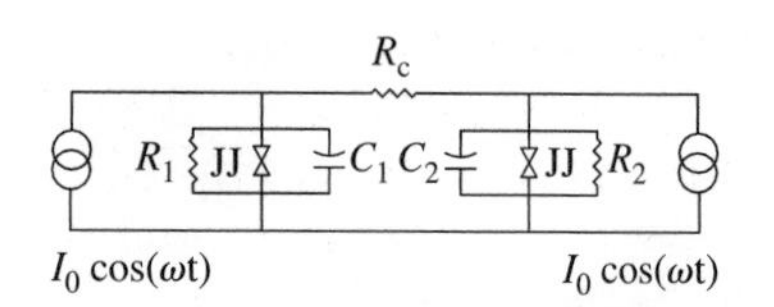

Fig. 9.24
Two Josephson junctions coupled through a resistor R_c and with identical ac bias.

The assumed arrangement is illustrated in Fig. 9.24, where the two Josephson devices are allowed to be nonidentical (different capacitances, resistances, and critical currents). The coupling resistor R_c allows current to pass between the two junctions. This can happen because of the difference between the voltage on junction no. 1: $(\hbar/2e)d\phi_1/dt$ and the voltage on junction no. 2: $(\hbar/2e)d\phi_2/dt$. Thus, accounting for all currents flowing through each junction,

$$\frac{\hbar C_1}{2e}\frac{d^2\phi_1}{dt^2}+\frac{\hbar}{2eR_1}\frac{d\phi_1}{dt}+I_{c1}\sin\phi_1=-\frac{\hbar}{2eR_c}\left[\frac{d\phi_1}{dt}-\frac{d\phi_2}{dt}\right]+I_0\cos(\omega t)$$
$$\frac{\hbar C_2}{2e}\frac{d^2\phi_2}{dt^2}+\frac{\hbar}{2eR_2}\frac{d\phi_2}{dt}+I_{c2}\sin\phi_2=\frac{\hbar}{2eR_c}\left[\frac{d\phi_1}{dt}-\frac{d\phi_2}{dt}\right]+I_0\cos(\omega t).\tag{9.43}$$

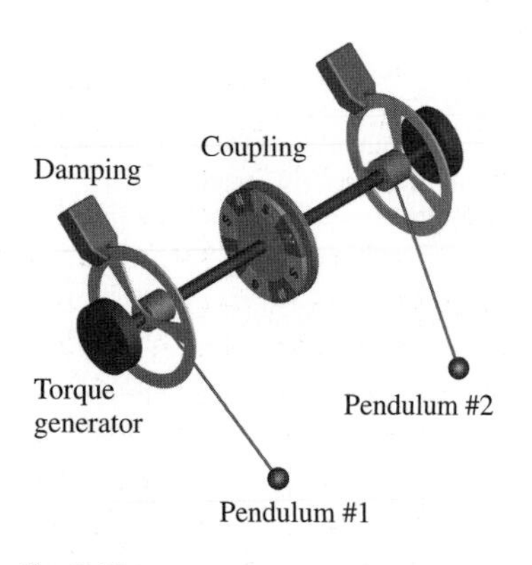

Fig. 9.25
Two driven, damped pendulums coupled via an eddy current induced torque. A ring shaped permanent magnet is attached to the shaft of one of the pendulums, while a conducting disc is fixed to the shaft of the other. The ring magnet consists of alternating north and south poles, as suggested in the drawing.

The generators, indicated by I_0 cos ωt, are meant to represent an applied *rf* field—an often employed model for junctions irradiated by microwaves.

The pendulum counterpart to the resistively connected Josephson devices is shown in Fig. 9.25. A ring shaped permanent magnet, or other similar structure, is attached to the shaft of one pendulum, while a solid conducting disc is fixed to the shaft of the other. The annular magnet, as depicted, has azimuthally alternating north and south poles. The pairs of north south poles provide for both symmetry and efficiency of the magnetic coupling. There is no mechanical contact between the ring magnet and metal disc—they are separated by an air gap. Any *difference* in the

[9] Solitons are nondispersive waves or pulses that can arise in nonlinear media. One realization of solitons occurs as the solution to the sine-Gordon equation when the potential has the same form, $(1 - \cos\phi)$, as the pendulum.

instantaneous angular rate of rotation of the pendulums produces relative slippage between the magnet and the conducting disc, and this in turn causes the induction of eddy currents which produce equal and opposite torques on the two pendulums. The governing equations for the two pendulums may be written

$$I_1\frac{d^2\theta_1}{dt^2}+b_1\frac{d\theta_1}{dt}+m_1gr_1\sin\theta_1=-A\left(\frac{d\theta_1}{dt}-\frac{d\theta_2}{dt}\right)+\Gamma\cos\omega t$$
$$I_2\frac{d^2\theta_2}{dt^2}+b_2\frac{d\theta_2}{dt}+m_2gr_2\sin\theta_2=A\left(\frac{d\theta_1}{dt}-\frac{d\theta_2}{dt}\right)+\Gamma\cos\omega t, \quad (9.44)$$

where it is assumed in this case that the forcing terms from the torque generators are equal. The constant A defines the proportionality between differential angular velocity and eddy current induced torque.

The correspondence between the equations for resistively coupled Josephson junctions (Eq. (9.43)) and those for the linked pendulums (Eq. (9.44)) is immediately apparent. This same system of coupled pendulums is discussed in Chapter 7.

These are nonlinear systems and their dynamical modes are numerous and complex. With realistic values of the junction parameters these coupled devices can be made to run chaotically just as is possible for pendulums. Hence the analysis mimics that found for coupled pendulums in Chapter 7 and we will not repeat that discussion here.

However, as an added illustrative example, the values of the various system constants were selected to place the coupled Josephson system in a very special synchronized mode (Blackburn et al. 2000). Typical results from a numerical simulation of Eq. (9.43), when the junctions are assumed to be identical, are presented in Fig. 9.26. In this case, the two Josephson oscillators are each running chaotically and the plot shows that they are actually synchronized for intervals called laminar times. This synchronization is evidenced by the fact that the difference between the junction phases becomes very small (i.e. modulo 2π). However, every so often, abruptly and unpredictably, the phase of one of the junctions briefly spins through one or more full rotations with respect to the other junction, after which they resume their synchronized chaotic dynamics. Figure 9.27 shows a plot—familiar from Chapter 7—of the probability distribution of these laminar times computed for several different values of the dimensionless coupling strength parameter α where

$$\alpha=\frac{1}{R_c}\sqrt{\frac{\hbar}{2eI_cC}}. \quad (9.45)$$

Four different levels of coupling are shown and for each of these the distributions show the characteristic exponential decay rate typical of intermittent synchronization, which, as shown earlier, suggests a Markov process.

Given an appropriate choice of parameters and constants, analogous dynamics would occur in the system of coupled pendula depicted in

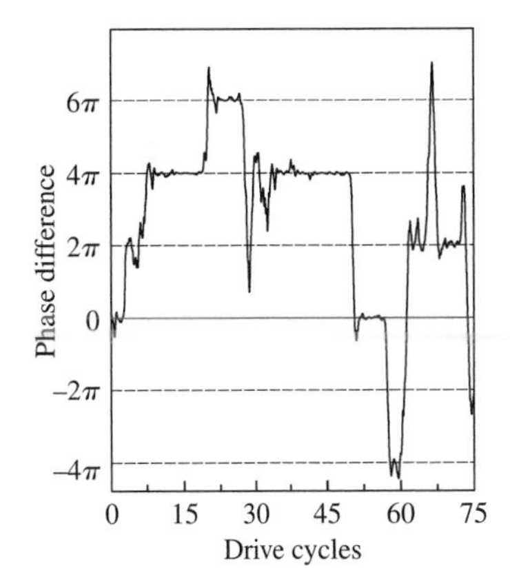

Fig. 9.26
Evolution in time of the difference in phase between two resistively coupled Josephson junctions. For this example, it can be seen that the oscillations are synchronized (phase difference modulo 2π) except during brief moments when there is loss of synchronization and the relative phase between the two junctions slips by a multiple of 2π.

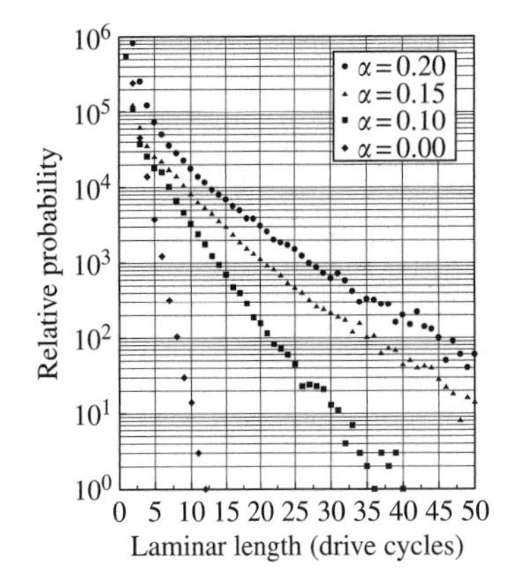

Fig. 9.27
Relative probability of the duration of laminar intervals for a driven pair of resistively coupled Josephson junctions. Results for four different coupling strengths are presented in these computer simulations. (Reprinted with permission from Blackburn et al. 2000, p. 5934). ©2000 by the American Physical Society.

Fig. 9.25. The motions of the pendula also would be chaotic in such a case, and synchronized most of the time. Occasionally one of the pendula would flip one or more times, lapping the other pendulum, after which they would again appear locked together in their chaotic motions.

9.6 Remarks

The connection between the driven damped pendulum and the Josephson junction is indeed remarkable. We have seen how superconductivity is a manifestation of long range quantum mechanical coherence, and how the leakage of the associated macroscopic wavefunction through a sufficiently thin barrier separating two superconductors can lead to the tunneling of Cooper pairs—the celebrated Josephson effect. The key role played in the Josephson tunneling current by the sine of the order parameter phase difference across a barrier is the link to the classical world of the driven pendulum, for which the gravitational restoring torque also has a sine dependence on the angular position of the pendulum.[10]

9.7 Exercises

1. Equation (9.20) can be made dimensionless if we define a dimensionless time $\tau = \omega_J t$ where $\omega_J = \sqrt{(2eI_c/\hbar C)}$ is the Josephson or "plasma" frequency.
   (a) Using this dimensionless time show that Eq. (9.20) becomes
   $$\frac{d^2\phi}{d\tau^2} + \beta_J \frac{d\phi}{d\tau} + \sin\phi = \frac{I}{I_c}. \tag{9.46}$$
   (Note that I is an applied bias current.)
   (b) Give the formula for β_J.
2. Equation (9.20) can be made dimensionless in a different way if we define a different dimensionless time $\tau = ((2_e I_c R)/\hbar)t$. (Again, the quantity $(2_e I_c R)/\hbar$ is an angular frequency.)
   (a) Starting from Eq. (9.20) show that the dimensionless equation is given by
   $$\beta_c \frac{d^2\phi}{d\tau^2} + \frac{d\phi}{d\tau} + \sin\phi = \frac{I}{I_c}. \tag{9.47}$$
   β is called the McCumber parameter and is related to the amount of hysteresis incurred in gradually increasing and decreasing the bias current. See McCumber (1968).
   (b) Give the formula for β_c.
   (c) Find a formula relating β_c and β_J.
3. The equation of motion for the pendulum,
   $$I\frac{d^2\theta}{dt^2} + b\frac{d\theta}{dt} + mgr\sin\theta = \Gamma, \tag{9.48}$$
   may be rendered dimensionless by techniques similar to those suggested in the previous problems.

[10] An exhaustive review of the ways in which pendulums represent mechanical counterparts of Josephson junction devices can be found in chapter 4, *Equivalent Circuits and Analogs of the Josephson Effect*, by T. A. Fulton in Schwartz and Foner (1977).

(a) Using the dimensionless time $\tau = \omega_{01} t$, where $\omega_{01} = \sqrt{(mgr)/I}$, show that Eq. (9.48) may be transformed to

$$\frac{d^2\theta}{d\tau^2} + \beta_1 \frac{d\theta}{d\tau} + \sin\theta = \frac{\Gamma}{mgr}. \qquad (9.49)$$

(b) Using the dimensionless time $\tau = \omega_{02} t$, where $\omega_{02} = (mgr)/b$, show that Eq. (9.48) may be transformed to

$$\beta_2 \frac{d^2\theta}{d\tau^2} + \frac{d\theta}{d\tau} + \sin\theta = \frac{\Gamma}{mgr}.$$

Note that the quantity mgr is the critical torque required to raise the pendulum to the horizontal position.

4. Start with Eq. (9.47) and consider the "overdamped" limit (in exercises 2 through 6) where β is very small and therefore $\beta(d^2\delta/d\tau^2) = 0$. Then the dimensionless equation simplifies to

$$\frac{d\phi}{d\tau} = \frac{I}{I_c} - \sin\phi.$$

In a phase space with axes $d\phi/d\tau$ and ϕ sketch phase plane diagrams (a) when $I/I_c < 1$, and (b) when $I/I_c > 1$.

5. The fixed points of the phase plane diagrams of $d\phi/d\tau = (I/I_c) - \sin\phi$ are where $d\phi/d\tau = 0$. These are the zeroes of the graphs drawn in Exercise 2. (a) Do both graphs have fixed points? (b) Give the equation that specifies the fixed points.

6. Consider the phase orbit of a general equation $d\phi/d\tau = f(\phi)$ near the fixed points. Near the fixed points the function $f(\phi)$ may be approximated by its linearization as $f(\phi) \approx \phi^* + \alpha\phi$ where ϕ^* is a fixed point. Therefore a small perturbation $\varepsilon = \phi - \phi^*$ from the fixed point is approximated by $\alpha\phi$. The rate of change of the perturbation is $d\varepsilon/d\tau = \alpha\varepsilon = d\phi/d\tau$. If α is negative then the orbit tends to the fixed point. In this case the fixed point is *stable*. If α is positive then the phase orbit tends away from the fixed point, which is now unstable. (This approach to the study of the stability of fixed points is called *linear stability analysis*.) (a) Examine phase plane diagrams for the case when $I/I_c = 0.3$ and determine the stable and unstable fixed points. (b) In view of your answers to (a) what is the long term value of the Josephson voltage V for $I < I_c$?

7. Now consider the overdamped case when $I/I_c > 1$. There are no fixed points but $d\phi/d\tau$ is still oscillatory. Calculate the period of the motion. Hint: Separate the variables in $d\phi/d\tau = (I/I_c) - \sin\phi$ and let $T = \int_{-\pi}^{\pi} d\phi/((I/I_c) - \sin\phi)$. Use the half angle substitution $u = \tan\phi$, $d\phi = 2/(1+u^2)du$, and $\sin\phi = 2u/(1+u^2)$ and show that $T = 2\pi/\sqrt{((I/I_c)^2 - 1)}$ in units of τ.

8. In the overdamped case where $I > I_c$ there is a long term voltage given by $V = (2e/\hbar)\langle d\phi/d\tau\rangle$. (a) Average this voltage over one period (using the result of exercise 5) to show that $V = (2e/\hbar)\sqrt{((I/I_c)^2 - 1)}$. (b) Draw a graph of V versus I/I_c on $[0, \infty)$ for $\beta = 0$, the overdamped limit.

9. Through comparison of the equation of motion for the damped nonlinear pendulum, Eq. (9.21), with that of the dynamical equation for the junction phase, Eq. (9.20), match up corresponding physical quantities in Column A with those in Column B.

A	B
Moment of Inertia, I	Conductance, $1/R$
Angular velocity, $d\theta/dt$	Capacitance, C
Damping factor, b	Voltage, V
Angle, θ	Source current, I
Maximum gravitational torque, mgr	Phase, ϕ
Applied Torque, Υ	Maximum zero voltage current, I_c

(a) Write an expression for the potential $V(\phi)$ and sketch its graph.
(b) The angle of the washboard and the size of the ripples dictate one of several possible behaviors of the system. Describe these possibilities. Hint: refer to page 179 of Van Duzer and Turner (1981).

10. The existence of Cooper pairs of electrons is required for superconductivity. These pairs form only when the temperature is lowered below the critical temperature T_c, and the pair have a binding energy/electron of $\Delta(T)$. At $T=0$ this energy has a value $\Delta(0)$ and it remains almost constant and then drops rapidly to zero as T approaches T_c. Near T_c, the binding energy is given approximately by $\Delta(T) \simeq 3.2kT_c\ (1 - T/T_c)^{1/2}$. (a) Sketch a graph of this function. (b) The gap energy at $T=0$ is related to the critical temperature by the equation $2\Delta(0) = \text{constant} \times kT_c$. For Mercury, the constant $= 3.52$. For Lead, the constant is 4.38 and the gap energy is $2\Delta(0) = 2.73 \times 10^{-3}$ eV. Find the temperature at which lead is superconducting.

11. Show that the restoring torque given in Eq. (9.32) leads to the parabolic washboard potential of Eq. (9.34).

The pendulum clock

10

10.1 Clocks before the pendulum

The spinning Earth is, of course, a timekeeper. The instantaneous position of the sun in the sky during the day, or of particular stars in the night sky, indicates a moment in time. Weather permitting, sundials could mark off daytime intervals as sequences of positions of the shadow cast by a stick or wedge (the gnomon).

While timekeeping with devices of various descriptions (water clocks, burning candles, flowing sand) has a long history, for millennia *accuracy* was not a high priority. Human affairs were mainly local and exactness was not essential. Ultimately, however, the need for public timekeeping led to a progression of significant advances and inventions.

The epoch of the first large mechanical clocks began around 1335 in Milan. Many towns and cities throughout Europe installed similar clocks in cathedrals and public squares. An example of an early English cathedral clock is the one at Salisbury (*c*.1275–1284) (Fig. 10.1).

Fig. 10.1
Mechanism of the clock at Salisbury Cathedral (1386). (Reproduced with permission.)

These timekeeping machines originally employed the so-called verge and foliot as their mechanical oscillators, as shown in Fig. 10.2, although, as pointed out in Ward (1970), they were later modified for use with pendulums.

The vertical rod, the *verge*,[1] is fitted with two small plates, the *pallets*. The cross beam at the top is the *foliot*. The interrelation between the unidirectional motion of the torque-driven crown wheel and the back-and-forth motion of the foliot beam can be understood with the aid of Fig. 10.3.

In this illustration, the system has been simplified visually to bring out the essentials. All wheel teeth have been deleted except for the two, which will engage the upper or lower pallets during this brief time interval. In the first frame, the system is observed at a moment when the leading edge of the top tooth is pushing on the upper pallet, causing a counterclockwise twist to the verge. At the mid-point of the sequence of illustrations, the right-moving top tooth hands off to the left-moving tooth at the bottom of the crown wheel. The leading edge of this bottom tooth now pushes on the

[1] From verge: *a Rod or Staff Carried as an Emblem of Authority or Symbol of Office*, The New Penguin English Dictionary, Penguin Books (2000).

Fig. 10.2
Verge and foliot mechanism. The weight and rope apply torque to the crown wheel. In this depiction, the upper pallet is being forced by contact with a tooth in the crown wheel, causing the beam (the *foliot*) to twist counter-clockwise when viewed from above. The lower pallet is hidden from view behind the vertical rod (the *verge*). Through a series of gears the hour hand is driven by the shaft from the crown.

lower pallet, causing the verge to begin twisting in a clockwise direction. At the end of the cycle, the following wheel teeth (not shown in the figure), top and bottom, will have moved into position and will then take their turns forcing the pallets. Thus the verge and the attached foliot beam oscillates, alternating between clockwise and counterclockwise rotations about the vertical axis.

In Fig. 10.2, the crown wheel has 21 teeth. This number was selected rather arbitrarily, for simplicity in the illustration, but whatever the tooth count, it must be odd for the alternation between pallets to work correctly. An even number would result in the system locking-up because both pallets would be engaged by oppositely moving teeth at the same moment.

The two weights hanging from the foliot, as shown in Fig. 10.2, are used to trim the period of oscillation. Moving them outwards slows the angular spin rate and so makes the period of oscillation longer; moving the weights inwards speeds up the oscillations. Thus, the basis of this mechanism is the adjustment of the moment of inertia of the swinging beam so as to serve as a speed regulator for the rotating crown wheel which repeatedly experiences angular acceleration, pause, angular acceleration, and so on as the drive is transferred back and forth between the upper and lower pallets.

The view (Fig. 10.4) of the clock from Dover Castle clearly shows the foliot mechanism. The large crown wheel is driven via a small pinion by the rope wound around the drum. The long yoke shaped balance beam and its hanging rectangular weights are visible at the top of the apparatus.

The verge-and-foliot does not make a very precise oscillator. Its performance is limited by the fact that there is no natural frequency to which it is drawn. By contrast, a good oscillator is highly tuned, possessing a sharp resonance that stabilizes the running frequency.

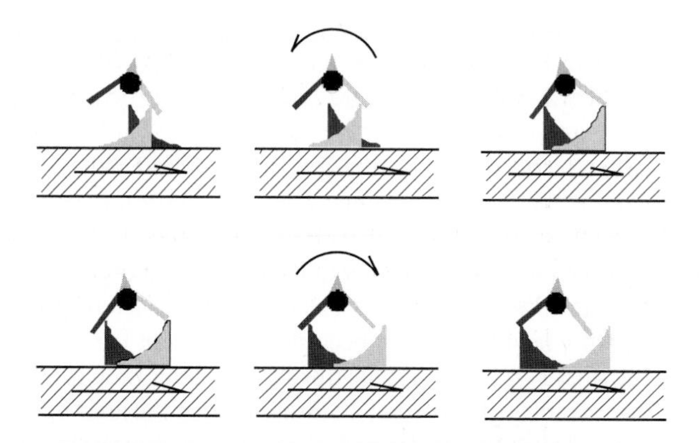

Fig. 10.3
Top view looking down on the crown wheel and verge (the foliot beam is omitted here). Portion of the sequence in which the verge is driven first counter-clockwise by a top tooth (light) forcing the upper pallet (light), then clockwise by a bottom tooth (dark) forcing the lower pallet (dark). For clarity, only a portion of the crown wheel together with the two teeth that participate at this stage in the sequence are shown. As the crown wheel rotates, the top tooth moves slowly to the right while the bottom tooth moves to the left.

Timekeeping in these early clocks was not better than within about a half or quarter hour per day. There were no minutes hands and the clock faces were marked only in hours.

Fig. 10.4
The clock from Dover Castle. This photograph clearly shows the large yoke-shaped foliot beam (at the top) with its two rectangular hanging trim weights. The design is essentially fourteenth century, but the date of the clock is likely early seventeenth century. (Reproduced with permission from the Science Museum/Science and Society Picture Library, London.)

10.2 Development of the pendulum clock

10.2.1 Galileo (1564–1642)

As already noted in Chapter 2, Galileo believed that the period of a swinging pendulum is independent of the amplitude. By 1637, he realized the possibility of using the pendulum as the central beating mechanism in a clock (Bedini 1991, pp. 24, 27). Eighteen years after the fact, his young assistant Vincenzio Viviani recounted in a letter to Prince Leopold de Medici (dated August 20, 1659), that sometime in the summer of 1641, "it came into his (Galileo's) mind that the pendulum could be adapted in place of the customary balance, to clocks driven by weight or spring, in the hope that the perfect natural equality of its motion would correct all the defects of the art of these clocks."

This is the definitive moment at which the history of the pendulum clock might be said to have commenced.

Galileo died in January of the next year, 1642, at the age of 78. At the end of the same year, on Christmas day, Isaac Newton was born.

With the help of Galileo's son Vincenzio, a design for a pendulum clock had been begun in 1641, but the work had then been set aside and was not resumed until April of 1649, well after Galileo's death (Bedini 1991, p. 29). A central idea was to employ a falling weight as a driving force to maintain the pendulum's motion, "and thus in a certain manner was perpetually maintained the to-and-fro motion of the pendulum, until the weight had run down to the bottom."[2]

Because there is always some energy loss, however small, in a pendulum, some means must be provided of supplying a replacement for this lost energy, or else the oscillator will inevitably run down and stop. Additionally, some means must be provided for counting the beats so as to derive and display the time. Both of these functions are provided by what is called an *escapement*.

Galileo's design used a form of escapement involving pivoted arms, notched wheels and so forth. It seems to have had some problems and a working model was left uncompleted at the time of Vincenzio's death in May of 1649 (Figs. 10.5 and 10.6).

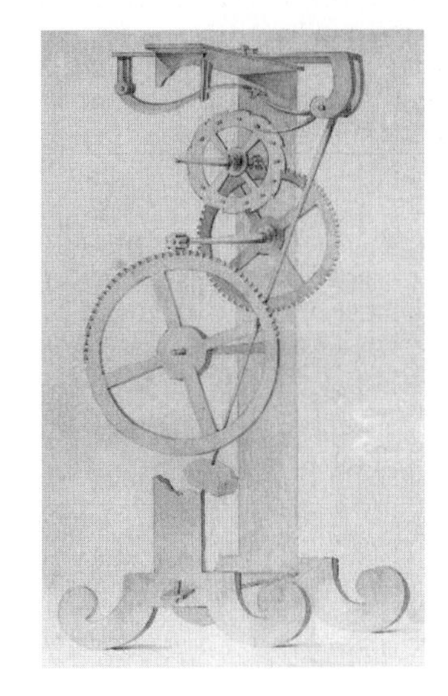

Fig. 10.5
Galileo's design for a pendulum clock. The drawing is dated to 1659 and is attributed to Galileo's friend Viviani. (Reproduced with permission from the Science Museum/Science and Society Picture Library, London.)

10.2.2 Huygens (1629–1695)

The next critical steps in the evolution of the pendulum clock were due to the Dutch scientist and mathematician, Christiaan Huygens (Fig. 10.7). In 1656–1657, Huygens invented what we now consider to be the first truly

[2] Letter from Viviani to Prince Leopold, August 20, 1659.

Fig. 10.6
A model of Galileo's pendulum clock design incorporating his escapement mechanism. (Reproduced with permission from the Science Museum/Science and Society Picture Library, London.)

practical design. His achievement lay in the realization of the escapement and its linkage to the swinging pendulum.

The clock was first described in Huygens' small book of 1658: *Horologium*. His major work, *Horologium Oscillatorium* (*The Pendulum Clock or Geometrical Demonstrations Concerning the Motion of Pendulums as Applied to Clocks*) was published fifteen years later, in 1673.[3] The title page from a facsimile edition (Huygens 1673) is shown in Fig. 10.8.

Part I of this treatise is devoted to the design of the pendulum clock as well as its possible application to the problem of determining longitude at sea. Huygens describes a test which was carried out during the voyage of a British ship in 1664. Two of his clocks, each with a pendulum length of nine inches and a weight of one-half pound,[4] were fitted out for sea and sent on a voyage to the west coast of Africa.[5] According to Huygens, the results were promising, and he went on to propose modifications to improve their stability with respect to the rolling and pitching motions of a ship. Part I of the *Horologium Oscillatorium* concludes[6] "Instruments which have been modified in this way have yet to be taken to sea and put to the test. But they offer an almost certain hope of success because, in experiments conducted so far, they have been found to hold up under every kind of motion much better than prior clocks."

It was not that easy. The problem of determining longitude with suitable accuracy had to wait for almost another century, and the solution lay not in pendulum clocks after all. In 1759 John Harrison perfected an extremely sophisticated marine chronometer which employed a mainspring and balance wheel (Sobel 1996, 1998).

The essential idea of Huygens for a clock was a very simple modification of the verge-and-foliot mechanism—just attach a pendulum directly to the verge and discard the foliot beam. This arrangement is shown in Fig. 10.9. Simple as this was, it was revolutionary. The design, as presented in the *Horologium Oscillatorium*, is shown in Fig. 10.10. Part I of the illustration shows a side view of the complete clock. The pendulum mass (X) hangs at the end of a rod from point (V), above which a double cord extends to the upper clamp, as shown in Part II of the figure. Other key mechanical components, also shown in Part I are the crown wheel (K) with 15 teeth, the horizontal verge (LM) with two pallets. A cord with a hanging weight is wrapped around the studded wheel (D) and provides gravitational drive to the crown wheel through intermediary gears and pinions ($CEFGHI$).

As the pendulum swings back and forth, the shaped rod (S), referred to as the crutch, follows the motion and causes the verge rod to turn correspondingly clockwise and counterclockwise (as viewed from its end). The pallets alternate between engaging a tooth on the right side of the crown wheel or a tooth on the left. Each time an engaged tooth is released (as the swinging pendulum twists the pallet out of the way), the crown wheel freely advances slightly until stopped when a tooth on the other side of the wheel

[3] For a superb translation, see Blackwell (1986). [4] Huygens (1986, pp. 28–32).
[5] See (Matthews, 2000), pp. 136–140. [6] Huygens (1986, p. 32).

Fig. 10.7
Portrait of Christiaan Huygens. (Reproduced with permission from the Science Museum/Science and Society Picture Library, London.)

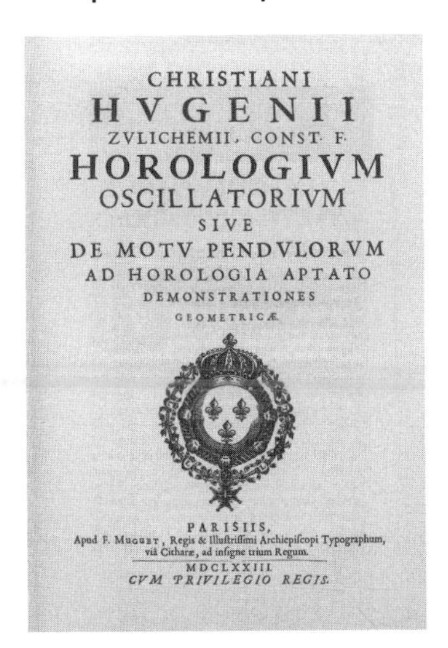

CHRISTIANI
HVGENII
ZVLICHEMII, CONST. F.
HOROLOGIVM
OSCILLATORIVM
SIVE
DE MOTV PENDVLORVM
AD HOROLOGIA APTATO
DEMONSTRATIONES
GEOMETRICÆ.

PARISIIS,
Apud F. Muguet, Regis & Illustrissimi Archiepiscopi Typographum, viâ Citharæ, ad insigne trium Regum.
MDCLXXIII.
CVM PRIVILEGIO REGIS.

Fig. 10.8
Title page from Horologium Oscillatorium (1673) by Christiaan Huygens.

Fig. 10.9
The basic arrangement for Huygens' pendulum clock escapement. The externally driven crown wheel rotates in only one direction (front edge to the right, back edge to the left), but does so in interrupted intervals paced by the pendulum driven to-and-fro motion of the verge.

hits the other pallet. The "tick" and "tock" are the clicks as crown wheel teeth alternately hit the pallets. Hence the operation of the escapement amounts to repeatedly blocking and releasing the crown wheel. Each time the forward rotation of the crown wheel is arrested, there is a slight recoil of the pallet.

Huygens pointed out in his discussion of the clock mechanism (Huygens 1986, p. 16).

For the small rod *S*, which is moved very slightly by the force of the wheels, not only follows the pendulum which moves it, but also helps its motion for a short time during each swing of the pendulum. It thus perpetuates the motion of the pendulum which otherwise would gradually slow down and come to a stop on its own, or more properly because of air resistance.

Huygens thus understood the dual requirements of a clock escapement: counting beats and making up for energy losses.

Note that in the verge and foliot discussed earlier, the crown wheel delivered alternating torques to the foliot beam, causing it to accelerate clockwise then counterclockwise. But, as was mentioned, the foliot had no natural rate of oscillation of its own. In contrast, Huygens' clock utilized the fundamental periodicity of the pendulum to regulate the beats.

Huygens' design incorporated one additional innovation. In Part II of Fig. 10.10 two curved cheeks are depicted. As the pendulum swings, the double suspension cord will wrap and unwrap along the inside faces of these cheeks. The reason Huygens included this feature, and a description of the precise shape of the cheeks are now outlined.

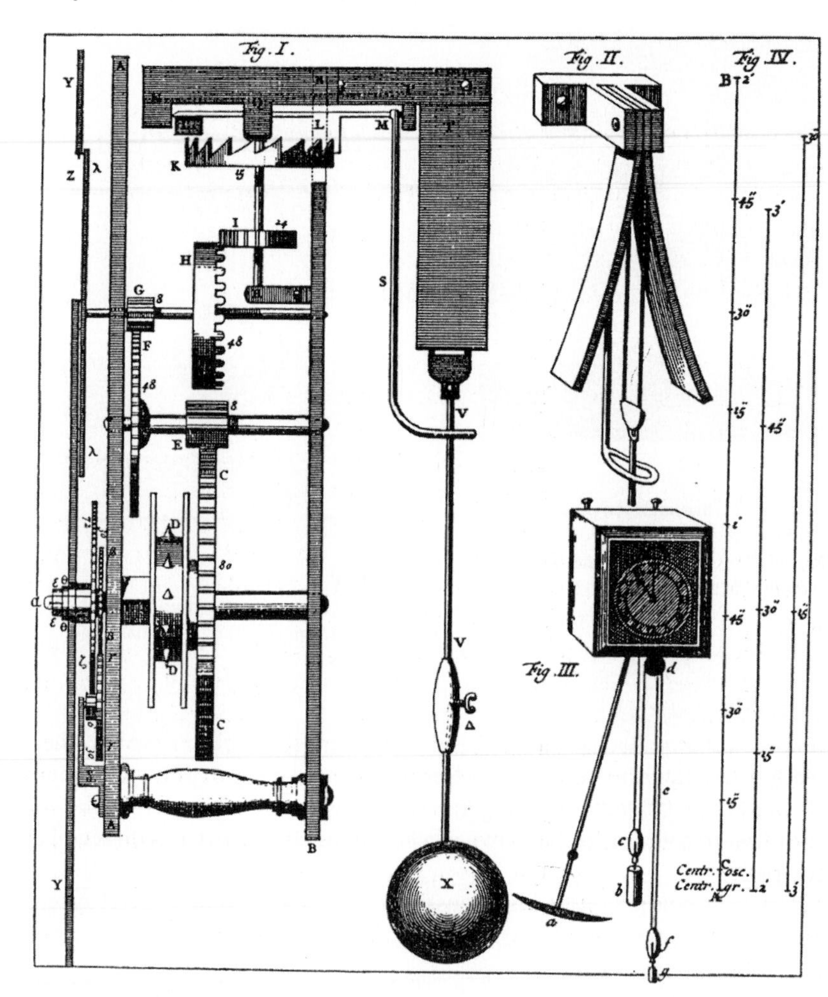

Fig. 10.10
Huygens' clock design, as illustrated in the *Horologium Oscillatorium* (1673).

Cycloids and tautochrones

In the *Horologium Oscillatorium* (page 19), Huygens remarked:

> We must now describe the form of the thin plates[7] between which the pendulum is suspended, and whose most important function is to guarantee the regular motion of the clock. For without them the simple oscillations of the pendulum will not be equal over a long period of time (even though some have thought otherwise), but rather oscillations through smaller arcs will occur in a shorter time.

In other words, Huygens realized that the simple pendulum was not, as Galileo had supposed, isochronous. The period of an oscillation was dependent on the size of the arc of the swing. The exact expression for the period is

$$T = \left[2\pi\sqrt{\frac{\ell}{g}}\right]\left\{\frac{2}{\pi}\int_0^{\pi/2}\frac{d\varphi}{\sqrt{1-k^2\sin^2\varphi}}\right\}, \tag{10.1}$$

[7] See Fig. 10.10, Fig. II. These plates are usually called *cheeks*.

where

$$k = \sin(\theta_m/2)$$

and θ_m is the maximum angular displacement on either side of the centre position. As discussed in Chapter 3, the integral in Eq. (10.1) is known as a complete elliptic integral of the first kind, $K(k)$. Since $K(0) = \pi/2$, in the limit of vanishingly small oscillations ($\theta_m \to 0$) this expression gives the familiar formula for the period

$$T = 2\pi\sqrt{\frac{\ell}{g}}. \tag{10.2}$$

Equation (10.1) can be used to plot the dependence of the period on the amplitude of the oscillation. This is shown in Fig. 10.11, which is an expanded version of part of Fig. 3.17. Clearly, the period increases as the size of the angular excursions increases. For example, with a maximum angular displacement of 30°, the period will be 1.7% longer than the value given by the expression in Eq. (10.2). A clock set to 30° would run 24 min slower per day than one running with a very small angular swing.[8] Even for just 3°, this effect amounts to 14.8 s lost per day.

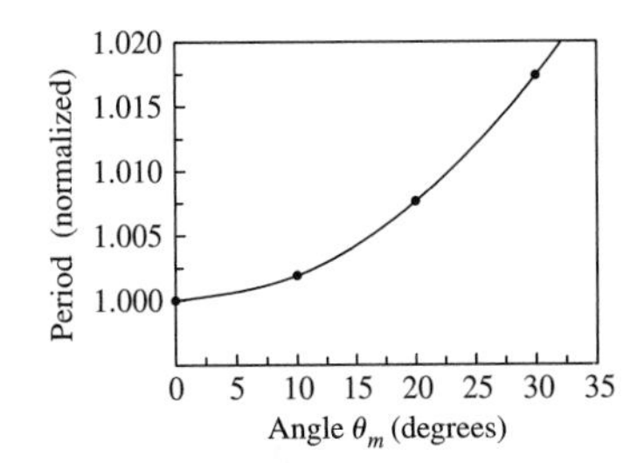

Fig. 10.11
Variation of the period of a simple pendulum with the oscillation amplitude, computed from Eq. (5.7). The vertical scale is normalized to the period in the limit of very small angles as expressed by Eq. (10.2).

The pendulum bob moves along a strictly circular arc. A completely equivalent oscillating motion would be exhibited by a small bead sliding without friction along a wire bent in the shape of a circular arc of the same radius. As with the pendulum, the time taken for the bead to complete a cycle (slide down the wire to the bottom of the arc, up the other side, and then reverse the trip, going back down the wire and finally up to the starting position) will depend on how far up the wire is the release point. A higher starting point leads to a longer round trip travel time (period of oscillation), as occurs with the pendulum.

This problem of increasing period with increasing amplitude of swing could be eliminated if there were some means to gradually and appropriately shorten the effective length of the pendulum for larger swings. This can be accomplished if the bob is constrained to follow a special form of curve (not quite a circular arc) called a *tautochrone*. A frictionless bead sliding down a wire bent in this shape will reach the lowest position in a time that is independent of the point of release on the wire (tautochrone: "same time"). In a gravitational field, the tautochrone is a *cycloid*.

The relationship of tautochrone and cycloid can be developed in either of two ways. One can start with a cycloid and then prove that it is indeed a tautochrone. Alternatively, by starting with the required property of isochronicity, the equation of the curve that possesses that property can be derived. Each of these two treatments is now presented.

The cycloid is a tautochrone

In parametric form, the equations defining a cycloid are

$$x = r(\phi + \sin\phi) \tag{10.3}$$

$$y = r(1 - \cos\phi). \tag{10.4}$$

[8] This is known as the *circular error*.

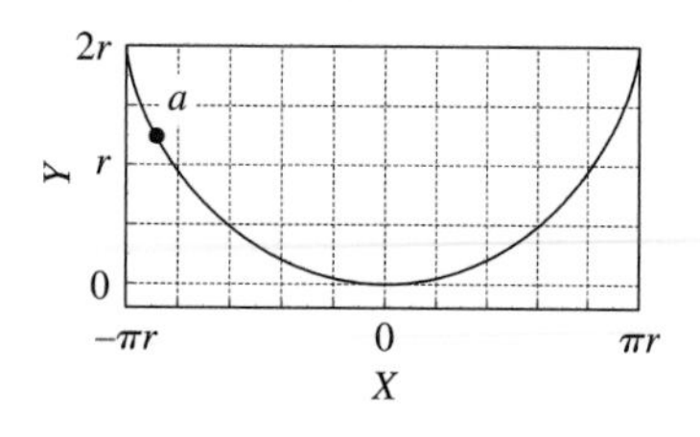

Fig. 10.12
A wire in the shape of a cycloid with a frictionless bead released from position a. As proved in the text, the time taken for the bead to reach the bottom of the loop is the same for all starting positions!

The constant r is the radius of the generating circle. By letting the parametric angle ϕ range from $-\pi$ to π, a full cycle of the cycloidal curve can be generated.[9] As shown in Fig. 10.12 there is a minimum at (0,0) and cusp maxima at $(-\pi r, 2r)$ and $(\pi r, 2r)$.

Let us suppose that a frictionless bead is positioned on a wire of this shape, and released. What is the time required for the bead to slide down to the bottom?

To answer this question, consider conservation of energy: $mgy_a = mgy + 1/2\,(mv^2)$, where the release point is (x_a, y_a) and v is the velocity at the general point (x, y). Hence, $v = \sqrt{2g(y_a - y)}$. The velocity is the speed along the wire, so

$$v = \frac{ds}{dt},$$

where ds is a differential element of arc length. Therefore,

$$\frac{ds}{dt} = \sqrt{2g(y_a - y)}$$

and so

$$dt = \frac{ds}{\sqrt{2g(y_a - y)}}.$$

But $ds^2 = dx^2 + dy^2$, hence

$$dt = \sqrt{\frac{dx^2 + dy^2}{2g(y_a - y)}}. \tag{10.5}$$

From the defining equations for the cycloid (10.3, 10.4), $dx = r(1 + \cos\phi)d\phi$ and $dy = r\sin\phi\, d\phi$. Substituting these in Eq. (10.5),

$$dt = \sqrt{\frac{r^2(2 + 2\cos\phi)}{2g[r(1 - \cos\phi_a) - r(1 - \cos\phi)]}}d\phi.$$

The time to go from (x_a, y_a) to the bottom (0,0) is then, in terms of the parametric variable ϕ,

$$t = \sqrt{\frac{r}{g}}\int_{\phi_a}^{0}\sqrt{\frac{1 + \cos\phi}{\cos\phi - \cos\phi_a}}d\phi. \tag{10.6}$$

Making use of the relations

$$\sin^2(u/2) = \frac{1 - \cos u}{2}$$

$$\cos^2(u/2) = \frac{1 + \cos u}{2}$$

[9] See MacMillan (1927, p. 320).

this becomes

$$t = \sqrt{\frac{r}{g}} \int_{\phi_a}^{0} \sqrt{\frac{\cos^2(\phi/2)}{\sin^2(\phi_a/2) - \sin^2(\phi/2)}} d\phi$$
$$= k\sqrt{\frac{r}{g}} \int_{\phi_a}^{0} \frac{\pm\cos(\phi/2)}{\sqrt{1 - k^2 \sin^2(\phi/2)}} d\phi, \quad (10.7)$$

where

$$k^2 = \frac{1}{\sin^2(\phi_a/2)}. \quad (10.8)$$

The integral in Eq. (10.7) is given in Gradshteyn and Ryzhik.[10]

$$\int \frac{\cos x}{\sqrt{1 - k^2 \sin^2 x}} dx = \frac{1}{k} \arcsin(k \sin x)$$

so

$$t = \pm k \sqrt{\frac{r}{g}} \frac{1}{k} 2 \arcsin k \sin(x)\Big|_{\phi_a/2}^{0} = \pi \sqrt{\frac{r}{g}}, \quad (10.9)$$

where the sign has been chosen which yields a positive transit time. The amazing thing about this result is that it does not depend on the point of release; this has happened because the specifics about the starting point were encapsulated in the factor k, and this has cancelled out in the final expression. In other words, no matter where the bead is let go, it will reach the bottom in exactly the same time. Thus it has been demonstrated that a cycloid is a tautochrone (McKinley 1979; Flores 1999).

A tautochrone is a cycloid

As a preliminary, consider the simplest equation of a one dimensional harmonic oscillator.

$$x = A \sin \omega t.$$

Note first that the frequency is constant, meaning that the period $T = 2\pi/\omega$ of a harmonic oscillator is independent of the amplitude of the oscillation, A.

The velocity and acceleration are $\dot{x} = A\omega \cos \omega t$ and $\ddot{x} = -A\omega^2 \sin \omega t$. This last expression is equivalent to $\ddot{x} = -\omega^2 x$. If the oscillating object is a mass m, then $F = m\ddot{x}$ and so the restoring force is exactly proportional to the displacement, with a proportionality constant $k = m\omega^2$.

Therefore, it can be concluded that a necessary and sufficient condition for the period of oscillations to be independent of the amplitude of those

[10] I. S. Gradshteyn and I. M. Ryzhik, *Table of Integrals, Series, and Products*, fifth edition, Academic Press (1994), formula 2.584, no. 3, p. 198.

oscillations is that there be a restoring force which is a linear function of the displacement from the equilibrium position.

Consider a frictionless bead on a curved wire. As a consequence of the necessary and sufficient condition just stated, the wire will be a tautochrone provided its shape is such that the bead experiences a restoring force *along the wire* that is proportional to the distance, *measured along the wire*, of the bead from the equilibrium (lowest) position. Consider the situation that is depicted in Fig. 10.13. The restoring force along the wire is $mg \sin\theta$. From the necessary and sufficient criterion for the wire to be a tautochrone, it is required that this restoring force be proportional to the distance s. So the condition $ks = mg \sin\theta$ is now *imposed*.

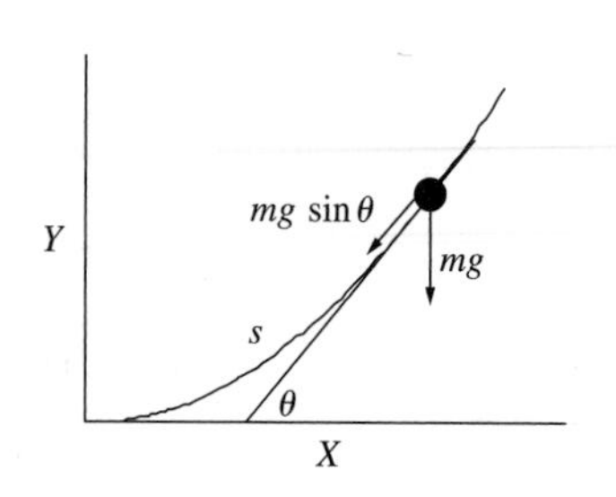

Fig. 10.13
A bead of mass m on a frictionless wire. The tangent to the wire at the location of the bead subtends an angle θ with the horizontal. The distance from the bead to the origin, measured along the wire, is s.

Therefore $dx = ds\cos\theta = k^{-1}mg\cos^2\theta\, d\theta$ and $dy = ds\sin\theta = k^{-1}mg \times \cos\theta \sin\theta\, d\theta$, or

$$\frac{dx}{d\theta} = \frac{mg}{k}\frac{1+\cos 2\theta}{2}$$
$$\frac{dy}{d\theta} = \frac{mg}{k}\frac{\sin 2\theta}{2}.$$

Integrating each of these two equations from $0 \to \theta$,

$$x = \frac{mg}{4k}[2\theta + \sin 2\theta] + C_1$$
$$y = \frac{mg}{4k}(1 - \cos 2\theta) + C_2,$$

where C_1 and C_2 are constants. If $x = y = 0$ at $\theta = 0$, then $C_1 = C_2 = 0$. Define $\phi = 2\theta$; the final result becomes

$$x = r(\phi + \sin\phi) \tag{10.10}$$
$$y = r(1 - \cos\phi) \tag{10.11}$$

with $r = mg/4k$. Equations (10.10) and (10.11) are just the defining expressions for a cycloid, as was given earlier in Eqs. (10.3) and (10.4). Thus it has been proved that the cycloid can be derived from the requirement of isochronicity.[11]

Returning to the matter of the oscillations on the wire, notice that Eq. (10.9) represents only one-quarter of a period. In a full cycle, the bead will go up the other side of the loop, then back down, and finally back up to the starting point. Therefore, the period of the oscillating bead on the cycloidal wire is

$$T = 4\pi\sqrt{\frac{r}{g}}. \tag{10.12}$$

[11] A similar argument is presented in "Development of an apparatus for two-dimensional collision experiment using a cycloidal slide" by M-h Ha, Y-k Kim, and S. B. Lee, *American Journal of Physics* **69** (2001), pp. 1187–1190, but for the case of a sphere rolling without slipping on a track rather than a bead sliding without friction on a wire. The result, however, is the same—the tautochrone is a cycloid.

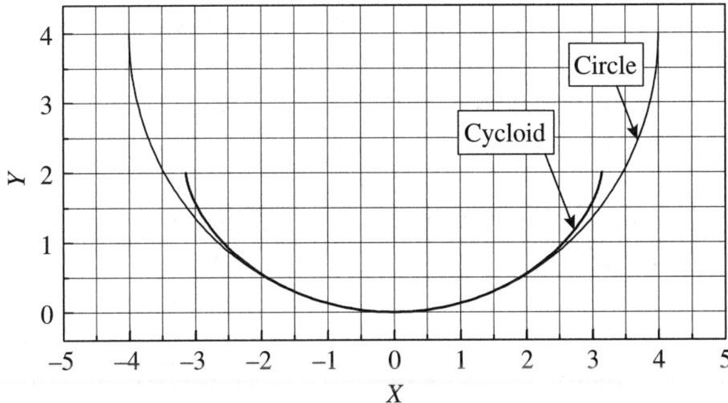

Fig. 10.14
Illustration of the relationship between the path of a cycloidal pendulum and the path of a conventional pendulum having the same curvature at the bottom of their swings.

Recalling Eq. (10.2), it is clear that a bead on a circular arc of radius $R = 4r$ will have the same period as the cycloid, but only for very small oscillations near the bottom of the loop. Otherwise the deviations illustrated in Fig. 10.11 will come into effect. The relationship between the circle and cycloid is illustrated in Fig. 10.14.

The proof just presented that a cycloid possesses an isochronous property was based on the principle of conservation of energy and then was worked through with the methods of calculus. But at the time of the *Horologium Oscillatorium*, Isaac Newton's *Principia* was not to appear for another fourteen years (1687). Huygens' methods of investigation were mainly an insightful combination of logical reasoning and geometrical constructions. He discovered the isochronous property and reported it in the *Horologium* as Proposition XXV in Part II.

On a cycloid whose axis is erected on the perpendicular and whose vertex is located on the bottom, the times of descent, in which a body arrives at the lowest point at the vertex having departed from any point on the cycloid, are equal to each other; and these times are related to the time of a perpendicular fall through the whole axis of the cycloid with the same ratio by which the circumference of a circle is related to its diameter.

Now reversing the argument, one could say that if the mass on the end of the string in a pendulum could somehow be *obliged* to follow a cycloidal path rather than a circular one, then the period of the pendulum would become independent of the size of the swing. Huygens set about addressing this point and found an interesting and, as it happened, useful property of cycloidal curves. As stated in Proposition VI of Part III of the *Horologium*, "By the evolution[12] of a semicycloid, starting at the apex, another semicycloid is described, which is equal and similar to its evolute, and whose base is a line which is tangent to the evolute cycloid at its apex."

In other words, the evolute of a cycloid is another cycloid.[13] Following the Proposition, Huygens went on to remark, "For it is clear that a

[12] Suppose one end of a string is fixed to a point on a curve in the plane and the remainder of the string is wrapped around the curve. The path traced by the free end of the string as it is peeled away while keeping the string taut is the *involute* of the original planar curve—the *evolute*.

[13] See Hildebrandt and Tromba (1996, p. 123). A geometrical proof in the style of Huygens may be found in appendix 4 of Rawlings (1948).

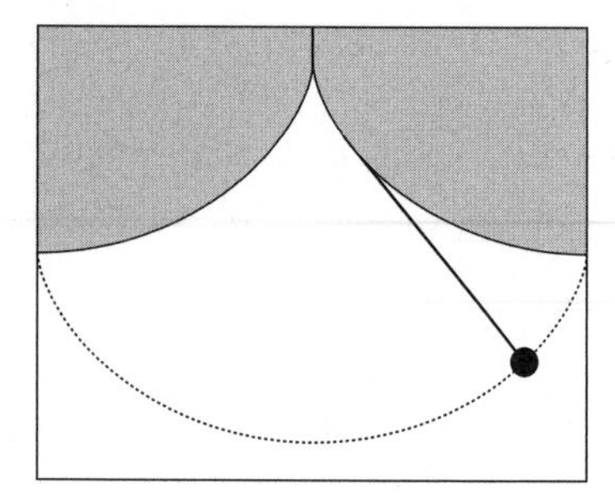

Fig. 10.15
A pendulum whose suspension cord is constrained between two cycloidal cheeks (shaded). By wrapping around the upper portion of a cheek, the cord is effectively shortened slightly for larger displacement angles, and the pendulum bob traces a cycloidal, not circular, path.

pendulum which is suspended and set in motion between two plates curved in the shape of a semicycloid describes an arc of a cycloid by its motion, and thus oscillations, whatever their magnitude, are completed in equal times."

Here then is the key point in Huygens approach to his pendulum clock: the period can be made independent of the amplitude of the swing by constraining the suspension between cycloidal cheeks, as shown in Fig. II of Fig. 10.10 and in Fig. 10.15.

10.2.3 The seconds pendulum and the meter: An historical note

An ideal simple pendulum, without any frictional losses and executing very small angle oscillations, will take exactly one second to swing from left-to-right or right-to-left provided its length is precisely[14] 99.357 cm (39.12 in.). (See Problem 2 in Chapter 2.) With this length, one has what has been historically referred to as a "seconds pendulum." In modern parlance, the period denotes the time for a complete dynamical cycle, so the period of a seconds pendulum actually is two seconds. However, the coincidental relationship between the half-period of the pendulum and the meter played an important, but short-lived role in the definition of the standard meter.

We now take for granted the existence of standards of measurement such as the precisely defined meter, second, and kilogram. Yet in the late 1700s and early 1800s the quantities used for measurement were anything but standardized. For example, in France prior to the revolution, a whole array of different length measurements were in common usage. These included the "toise" which was defined as six royal feet or about two meters. Or there was the "parish" measuring rod which consisted of the sum of the lengths of the feet of sixteen randomly sized men, measured as they left the parish church. Aside from the plethora of imprecisely defined units, the same unit could also mean different measures in different places or different circumstances. For example, a pint of wine in Paris was relatively small compared to a pint of wine in Précy-sous-Thil that was more than three times the Parisian pint (Heilbron 1989).

The French system also had a feudal component in that the landowner would always want his tenants to pay him in the largest size of a particular unit of measurement whereas he would want to sell the produce in the smallest size of the same unit. In England, the situation was somewhat less confusing and there were even laws that mandated a national standard for weights and measures. Nevertheless standard measures of length and volume contained inconsistencies. For example, a gallon typically contained less, by a half pint, than four quarts.

Contemporary with these domestic issues, scientists were swinging pendulums in various parts of the world to determine the shape of the world, as noted in Chapter 2. Momentum grew on both sides of the Channel for using the length of the "seconds" pendulum as the standard

[14] These numbers are for sea level at latitude 45°.

meter. Christopher Wren is believed to have been the first to suggest this standard, but it also received early support from Christiaan Huygens and Jean Picard. Of course, because of deviation of the earth's shape from sphericity, variable topography and density, the standard pendulum would have to be swung in a specified location.

After the French revolution the National Assembly turned to the question of a single standard. A Committee of Weights and Measures was appointed. In 1790, Talleyrand[15] suggested that the unit of length be that of a seconds pendulum at 45° latitude. And in 1792, new measurements were made on the seconds pendulum at the Paris Observatory. But by 1791 events had moved away from a pendulum standard. The National Academy, upon which the Assembly relied in matters of science, had another agenda, namely their long-standing project of measuring the meridian through Paris. Therefore the Assembly voted to define the meter as one ten-millionth of the distance from the equator to the north pole, measured along the meridian of longitude passing through Paris. Monies were appropriated, new surveys along the meridian were funded, and in 1799 these measurements led to a brass meter bar as the contemporary standard meter.

In England, the pendulum seemed to fare somewhat better. For example, Henry Kater, of Kater's pendulum fame, was appointed a member of the 1819 royal commission on weights and measures. Yet, there was still the matter of where to swing the pendulum and as pendulum measurements became increasingly more precise this issue was increasingly more urgent. In 1824, the pendulum was given a legal niche in the measurement of a meter in a rather unique way.(Simpson 1993). The Weights and Measures Act of 1824 actually mandated an Imperial standard yard based upon a length previously constructed in 1760. However, the act stipulated that, in the event that this physical yard was destroyed or damaged, a new standard length would be created using a pendulum swung in a place with the *same latitude as London*. (This lack of precision did not help the pendulum's cause.) Remarkably, on October 16, 1834, a fire in Westminster burnt the Commons and damaged the standard yard beyond repair. It would seem that the pendulum was about to create a new standard of length. But the promise was not to be realized. Because of delays, lack of agreement as to where the pendulum should be swung, advances in other mensuration techniques, the British pendulum standard was abandoned in 1851. Thus the pendulum's legal celebrity as a standard of length was relatively short-lived.

The definition of the meter has been changed significantly. Now it is the distance travelled by light in a vacuum within a time of 1/299, 792, 458 of a second. Thus the length definition is derived from the universality of the speed of light, and the definition of the second (which is now based on a resonant frequency of the cesium atom: 9, 192, 631, 770 Hz).

[15] Charles Maurice de Talleyrand-Périgord 1754–1838. French statesman and diplomat during the turbulent years before and after the Revolution, member of the Estates-General and of the National Assembly 1789, Foreign Minister under Napoleon Bonaparte 1799–1807.

10.2.4 Escapements

Huygens' clock used the verge and crown wheel (now more generally referred to as an escape wheel or as the scape wheel). The angle between the pallets was typically 90° or slightly more. For correct block and release of the teeth in the escape wheel, the verge has to rock back and forth through a considerable angle—perhaps 20° or so (the rocking of the foliot beam in Fig. 10.3 shows this same effect). Thus the pendulum must swing with a rather large amplitude of oscillation for the verge to work properly. But of course, large swings introduce circular errors which could in turn be rectified with cycloidal cheeks.

Within a short time significant technical advances took place in escapement design. About 1671, the *anchor* escapement appeared. The originator of this innovation is not known for certain, but it is variously attributed to the brilliant scientist Robert Hooke or the clockmaker William Clement. A drawing of this escapement is shown in Fig. 10.16. As the pendulum swings back and forth, the anchor rocks to and fro alternately blocking and releasing the escape wheel, which is powered by weights or a spring so as to turn clockwise as viewed in the illustration. One notable change in this arrangement is that whereas in the verge escapement, the shaft of the verge and the shaft of the crown wheel are perpendicular to one another (see Fig. 10.9), but in the anchor escapement, the shaft of the anchor is parallel to the shaft of the escape wheel. That is, for the anchor escapement, all rotating components (pendulum, escape wheel, anchor) move in parallel planes. This means that a thin clock mechanism can be produced.

Fig. 10.16
The anchor escapement consisting of the escape wheel and pallet.

The anchor escapement can accommodate quite small pendulum oscillations, in contrast to the verge and crown wheel. This improvement in performance immediately made possible long pendulum clocks and reduced the potential usefulness of cycloidal cheeks. Rawlings (1948) points to the problem of scale. He takes as a working example, a pendulum of length 100 cm. For a maximum angular displacement of $\pm 3°$, he calculates that the portion of the cheeks that must be in contact with the suspension cord is only a little over a millimeter in size. Producing perfectly shaped cheeks on this scale might be difficult, but more problematical is the requirement of maintaining perfect contact between the suspension fiber and cheek over such a tiny area. Cycloidal cheeks are no longer a part of the strategy for making high performance pendulum clocks.

The *dead-beat* escapement was invented by George Graham in 1715. Its design eliminated an undesirable aspect of the anchor escapement—recoil. In the anchor movement, just after a tooth of the escape wheel is intercepted by one of the pallets, there is a slight recoil as the pallet pushes further in and forces the escape wheel slightly backwards. In clocks which employ this form of escapement, this results in a slight observable shiver in the seconds hand at each tick.[16]

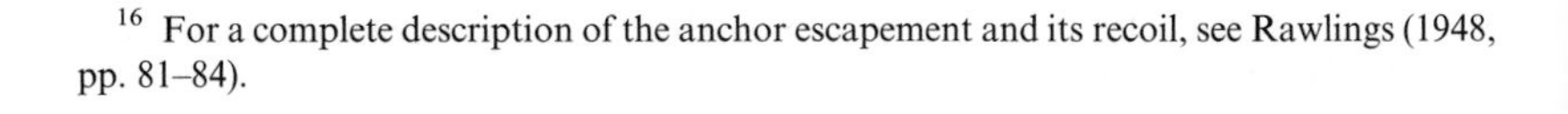
[16] For a complete description of the anchor escapement and its recoil, see Rawlings (1948, pp. 81–84).

The dead-beat design is shown in Fig. 10.17. Each pallet has a relatively large stopping face and a smaller, angled impulse face. The stopping faces are on the outside of Pallet A and on the inside of Pallet B. Here the pendulum, in its swing to the right, has carried pallet B just away from a tooth of the escape wheel. But notice that the tip of the tooth now pushes slightly on the angled impulse face of the pallet. This action gives a slight helping torque (counterclockwise) to the pendulum. When the pendulum rotates a little more, pallet B will swing clear and the escape wheel will freely rotate until one of its teeth is caught by the stopping face of pallet A. The pendulum then completes it swing to the right and reverses itself now slowly lifting pallet A until it starts the impulse sequence already seen for pallet B, as in the figure. This time the helping torque from the angled pallet face is clockwise. These helping torques from the escapement act to make good the frictional losses[17] and keep the pendulum running. No recoil is generated as a wheel tooth slides along a stopping face because the stopping faces are arcs with the pallet axis as their centre.

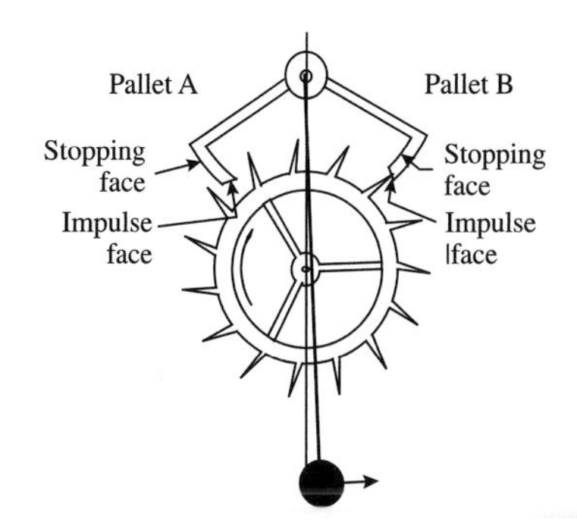

Fig. 10.17
The dead-beat escapement.

A close-up view of a wheel tooth and pallet is shown in Fig. 10.18. The various proportions in the design of a dead beat escapement seem to be partly science and partly art. For example, Kesteven (1978) states that the slope angle of the impulse face δ is "generally of order 80°," whereas Rawlings (1948) gives 45° as a good value. On the matter of pallet geometry, Rawlings[18] says: "The comparative advantages of long and short pallet arms have been debated for a century."

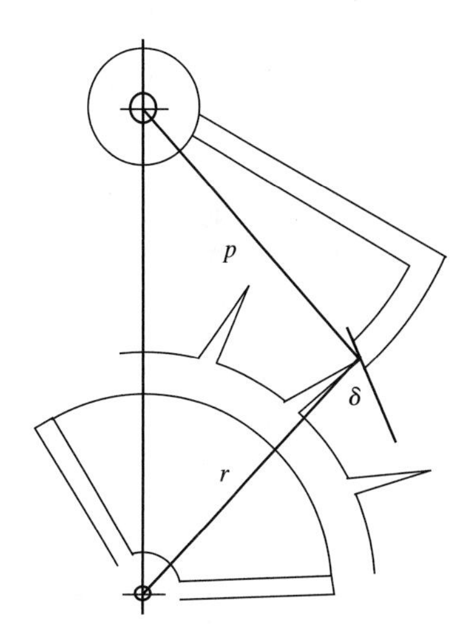

Fig. 10.18
Geometry of the dead beat escapement showing the contact between an escape wheel tooth and the impulse face on the pallet.

There are many designs for escapements, of which could be mentioned the pin-wheel, Brocot pin pallet, Mudge gravity, Grimthorpe three-legged gravity, Riefler, and Bloxham. They can be quite complex and ingenious in their attempts to refine the art and enable more and more accuracy in clocks. The books by Rawlings (1948) and Smith (1975) are recommended on this topic. An interesting paper by Kesteven (1978) discusses the mathematics of the design and performance of the dead-beat and Grimthorpe escapements.

Simple model of the escapement

In 1872, George Airy, the seventh Astronomer Royal, published a mathematical analysis of the mechanics of escapements in pendulum clocks (Kesteven 1978). Kesteven's paper is a condensation of Airy's original work. Some of the flavor of these discussions is now outlined.

The purpose of the escapement is to provide enough torque on the pendulum to overcome the effect of friction on the pendulum and therefore keep the pendulum in a steady state oscillatory motion. But there is another benefit provided by the escapement. We recall that one of Huygens' contributions to the design of the pendulum clock was the pair of cycloidal cheeks which ensured that the pendulum's oscillations were isochronic. It

[17] Rawlings (1948) points out (p. 67) that about four fifths of the energy of the escapement is used to overcome the effects of air friction. While only about one fifth compensates for losses associated with the pivot, the suspension, and friction in the escapement itself.

[18] Rawlings (1948, p.85).

was noted earlier that the region of contact with the cycloidal cheeks is quite small and that they are not easily utilized to assure isochronicity. Modern pendulum clocks do not have these leaves. Instead, the escapement provides some of the needed compensation. Through its reaction back on the pendulum the escapement slightly affects the period of the pendulum, and it happens that this effect can be made to counter the lack of isochronicity.

Let us introduce two quantities that specify the deviation of the pendulum's period from its zero displacement value of $T_0 = 2\pi\sqrt{l/g}$. The first deviation is due to the finite amplitude α of the pendulum swing. This effect is called the *circular error* and is defined as $E_C = (T_R - T_0)/T_0$, where T_R is the period of the nonlinear pendulum. Using the power series expansion for T_R given in Chapter 3 it is found that $E_R \approx \alpha^2/16$, where α is the angular amplitude of the pendulum. The other deviation is called the *escapement error* and is defined as $E_E = (T_E - T_0)/T_0$, where T_E is the period of the pendulum when the escapement is connected to the pendulum. Under some fairly general assumptions the escapement error can be shown to be $E_E = S_1/(2mgl\alpha)$, where S_1 is the amplitude of the fundamental frequency of the escapement forcing term and the other quantities are the parameters of the equivalent simple pendulum. From this relationship it is clear that a stronger reaction of the escapement on the pendulum creates a stronger effect on the period of the pendulum, but the effect diminishes with increasing amplitude of oscillation. Remarkably, it is found that E_E and E_R have opposite signs and therefore it is theoretically possible to achieve isochronism without the use of the cycloidal leaves. In practice perfect balance is not likely, but the addition of the cheeks is still not sufficiently helpful to justify the effort.

The calculation of S_1 is specific to a particular escapement. Such calculations can suggest optimal angles of pendulum displacement at which the escapement should contact the pendulum. However, any reasonably tractable model is only an approximation to reality and therefore adjustment of the escapement for optimal performance must be done empirically.

While the paper by Kesteven (1978) provides a mathematical analysis peculiar to each of the dead-beat and Grimthorpe escapements, it is possible to get some idea of the effect of the escapement from a generic model. For all escapements the pendulum receives an impulse when it is positioned on either side of the vertical. The thrust or drive function is antisymmetric although the two pulses are not spaced symmetrically about the origin. In other words the pendulum may be at some angle δ_1 for one contact with the escapement and at an angle $-\delta_2$ for the other contact, but the values of these angles may differ only slightly. For the dead-beat escapement Kesteven uses a square wave drive whereas for the Grimthorpe escapement the drive is a square wave with a piece of one corner of the square being cut off. For the purposes of a considerably simpler model, an idealized escapement can be imagined as contacting the pendulum equally on either side of the vertical with a very short sharp pulse. In dimensionless units the equation of motion of the linearized pendulum is

$$\frac{d^2\theta}{dt^2} + 2\gamma\frac{d\theta}{dt} + \omega_0^2\theta = F(t) \tag{10.13}$$

with the simple initial conditions that $\theta(0) = 0$ and $\dot{\theta}(0) = 0$ The dimensionless torque $F(t)$, provided by the escapement, is approximated by a series of Dirac delta functions, introduced in Chapter 3, which approximate sharp pulses given at $T/4$ and $3T/4$, where T is the period of the pendulum. Exactly where the pulses are delivered does not materially affect the motion, but the choice of $T/4$ and $3T/4$ is one that simplifies the mathematics. The small amplitude period depends on the friction γ and the natural frequency ω_0 according to $T = 2\pi/\omega$, where $\omega = \sqrt{\omega_0^2 - \gamma^2}$. Thus, in the first two periods of the pendulum the applied torque is

$$F(t) = A[\delta(t - T/4) - \delta(t - 3T/4) + \delta(t - 5T/4) - \delta(t - 7T/4)]. \tag{10.14}$$

Therefore the full equation of motion is

$$\frac{d^2\theta}{dt^2} + 2\gamma\frac{d\theta}{dt} + \omega_0^2\theta = A\sum_{n=0}^{k}[\delta(t - (4n+1)T/4) - \delta(t - (4n+3)T/4)], \tag{10.15}$$

where k is the number of periods the pendulum is in contact with the escapement. Using the Laplace transform method (introduced in Chapter 3 for the pushed swing), the transform of the equation of motion is

$$\Theta(s) = \frac{A}{(s+\gamma)^2 + \omega^2}\left[\sum_{n=0}^{k}\left[e^{-s(4n+1)T/4}\right] - \left[e^{-s(4n+3)T/4}\right]\right], \tag{10.16}$$

which yields a solution

$$\theta(t) = -\frac{A}{\omega}\sum_{n=0}^{k}[\cos\omega t(e^{-\gamma(t-(4n+1)T/4)}U(t - (4n+1)T/4)$$
$$+ e^{-\gamma(t-(4n+3)T/4)}U(t - (4n+3)T/4)]. \tag{10.17}$$

This solution looks very complex, but it is quite similar to that for the pumped swing. The effect of the escapement is to provide a slow buildup of a sinusoidal motion as shown in Fig. 10.19. The only significant difference is that while the swing model was pumped once during a cycle, the clock pendulum is pumped twice by the escapement.

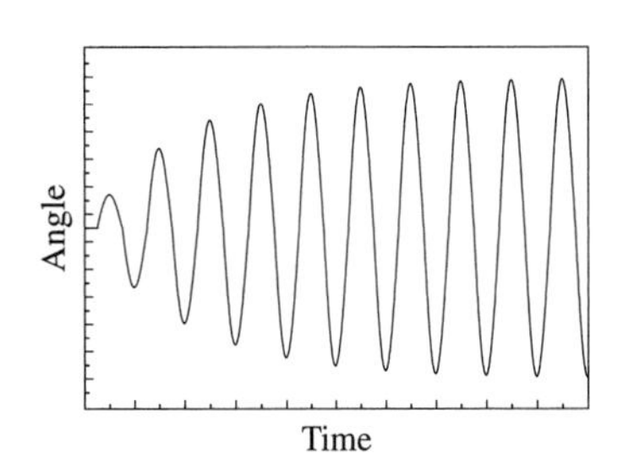

Fig. 10.19
Evolution of the pendulum angle as a result of the action of the escapement.

10.2.5 Temperature compensation

If a pendulum consists, as is often the case, of a mass on the end of a metal rod, and the temperature of the metal rod shifts, then thermal expansion will change the length of the pendulum. The change in length from l_0 at some temperature T_0 to a new value l at temperature T is governed by the linear relationship

$$l = l_0(1 + \alpha\delta T), \tag{10.18}$$

where $\delta T = (T - T_0)$ and α is the coefficient of thermal expansion. Expressed another way, the length change is

$$\Delta l = \alpha l_0 \Delta T. \tag{10.19}$$

As an example, suppose the rod is made from steel, for which $\alpha =$ 0.000012 /°C. For a temperature rise of, say 5°C, a 1 m rod would grow longer by 0.06 m (a relative change of 6 parts in 100,000). This may seem negligibly small, but consider the following facts:

1. There are 86,400 seconds in a day.
2. A pendulum with a half-period of 1 s is almost a meter in length.
3. The period varies as the square root of the length. Thus the *proportional* change in period is half of any small *proportional* change in length.

For the example just proposed, the lengthened steel rod would result in a period change of 3 parts in 100,000, enough to produce an error of more than a second per day. This demonstrates the practical significance of thermal expansion in the operation of pendulum clocks.

In 1729, John Harrison invented the so-called *gridiron* pendulum which cleverly compensated for temperature variations. The underlying principle of the gridiron compensator is illustrated in Fig. 10.20. Suppose the desired suspension length of a pendulum is L_0. If its temperature rises, a rod of this initial length will expand by an amount Δ_0, as suggested in the diagram. With this new total length, $L_0 + \Delta_0$, the pendulum period will increase and the clock will run slow.

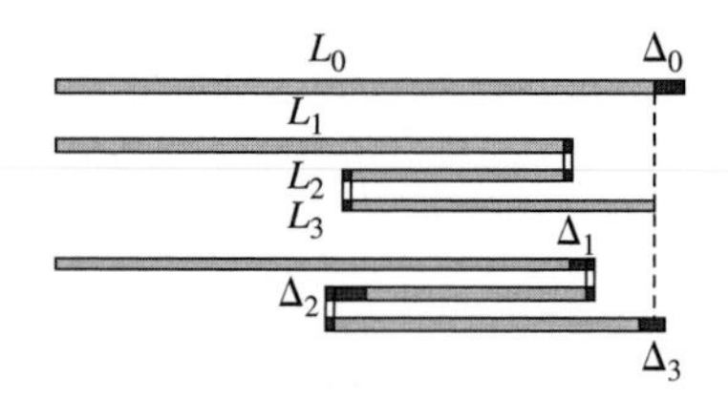

Fig. 10.20
Principle of the gridiron temperature compensator. In each case, the original lengths of the segments are designated L_0, L_1, L_2, L_3 and the expansions are $\Delta_0, \Delta_1, \Delta_2, \Delta_3$.

Now suppose that a new rod is made from three segments, connected as shown, with the first and last segments formed from the same metal as in the original, but with a different material employed for the middle segment. When this composite structure is heated, the first and last segments increase in length by Δ_1 and Δ_3, while the mid section grows by Δ_2. From the illustration it can be seen that the free end is carried to the right by the expansions of the first and last segments, but is carried to the left by the expansion of the middle segment.

Measured from the left-hand end, the new length will be $(L_1 + \Delta_1) - (L_2 + \Delta_2) + (L_3 + \Delta_3)$. The goal here is to find the necessary condition for zero net displacement of the free end. Therefore set

$$(L_1 + \Delta_1) - (L_2 + \Delta_2) + (L_3 + \Delta_3) = L_0$$

or

$$(L_1 - L_2 + L_3) + (\Delta_1 - \Delta_2 + \Delta_3) = L_0.$$

But of course, $L_1 - L_2 + L_3 = L_0$ since these are the unexpanded lengths as indicated in the middle figure. Hence we require

$$(\Delta_1 - \Delta_2 + \Delta_3) = 0. \tag{10.20}$$

It is quite feasible to make the first and last segments equal in length, that is, $L_1 = L_3$, in which case $\Delta_1 = \Delta_3$.. Then the requirement for a constant length of the composite rod is

$$\Delta_2 = 2\Delta_1. \tag{10.21}$$

The example just discussed was made up of three segments. The scheme can be generalized to any odd number of segments. The next possibility would be composed of five pieces. The condition for zero net length change then becomes

$$(\Delta_1 - \Delta_2 + \Delta_3 - \Delta_4 + \Delta_5) = 0.$$

Segments 1,3,5 are made from one material, while segments 2 and 4 are from the alternate metal. Then by choosing to have $L_1 = L_3 = L_5$ and $L_2 = L_4$,

$$\Delta_2 = \frac{3}{2}\Delta_1. \tag{10.22}$$

For seven segments, the necessary condition is

$$\Delta_2 = \frac{4}{3}\Delta_1. \tag{10.23}$$

so the pattern for Δ_2/Δ_1 is $2/1, 3/2, 4/3, 5/4, \ldots$, depending on the number of back and forth stages in the system.

Suppose now that the compensator is designed so that *all* rods are initially of the same length L. Then from Eq. (10.19) the length changes are $\Delta_1 = \alpha_1 L \Delta T$ and $\Delta_2 = \alpha_2 L \Delta T$. Therefore, $\alpha_1/\alpha_2 = \Delta_1/\Delta_2$. Using Eqs. (10.21), (10.22), or (10.23), the condition for zero displacement of the free end becomes a required ratio of thermal expansion coefficients:

$$\frac{\alpha_2}{\alpha_1} = 2, \text{or} \frac{3}{2}, \text{or} \frac{4}{3}, \text{or} \ldots \tag{10.24}$$

The expansion coefficients for some common metals are given in the following table.

Metal	Expansion coefficient /°C
Red brass	0.0000189
Carbon steel	0.000012
Aluminum	0.0000255
Copper	0.000016

Notice that the ratio of brass to steel is 18.9/12 = 1.575, a value quite close to the ideal of 1.5 for a 5 segment design. This fortuitous relationship between the expansion coefficients of these two common and easily worked metals means that, with some slight refinements, a five segment temperature compensator is a practical possibility.

Harrison incorporated these ideas in the gridiron. The actual arrangement was symmetrical, as shown in Fig. 10.21. There are nine rods in total,

Fig. 10.21
General arrangement of metal rods in a gridiron pendulum. Where a pin in a cross beam is absent, the rod passes freely through. The array is symmetric with a total of five steel and four brass rods.

but the configuration is equivalent to the five segment case, mirrored about the centre rod. Gridirons are relatively common in quality pendulum clocks, even nowadays.

Other strategies have been devised for temperature compensation. In 1721 Graham placed a container with mercury at the end of the steel pendulum rod. When warmed, the steel rod lengthens, but the mercury rises, just as in a thermometer, counteracting the effect. In 1895 C. E. Guillaume produced an alloy of iron and nickel, known as Invar, that has a coefficient of expansion about one hundred times smaller than normal metals. Pendulum rods made of Invar avoid the need for compensation from a gridiron or other techniques.

10.2.6 The most accurate pendulum clock ever made

By the end of the nineteenth century, the state of the art in pendulum clocks, as typified by those from Siegmund Riefler of Munich, was a daily error of around 0.1 s. Riefler's clock took the additional step of providing a partial vacuum for the pendulum by enclosing the clock mechanism within a metal cylinder. Riefler regulators were the standard for timekeeping for about thirty years, until the final masterpiece of pendulum clock design appeared in 1921—William H. Shortt's subtle and intricate dual pendulum arrangement.

The idea behind the Shortt clock is to somehow allow a pendulum to oscillate alone and without any interference (free pendulum), and only to supply it with enough energy to keep it going. This master pendulum subtly communicates with the slave pendulum whose task is to count pulses and display time. Thus the complete system was composed of a pair of pendulum oscillators, a *master* and a *slave*.

The genesis of this arrangement was to be found in earlier systems for electrically synchronizing networks of clocks such as were located in schools and railways. A nice discussion of early synchronizing schemes and of the operating principles of the Shortt clock may be found in Rawlings (1948); a figure from that book is reproduced in Fig. 10.22. A very detailed description of the action and role of every part is provided in a set of manuals for building a working copy of the Shortt clock.[19] A photograph showing both master and slave units is given in Fig. 10.23. Note that this system is electro-mechanical, in contrast to all of the preceding clocks which were purely mechanical. Only electrical wires connect master and slave, a fact apparent from Fig. 10.22. Close up photographs of the master and slave mechanisms are shown in Fig. 10.24. The components in the schematic drawing of Shortt's design, Fig. 10.22, are clearly recognizable in the two photographs.

Briefly, the operation of the clock may be described as follows:[20] The time-keeping element is a pendulum that swings freely, except that every

[19] ModelPlans no. 130: *Shortt Hit and Miss Synchronometer* by Michael Adler, available from M. W. Models, 4 Grays Road, Henley-on-Thames, England.
[20] Encyclopedia Brittanica Online, "clock."

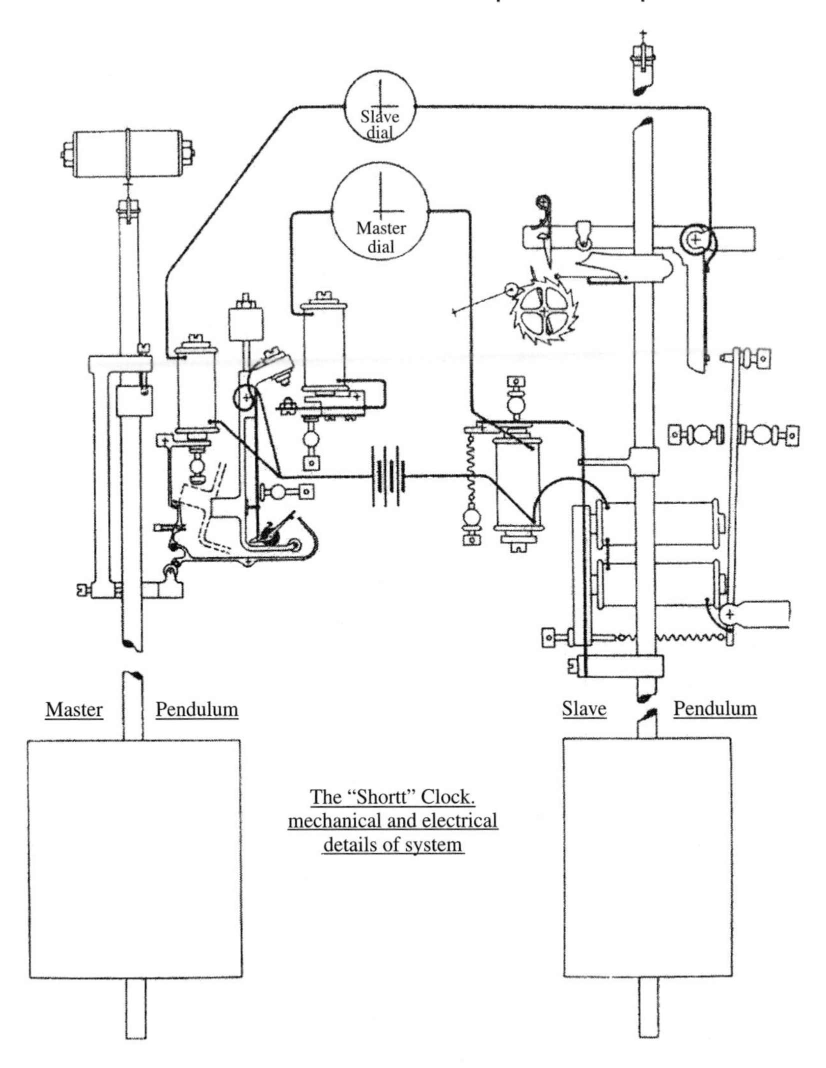

Fig. 10.22
The Shortt clock, showing both *master* and *slave* mechanisms.

half-minute it receives an impulse from a gently falling lever. This lever is released by an electrical signal transmitted from its slave clock. After the impulse has been sent, a synchronizing signal is transmitted back to the slave clock that ensures that the impulse to the free pendulum will be released exactly a half-minute later than the previous impulse.

About one hundred Shortt clocks were manufactured and installed mainly in astronomical observatories around the world. They performed exceedingly well, accumulating errors of only about a second a year.

The pursuit of perfection in mechanical clocks seems neverending. In the 1950s and 1960s, in Russia, F. M. Fedchenko (1957, 1962), Fedchenko and Fleer (1966) designed a single pendulum clock to rival the Shortt master–slave system. In appearance, Fig. 10.25, the clock was similar to the Shortt master. This design is unique in two respects: the suspension system and the electromagnetic impulsing.

Fig. 10.23
Photograph of the Number 16 Shortt clock, master on the left, on display at the National Maritime Museum, Greenwich (detail). ©National Maritime Museum, London.

The standard suspension springs for pendulum clocks are thin rectangular metal ribbons, typically about an inch in length and a half-inch in width, clamped between thicker plates at the top and bottom. The pendulum rod is fixed to the lower of the clamping plates. When the pendulum swings, the spring flexes, as suggested in Fig. 10.26

The conventional idea behind these suspensions is that the bending of the spring into a curve effectively shortens the pendulum in a manner reminiscent of the cycloidal cheeks of Huygens, thus producing isochronism (period independent of amplitude of swing). However Fedchenko discovered through careful experimental observation that something rather different actually was happening. He found that the flexing actually *raised* the point where the extension of a line from the pendulum bob along the rod intersected a vertical line descending from the upper clamp (these intersections are shown in the figure as small circles); for cycloidal cheeks, the intersection point would be *lowered* as the angle increased. Fedchenko concluded that "a normal spring suspension does not improve in principle the isochronism of pendulum oscillations; on the contrary, it makes them worse" (Fedchenko and Fleer 1966). Fedchenko's refinement of the standard suspension was to introduce a double spring configuration that in combination did produce the required slight lowering of the effective rotation axis and so achieved the shortening of the equivalent pendulum length that was needed for isochronism.

The Fedchenko clock used electromagnetic pulsing as a means of injecting energy sufficient to keep it going. A permanent magnet was attached at the bottom of the pendulum . As the pendulum swung to the centre-down position, the magnet passed quickly into a pair of concentrically wound coils. One of these, a pick-up coil, responded to the moving

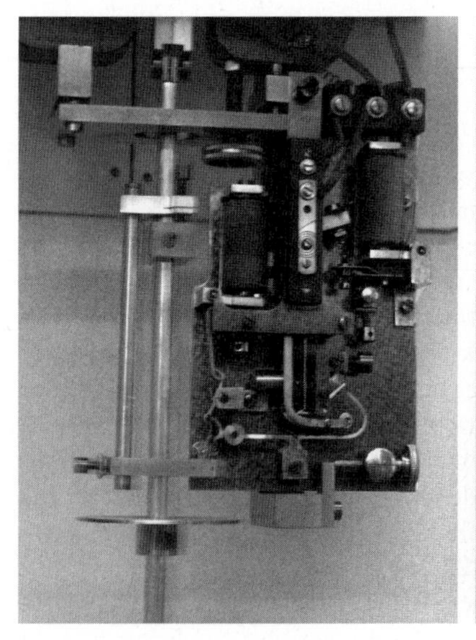

Fig. 10.24
Photographs by one of the authors (JAB) of the Shortt pendulum clocks in the Science Museum, London showing the mechanism of the Master (left) and Slave (right). (Reproduced with permission from the Science Museum, London.)

magnet by generating an induced signal. This signal was then fed to a transistorized amplifier which immediately produced a drive current for the second drive coil. The polarity and timing of this pulser was arranged so that the magnetic field created by the drive coil would repel the permanent magnet on the pendulum just as the pendulum swung past the lowest position, giving it a slight "kick."

Horologists might debate the point, but it would seem that the Shortt and Fedchenko clocks performed at about the same (impressive) level.

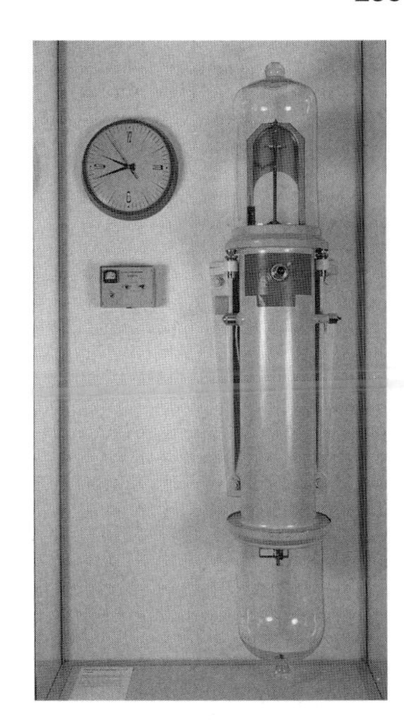

Fig. 10.25
The Fedchenko clock AChF-3 displayed in the National Maritime Museum, Greenwich. ©National Maritime Museum, London.

10.3 Reflections

It took nearly six centuries to progress from the verge and foliot to the Shortt pendulum clock. Over this span of time, Western Europe has moved from the Middle Ages, through the Renaissance, the Enlightenment, the Industrial Revolution, and into the twentieth century. The error in the early mechanical clocks of about a half-hour per day was ultimately reduced to about one second per year for the very best pendulum based devices. The Shortt master–slave clock was the pure essence of technical craft and ingenuity. It was an accurate, albeit delicate, timekeeping instrument.

The accelerating rate of technological change caught up with the just invented Shortt clock. Not long after its introduction in 1921, a remarkable advance in technology sealed the fate of pendulum clocks as the ultimate reference timekeepers. It was discovered that resonant quartz crystals could serve as the frequency stabilizing elements in electronic oscillators; by 1929 this principle had been applied to clocks. The performance of these quartz clocks was extraordinary, bringing errors down to less than a second in a decade. Beginning in the 1970s, the microelectronics revolution created the integrated circuit. Now a relatively inexpensive plastic watch with a small silicon chip and a tiny quartz crystal can rival the best products of the clockmakers art, at least insofar as accuracy is concerned.

But pendulum clocks persist. Their mechanisms continue to fascinate and they continue to tick away in their outdated mechanical elegance. It is an almost curious fact that a pendulum whose length is comparable in scale to the size of a person, oscillates in the gravitational field of the earth at a rate that is so sympathetic to human nature.

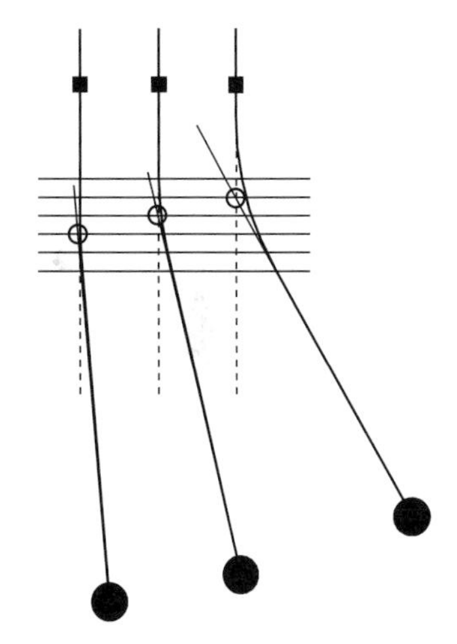

Fig. 10.26
Geometry of a conventional metal ribbon pendulum support. The upper end of the spring, which is viewed edge on, is clamped at the point indicated by the square. As the pendulum oscillates to larger angles, the spring flexes as illustrated.

10.4 Exercises

1. In a typical pendulum clock the slow downward motion of weights provides the energy to overcome friction in the clock's movement. If a weight of a 5 kg mass descends through 0.1 m in one week, how much energy, on average, does the clock dissipate in 1 s?
2. Studies (p. 67 of Rawlings (1948)) have shown that about 80% of the power loss in a pendulum clock is due to friction of the pendulum with air. Therefore observation of the decay of the pendulum's motion, when disconnected from the clock mechanism, can give a reasonable estimate of the rate at which the

pendulum loses energy. In this exercise we show that the instantaneous rate of energy loss (energy loss per second) is not constant, but is proportional to the square of the angular amplitude of the pendulum, θ.

(a) For a pendulum of length l, mass m, and angular displacement θ, write an expression for the potential energy of the pendulum.
(b) Assume that θ is small and rewrite you answer to (a) using a first order approximation for the trigonometric function.
(c) In time the amplitude of the undriven pendulum decays exponentially according to $\theta(t) = \theta_0 e^{-\alpha t}$ where α is a constant that depends upon the pendulum configuration. Use this fact and your answer to (b) to determine the rate of energy loss. Thereby show that, the energy loss per second is proportional to the square of the pendulum's amplitude during a given period.

3. A vehicle, having a spoked wheel of radius R, travels with a velocity v. Suppose that a spot of paint is put on a spoke at a distance $a < R$ from the axle.

(a) Find the parametric equations, $\{x(t),y(t)\}$, of the spot of paint. Initially the paint spot is at its lowest point such that $x(0) = 0$, $y(0) = R - a$. Hint: first consider the motion of the paint spot relative to the axle of the wheel and then consider the motion of the axle relative to the ground.
(b) Let $R = 1$, $v = 1$, and $a = 0.8$ and plot a graph of the motion of the spot on the interval $t \in [0, 2\pi]$.
(c) Let the paint spot be located at the circumference such that $a = R$. Then change the parameter of the equations to $\phi = (v/R)t$. Your final equations should match Eq. (10.3) and Eq. (10.4).

4. Recreate in detail the steps needed to go from the relation $dt = ds/\sqrt{2g(y_a - y)}$ to Eq. (10.9).

5. In this exercise we show that the claim in the text that $\phi = 2\theta$ is true for a cycloid.

(a) Start with the parametric equations for the cycloid and determine the slope, dy/dx.
(b) Note that in Fig.10.13 $\tan\theta$ = the slope of the curve $= dy/dx$. Equate the two expressions.
(c) Use the identity $\tan 2u = (2\tan u)/(1 - \tan^2 u)$ to show that $\tan\phi/2 = (dy/dx)_{\text{cycloid}}$ and therefore that $\phi = 2\theta$ for a cycloid.

6. The treatment of the issue of temperature compensation in the text assumes that the massive part of the pendulum is in the bob and therefore the pendulum in the clock is considered to be a simple pendulum. Let us take this same approach with the temperature compensated pendulum clock (invented by Graham in 1721) that uses Mercury. Assume that the pendulum consists of a long thin, relatively light, steel rod. Affixed to the bottom of the rod, and concentric with it, is a round glass container that holds mercury. When the clock is calibrated at temperature T_0, the length of the steel rod is L_0 and the height of the Mercury in the glass container is l_0. Ignoring the mass of the steel rod we treat the pendulum as if all the mass were concentrated at the center of mass of the Mercury.

(a) What is the effective length of the equivalent simple pendulum, X_0?
(b) The volume coefficient of expansion of the Mercury is $\Delta V/(V\,dT) = \beta$. Assume that the glass container expansion is negligible and that the expansion of the Mercury only occurs in the direction of its height l. Derive an expression for $\Delta l = (l - l_0)$ in terms of β, l_0 and ΔT.
(c) Given that the coefficient of linear expansion for steel is α, derive an expression for the correct ratio of the rod's length to the height of the Mercury, that is, find L_0/l_0.
(d) For $\alpha = 0.000018$ and $\beta = 0.00018$ calculate L_0/l_0.

7. The pendulum described in the previous question can be modelled more closely by considering the pendulum to be a compound pendulum consisting of steel rod of finite mass M_{Fe} and the Mercury as having a mass M_{Hg}. Now, as shown in Chapter 3, the period of the pendulum depends on the ratio I/d where I is the pendulum's moment of inertia, and d is the distance between the pivot point, at the top of the rod, and the center of mass. In order to properly compensate for temperature this ratio must remain unchanged as temperature changes. In this exercise we make the simplifying assumption that the moment of inertia will remain approximately constant with temperature but that the compensation calculation must be done in such a way that $d_0 = d(T)$.

(a) Determine the formula for the center of mass in terms of the nomenclature of the last problem.
(b) Find the differential of this formula treating L, l and T as variables.
(c) Using the condition given above find an expression for L_0/l_0 in terms of M_{Fe}, M_{Hg}, α and β.
(d) Show that in the limit as M_{Fe} gets small compared to M_{Hg}, this formula reduces to the result found in part (c) of the previous question.
(e) Express the mass of each component in terms of its density, cross-section and length; that is, $M_i = \rho_i l_i A_i$, and substitute into the formula found in (c) to derive a quadratic expression for L_0/l_0 and solve the expression for L_0/l_0.
(f) The rod is thin compared to the glass container, $\alpha = 0.000018$, $\beta = 0.00018$, $\rho_{Hg} = 13.6\ g/cm^3$, and $\rho_{Fe} = 8.5\ g/cm^3$. For the following values of A_{Fe}/A_{Hg} $\{0.1, 0.001, 0.0001, 0.00001\}$, compute the ratio L_0/l_0 and compare your answer to that found in exercise 6(d).

8. In the discussion of the generic model of the escapement, only the main equations for the derivation of the escapements effect upon the pendulum motion were stated. Beginning with the equation of motion Eq. (10.15) fill in all the steps necessary to arrive at Eq. (10.17). Use a computer package such as Maple or Mathematica to draw the graph of $\theta(t)$ for the following parameters: $t_i = 0$, $t_f = 10$, $T = 1$, $\omega = 2\pi$, $\gamma = .4$, $A/\omega = 0.2$, and $\kappa = 10$.

Appendix A: Pendulum Q

The pendulum is a mechanical oscillator. Any real pendulum suffers inevitable energy loss through a variety of mechanisms, including bearing friction and air friction, as well as more subtle effects such as stretching and flexing (Nelson and Olsson 1986). The "Q" of an oscillator is a specification that reflects its selectivity, that is the degree to which the system will prefer to oscillate at a particular value of frequency. The higher the Q, the sharper the selectivity and the better the oscillator. The definition of is

$$Q = 2\pi \frac{\text{energy stored in the oscillator}}{\text{average energy dissipated in one period}}. \tag{A.1}$$

The denominator can be transformed as follows. Let P be the rate at which energy is dissipated. Then in one period of the motion $2\pi/\omega$, the energy dissipated will be $2\pi P/\omega$ and so $Q = E/(P/\omega)$. But $1/\omega$ is the time to rotate through one radian, and thus an equivalent definition of Q is

$$Q = \frac{\text{energy stored in the oscillator}}{\text{average energy dissipated per radian}}. \tag{A.2}$$

A.1 Free pendulum

The linearized (small oscillation amplitude) undriven pendulum equation was

$$\frac{d^2\theta}{dt^2} + (2\gamma)\frac{d\theta}{dt} + \left(\frac{g}{l}\right)\theta = 0. \tag{A.3}$$

In the underdamped regime, $\gamma < \omega_0$ where $(\omega_0)^2 = g/l$, the solution of the equation of motion is a decaying oscillation:

$$\theta(t) = Ce^{-\gamma t}[\cos(\omega_1 t + \alpha)], \tag{A.4}$$

where

$$\omega_1 = \sqrt{\omega_0^2 - \gamma^2}. \tag{A.5}$$

The two constants of integration, C and α, can be determined from specified initial conditions. It is evident from this expression that the presence of damping causes the oscillation frequency to shift from ω_0 to the smaller value ω_1.

Using $E = (1/2)ml^2(d\theta/dt)^2 + \frac{1}{2}mg\ell\theta^2$ for the total energy (kinetic plus potential), together with Eq. (A.4), one obtains

$$E(t) = \frac{1}{2}ml^2C^2e^{-2\gamma t}[\gamma\cos(\omega_1 t + \alpha) + \omega_1\sin(\omega_1 t + \alpha)]^2 + \frac{1}{2}mg\ell C^2e^{-2\gamma t}\cos^2(\omega_1 t + \alpha).$$

This has the form: $E(t) = \text{const} \times \exp(-2\gamma t) \times f(t)$, where $f(t)$ is periodic with $T = 2\pi/\omega_1$. Hence,

$$E(t + T) = E(t)e^{-2\gamma T}, \tag{A.6}$$

which says that over a time interval equal to one period, the energy will decrease by the indicated exponential factor. Using this result in the first definition of Q, we have

$$Q = 2\pi\frac{E(t)}{E(t) - E(t + T)} = \frac{2\pi}{1 - e^{-2\gamma T}} \approx \frac{\pi}{\gamma T} \tag{A.7}$$

and, finally,

$$Q = \frac{\omega_1}{2\gamma} \approx \frac{\omega_0}{2\gamma}, \tag{A.8}$$

where the final approximation is for light damping. Clearly, the smaller the dissipation γ, the larger the Q.

A.2 Resonance

When a linearized pendulum is driven by a sinusoidal torque of amplitude Γ and frequency ω, Eq. (A.3) changes to

$$\frac{d^2\theta}{dt^2} + (2\gamma)\frac{d\theta}{dt} + \left(\frac{g}{l}\right)\theta = \frac{\Gamma}{ml^2}\cos\omega t. \tag{A.9}$$

The well-known solution to this inhomogeneous differential equation is

$$\theta(t) = Ce^{-\gamma t}[\cos(\omega_1 t + \alpha)] + B\cos(\omega t - \varphi), \tag{A.10}$$

where

$$B = \frac{\Gamma/ml^2}{\sqrt{(\omega_0^2 - \omega^2)^2 + (2\gamma)^2\omega^2}} \tag{A.11}$$

and the phase lag is given by

$$\tan\varphi = 2\gamma\frac{\omega}{(\omega_0^2 - \omega^2)}. \tag{A.12}$$

The first term in Eq. (A.10) is just Eq. (A.4), which is the general solution of the homogeneous reduction of Eq. (A.9) in the underdamped regime.

The second term in this expression for $\theta(t)$ is a so-called particular solution of the full inhomogeneous equation, obtained from a trial function such as $\theta = c\cos\omega t + d\sin\omega t$. Note that Eq. (A.10) contains only two arbitrary constants, C and α, as expected of a second order differential equation. Observe also that the first term is a transient that decays away, leaving only a long term oscillation embodied in the second term. This sustained oscillation runs at the forcing frequency ω but is phase shifted from the drive by φ.

Using Eq. (A.8), B can be re-expressed

$$B = \frac{\Gamma}{mg\ell}\frac{\omega_0/\omega}{\sqrt{((\omega_0/\omega)-(\omega/\omega_0))^2+\frac{1}{Q^2}}} \tag{A.13}$$

and, as well, the phase lag is

$$\tan\varphi = \frac{1}{Q}\frac{1}{((\omega_0/\omega)-(\omega/\omega_0))} \tag{A.14}$$

The amplitude B of the $\cos(\omega t - \varphi)$ term peaks at a particular drive frequency, the so-called "resonant" frequency

$$\omega_R = \omega_0\sqrt{\left(1-\frac{1}{2Q^2}\right)} \tag{A.15}$$

at which the amplitude of the forced oscillation is

$$B_R = \frac{\Gamma}{mg\ell}\frac{Q}{\sqrt{1-(1/4Q^2)}}. \tag{A.16}$$

For large Q (weak damping) this amplitude is $\approx Q(\Gamma/mg\ell)$. Figure A.1 shows plots of Eq. (A.13) and Eq. (A.14). These are the standard views of resonance in the case where the system is subjected to a sinusoidal forcing of fixed strength but variable frequency. It will be noticed that under these

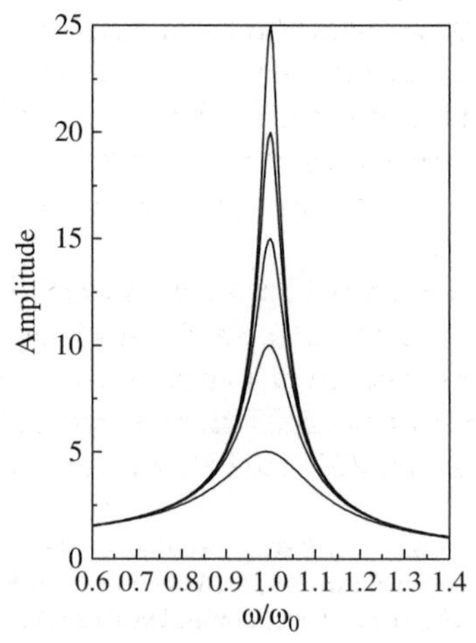

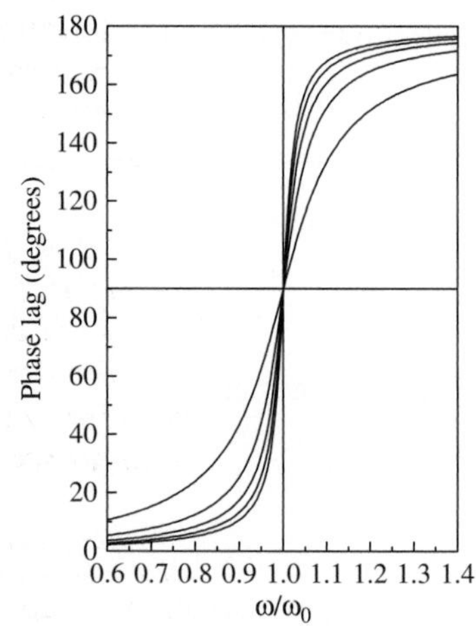

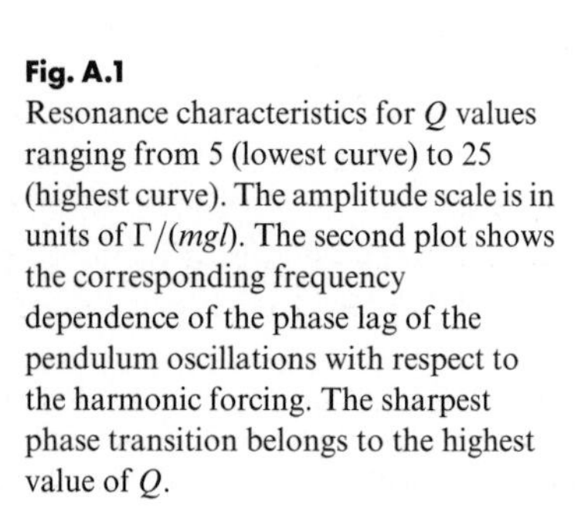

Fig. A.1
Resonance characteristics for Q values ranging from 5 (lowest curve) to 25 (highest curve). The amplitude scale is in units of $\Gamma/(mgl)$. The second plot shows the corresponding frequency dependence of the phase lag of the pendulum oscillations with respect to the harmonic forcing. The sharpest phase transition belongs to the highest value of Q.

conditions, the higher the Q, the larger the response at resonance. This is simply because lower damping at constant forcing strength leads to larger oscillations. However, the pendulum Eq. (A.9) was linearized, so small angle oscillations were assumed. Clearly, it would not be valid to take this resonance analysis outside the domain of small angles.

A useful alternate approach would be to change both Q and the drive amplitude $\Gamma/mg\ell$ in such a way that the height of the resonant peak does not change. This amounts to reducing the forcing strength when Q is increased—lower friction, less applied torque. Figure A.2 illustrates the resulting characteristics. Clearly the effect of increased Q is a sharpening of the response around the resonant frequency.

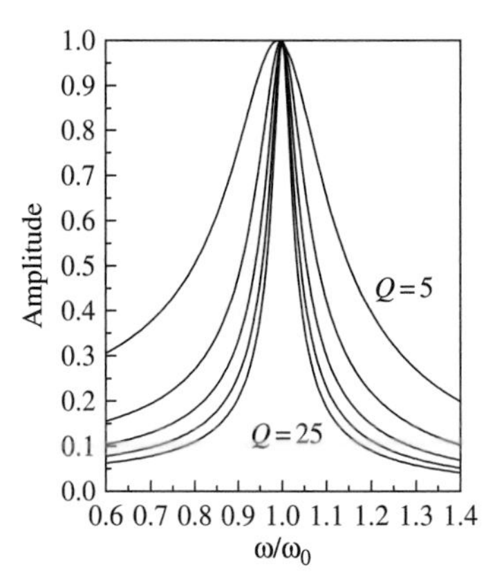

Fig. A.2
Resonant response of a harmonically forced pendulum when the drive strength is adjusted so that the peaks have the same height for all values of damping ($Q = 5, 10, 15, 20, 25$).

Suppose we ask the question: At what frequencies will the magnitude of the oscillations decrease to $1/\sqrt{2}$ of the resonant amplitude? This can be addressed by considering Eq. (A.13); the answer, for large Q, is

$$\omega = \omega_0\left(1 \pm \frac{1}{2Q}\right). \tag{A.17}$$

At these two frequencies lying on either side of the peak, the oscillation amplitude drops to 0.707 of its maximum value at resonance. Equation (A.17) shows that the ratio of the resonant frequency to the width of the resonant peak at a reduced height of $1/\sqrt{2}$ is equal to $\omega_0/(\omega_0/Q) = Q$. This demonstrates that an equivalent definition of Q is:

$$Q = \frac{\text{resonance frequency}}{\text{width of resonance}}. \tag{A.18}$$

A high Q means a sharp resonance.

This discussion about Q applies equally to any type of oscillator, and thus can be applied to torsion pendulums, mechanical watches, quartz oscillators, atomic clocks, and the like. In each instance, the idea is to lock on to a sharply defined resonant frequency by placing the resonator in an appropriate feedback loop.

A.3 Some numbers from the real world

According to Woodward,[1] "The Q of the balance and hairspring in a good mechanical watch might be no more than 100, whilst that of a good pendulum in air might be 10,000. (In a vacuum, it can be 100,000.)" The master pendulum in a Shortt clock had a weight of 10 pounds. It was reported[2] to lose energy at the rate of 3.8 erg/s. Supposing that the pendulum period was of the order of 2s and the swing amplitude was 1°,we can infer that the Q of a Shortt clock was about 15,000.

Torsion pendulum apparatus for gravitational law and equivalence principle experiments have Q values of around 1500 (Hu and Luo, 2000; Wang et al, 2001). Measurements suggesting a first observation of a supersolid phase in Helium-4 were carried out with a torsion pendulum

[1] See Woodward (1995, pp. 11, 12, 91).

[2] See Rawlings (1948, p. 78, 125).

whose resonant period was of the order of 1 ms and whose Q was stated to be 10^6 (Kim and Chan 2004). In the field of gravity wave detection, pendulums composed of fused silica masses suspended at the end of fused silica fibers have achieved Q values as large as 2×10^7 (Cagnoli et al. 2004). Microwave cavity resonators have Q values of around 10,000. For the quartz crystals used in clocks and watches $Q \approx 10^5$. Atomic clocks have $Q \gtrsim 10^8$.

Appendix B: The inverted pendulum

An undriven simple pendulum, left to itself, hangs in the "down" position. If the bob is manually displaced very slightly and released, it returns to the down position after a number of diminishing oscillations. This noninverted state is stable. However, the "up" position, with the mass poised vertically above the pivot, is patently unstable. Practically speaking, it is impossible to position the bob so perfectly in this inverted position that when released it will remain in that state. Any infinitesimal perturbation will cause the bob to fall down, swinging either to the left or to the right.

An historically famous and surprising result is that the inverted position of a pendulum can be stabilized if the pivot point is suitably oscillated up and down.[1] This type of transformation of a normally unstable state to a stable state is known as *dynamic stabilization* (Butikov 2001). To see how this comes about, we begin with the equation of motion for a pendulum, without any explicit external torque, as depicted in Fig. B.1.

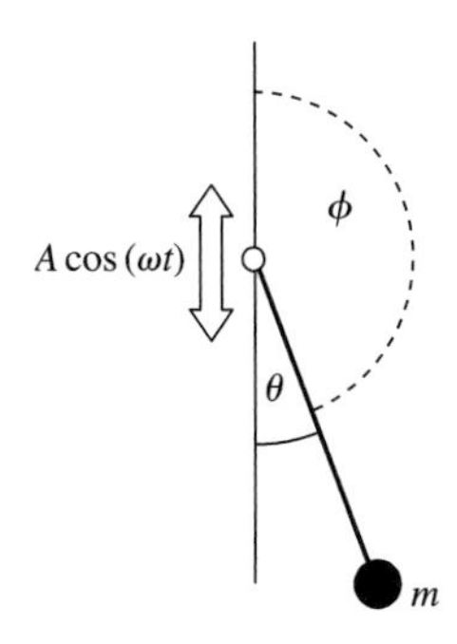

Fig. B.1
Basic geometry of a simple pendulum with a vertically oscillating point of suspension. The length of the rod is ℓ.

$$I\frac{d^2\theta}{dt^2} + b\frac{d\theta}{dt} + m\ell[g - A\omega^2 \cos(\omega t)] \sin\theta = 0. \tag{B.1}$$

The angular coordinate of the pendulum bob, θ, is measured counterclockwise from the down position, b is the damping coefficient, and I is the total moment of inertia of all rotating components of the pendulum (for the simplest pendulum consisting of just a bob on the end of a massless rod, $I = m\ell^2$, otherwise $I \neq m\ell^2$). The up and down pivot oscillation $y = A\cos\omega t$ contributes the additional modulating term in brackets which creates a new *effective* gravitational acceleration.

A natural time scale is provided by the reciprocal of the forcing frequency ω^{-1}. For reasons that will become apparent, we choose the new time unit to be $t^* = \frac{1}{2}\omega t$,

$$\ddot{\theta} + \frac{2b}{\omega I}\dot{\theta} + \frac{4m\ell}{\omega^2 I}[g - A\omega^2 \cos(2t^*)] \sin\theta = 0. \tag{B.2}$$

Define

$$Q = \frac{\omega_0 I}{b}, \tag{B.3}$$

[1] See Smith and Blackburn (1992).

where ω_0 is the natural frequency of the undriven pendulum: $\sqrt{mg\ell/I}$. Let $\Omega = \omega/\omega_0$ be the normalized pivot drive frequency; then the dimensionless equation of motion may be written

$$\ddot{\theta} + \left[\frac{2}{\Omega Q}\right]\dot{\theta} + 4\left[\frac{1}{\Omega^2} - \frac{A}{\ell}\frac{m\ell^2}{I}\cos(2t^*)\right]\sin\theta = 0. \quad \text{(B.4)}$$

If, instead of θ, we choose the complementary angle ϕ as indicated in Fig. B.1, then the equation of motion becomes

$$\ddot{\phi} + \left[\frac{2}{\Omega Q}\right]\dot{\phi} - 4\left[\frac{1}{\Omega^2} - \frac{A}{\ell}\frac{m\ell^2}{I}\cos(2t^*)\right]\sin\phi = 0. \quad \text{(B.5)}$$

At this point it is convenient to define the dimensionless parameters: $\delta = \Omega^{-2}$ and $\varepsilon = (A/\ell)\,(m\ell^2/I)$. Evidently, δ is the square of the ratio of the period of the pivot oscillation to the period of the natural pendulum oscillation. In the simplest case of a bob at the end of a massless rod, $I = m\ell^2$ and thus $\varepsilon = A/\ell$. More generally, ε will be A/ℓ times a factor somewhat less than unity; A/ℓ is a measure of the magnitude of the pivot excursions as a fraction of the pendulum length.

When damping is negligible and for small angles,

$$\ddot{\theta} + [(4\delta) - 2(2\varepsilon)\cos(2t^*)]\theta = 0 \quad \text{(B.6)}$$

$$\ddot{\phi} + [(-4\delta) - 2(-2\varepsilon)\cos(2t^*)]\phi = 0. \quad \text{(B.7)}$$

Equation (B.6) applies to small displacements around the straight down position, whereas Eq. (B.7) describes motion about the straight up orientation.

The earlier choice of a time scale, which produced the factor $\cos(2t^*)$ together with the form of bracketing of the terms in Eqs. (B.6) and (B.7) was motivated as a way of emphasizing the exact correspondence between the pendulum equations and the canonical form of the Mathieu equation (Mathieu 1868): [2]

$$\frac{d^2y}{dv^2} + (a - 2q\cos 2v)y = 0. \quad \text{(B.8)}$$

Comparing terms in the equations, it is clear that, for the conventional hanging-down orientation,

$$a = 4\delta \quad \text{(B.9)}$$

$$q = 2\varepsilon \quad \text{(B.10)}$$

and that for the inverted pendulum,

$$a = -4\delta \quad \text{(B.11)}$$

$$q = -2\varepsilon. \quad \text{(B.12)}$$

Note that, as defined, the parameters δ and ε are both positive quantities.

[2] See Corben and Stehle (1977, pp. 67–69).

A myriad of mathematical properties of the Mathieu equation have been examined in clinical detail (McLachlan 1947; Abramowitz and Stegun 1972). A significant benefit with respect to an understanding of the present system is provided by well known results concerning the domains of stability for solutions of the Mathieu equation.[3] The intriguing geometric structure of these stability zones in the parameter space (a, q) is illustrated[4] in Fig. B.2.

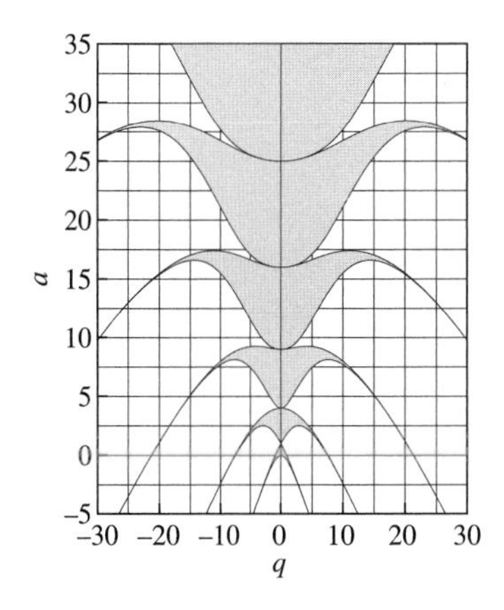

Fig. B.2
Stability zones (shaded) for solutions of the Mathieu equation.

Our focus here is on the inverted state, for which a and q are both *negative*. Only a small portion on the left-hand side of the lowest stability zone in Fig. B.2 applies to this case. Furthermore, since it is unlikely that very large up and down pivot excursions would be desirable, the parameter ε might reasonably be limited to a range of, say, (0,1); a value of 1 corresponds approximately to a vertical pivot displacement amplitude equal to the pendulum length ℓ. Therefore we concentrate our attention on a restricted portion of Fig. B.2, now magnified and replotted in Fig. B.3.[5]

The noninverted state of the pendulum (marked "down" in the figure) is confined to the region with both $a > 0$ and $q > 0$.

Figure B.3 contains a great deal of information. For example, when the amplitude of the pivot oscillations is zero (q and ε both equal to 0), then all possibilities are restricted to the vertical axis in the graph. Of course, an oscillation of zero amplitude is actually no oscillation; the pivot does not move. All points along the vertical line above $a = 0$ belong to stable zones for the noninverted state, while all points along the vertical axis below $a = 0$ belong to the lowest unstable zone. This says that for a stationary pivot point, the inverted state is unstable, but the noninverted (hanging down) state is stable—the expected result.

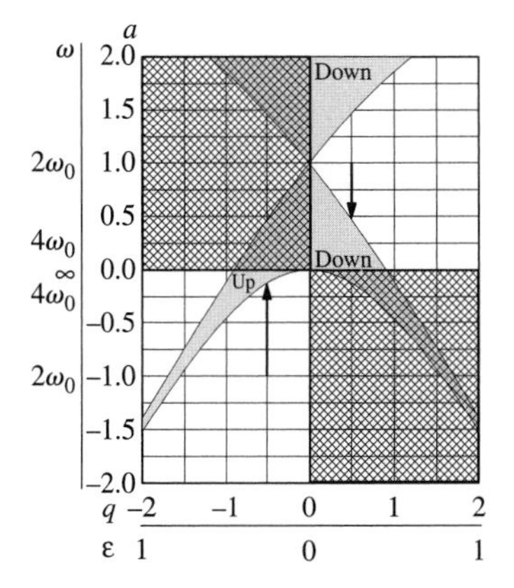

Fig. B.3
Blow up of the stability zones for the Mathieu equation showing the first stability regions for the noninverted state and the single stability region for the inverted state of the pendulum. The two hatched quadrants are not physically admissible for this system. The magnitude of a is 4δ, so equivalently, $\omega = 2\omega_0\sqrt{1/|a|}$.

But what if the pivot really is vibrating with finite amplitude ($\varepsilon > 0$)? As an example, take $\varepsilon = 0.25$ and suppose the frequency of the pivot oscillation is slowly increased beginning at a value twice the natural frequency of the pendulum: $\omega = 2\omega_0$. At the starting frequency, as Fig. B.3 shows, both inverted and noninverted states are unstable, so the pendulum bob would not be stationary in either up or down positions; it would instead execute some kind of rotational motion. As the frequency is turned up, the noninverted state first stabilizes at a just below 0.5, as indicated by the down-pointing arrow. The upward pointing arrow shows that stabilization of the inverted state occurs at a higher drive frequency ($a \approx 0.1$).

Note too, that for $\varepsilon > 0.45$ the down state cannot be stabilized at high frequencies, whereas the inverted state still can become stable, but only within an *interval* of pivot frequencies. Just at $\varepsilon = 0.45$, the inverted state is stable from $a = 0$ to $a = -0.38$, or equivalently, over a frequency range $(3.23\omega_0 \to \infty)$.

[3] See Abramowitz and Stegun (1972, Fig. 20.6, p. 728) and Corben and Stehle (1977). The matter of the stability of solutions hinges on a determination of the nature of the exponent (real or imaginary) in Floquent solutions of the equation.

[4] See McLachlan (1947, fig. 8(B), p. 41).

[5] The modifications that occur to this plot when there is weak dissipation are detailed in Leiber and Risken (1988).

Fig. B.4
Photographs of an experimental apparatus exhibiting the noninverted state, the inverted state, and an unstable (rotating) state. The small cylindrical pendulum bob is at the end of a short rod. The entire pendulum assembly is attached to the top of a steel shaft which is vibrated up and down by a loudspeaker driver (not shown). What appears to be the base is actually a triple bearing guide for the shaft. The total height of the components in the photograph is about 3 cm.

A demonstration of this dynamic stabilization was described in (Smith and Blackburn, 1992). As can be seen in the photographs, Fig. B.4, the pendulum was a small aluminum cylinder at the end of a short rod. The supporting frame was attached to a steel shaft which was driven up and down by a modified loudspeaker transducer. The circular plate at the bottom of the photos is an assembly that centers and guides the vertically oscillating rod. The output of a signal generator was amplified and fed to the voice coil of the loudspeaker. For this apparatus, the bob mass was 0.19 g and the natural frequency was $\omega_0 = 39$ rad/ sec. The forcing frequency was adjustable over the range $1.6\omega_0 - 18\omega_0$. Other reports of experimental studies of the inverted pendulum have been published, often with a focus on chaotic motions (Levin et al. 1985; Starrett and Tagg 1995).

We conclude this discussion by drawing attention to an interesting contribution to the topic by A. B. Pippard (Pippard 1987). In his paper, Pippard shows that by replacing the sinusoidal pivot displacement with alternating parabolic arcs or a sawtooth, a simpler mathematical analysis of stability follows, sidestepping the complexities of the Mathieu equation. The possibility that stable oscillations about the inverted state might occur for certain choices of *direct excitation* (that is, a properly chosen sinusoidal applied torque), without the need for vertical oscillations of the pivot as discussed above, has been considered by Miles (1988*a*).

Appendix C: The double pendulum

The double pendulum is one realization of the compound pendulum, and a simple version of a double pendulum is shown in Fig. C.1. Shinbrot et al. (1992) studied the chaotic behavior of this system and showed that the largest Lyapunov exponent was distinctly positive—a primary characteristic of chaotic motion. This group also constructed a metal double pendulum whose motion is approximated by the mathematical model. The device is energized by a hand crank and, for short times, effectively demonstrates chaotic motion.

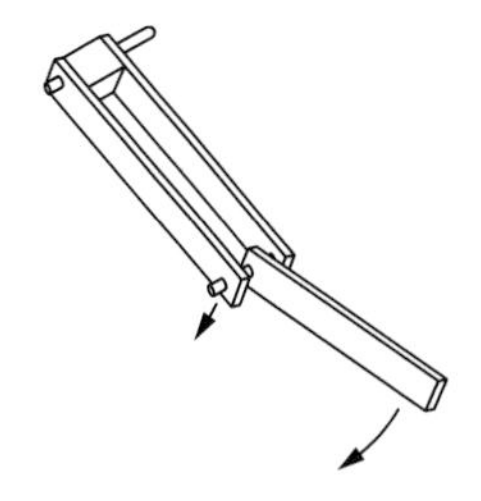

Fig. C.1
A hand cranked double pendulum. (Reprinted with permission from Shinbrot et al. (1992), p. 491. ©1992, American Association of Physics Teachers.)

An approximately equivalent model of this pendulum is shown in Fig. C.2. The two angular degrees of freedom, θ_1 and θ_2, suggest that its motion will be more complex than that of the simple pendulum. Not only will periodic motion be somewhat complex, but the extra degree of freedom allows for the possibility of chaotic motion. Our model does not include damping or forcing and therefore even in a chaotic state, there is no attractor. However, the system can exhibit sensitivity to initial conditions. Let us construct the equations of motion using the Lagrangian formulation of the problem. This treatment is similar to that of Shinbrot et al. (1992).

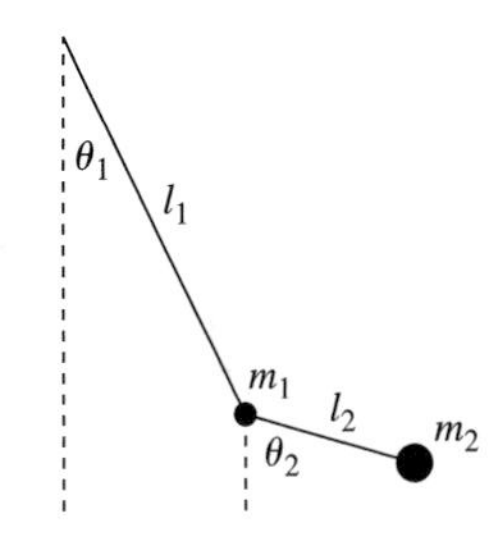

Fig. C.2
Schematic diagram of the mathematical model of the double pendulum.

The zero of potential energy is taken as the position of rest for the pendulum. Therefore the potential energy of the raised pendulum given by

$$V = m_1 g l_1(1 - \cos\theta_1) + m_2 g[l_1(1 - \cos\theta_1) + l_2(1 - \cos\theta_2)], \qquad \text{(C.1)}$$

where the various quantities are as indicated in Fig. C.2. Calculation of the kinetic energy is slightly more involved. Consider the x and y coordinates of m_2. These are given by

$$\begin{aligned} x_2 &= l_1 \sin\theta_1 + l_2 \sin\theta_2 \\ y_2 &= -l_1 \cos\theta_1 - l_2 \sin\theta_2 \end{aligned} \qquad \text{(C.2)}$$

as measured from a coordinate system located at the top of the pendulum. The kinetic energy of m_2 is then

$$\frac{1}{2} m_2(\dot{x}_2^2 + \dot{y}_2^2) = \frac{1}{2} m l_1^2 \dot{\theta}_1^2 + \frac{1}{2} m_2 l_2^2 \dot{\theta}_2^2 + m_2 l_1 l_2 \cos(\theta_1 - \theta_2)\dot{\theta}_1\dot{\theta}_2 \qquad \text{(C.3)}$$

and the combination of this particle's kinetic energy with that of m_1 leads to a total kinetic energy of

$$K = \frac{1}{2}m_1 l_1^2\dot{\theta}_1^2 + \frac{1}{2}m_2 l_1^2\dot{\theta}_1^2 + \frac{1}{2}m_2 l_2^2\dot{\theta}_2^2 + m_2 l_1 l_2 \cos(\theta_1 - \theta_2)\dot{\theta}_1\dot{\theta}_2. \quad \text{(C.4)}$$

The energy expressions are now combined into the Lagrangian which, in turn, is the basis for calculation of the equations of motion in the usual way:

$$\frac{d}{dt}\left(\frac{\partial L}{d\dot{\theta}_i}\right) - \left(\frac{\partial L}{\partial \theta_i}\right) = 0.$$

Careful use of the Lagrangian equations leads to the following coupled equations of motion:

$$\begin{aligned} \ddot{\theta}_1 &= -\frac{l_2}{\mu l_1}\ddot{\theta}_2 \cos(\theta_1 - \theta_2) - \frac{l_2}{\mu l_1}\dot{\theta}_2^2 \sin(\theta_1 - \theta_2) - \frac{g}{l_1}\sin\theta_1 \\ \ddot{\theta}_2 &= -\frac{l_1}{l_2}\ddot{\theta}_1 \cos(\theta_1 - \theta_2) + \frac{l_1}{l_2}\dot{\theta}_1^2 \sin(\theta_1 - \theta_2) - \frac{g}{l_2}\sin\theta_2, \end{aligned} \quad \text{(C.5)}$$

where $\mu = 1 + (m_1/m_2)$. To facilitate numerical solution, the angular acceleration terms can be decoupled by substitution of the second equation into the first equation and vice-versa. Then

$$\begin{aligned} \ddot{\theta}_1 &= \frac{g(\sin\theta_2\cos(\theta_1-\theta_2) - \mu\sin\theta_1) - (l_2\dot{\theta}_2^2 + l_1\dot{\theta}_1^2\cos(\theta_1-\theta_2))\sin(\theta_1-\theta_2)}{l_1(\mu - \cos^2(\theta_1-\theta_2))} \\ \ddot{\theta}_2 &= \frac{g\mu(\sin\theta_1\cos(\theta_1-\theta_2) - \sin\theta_2) + (\mu l_1\dot{\theta}_1^2 + l_2\dot{\theta}_2^2\cos(\theta_1-\theta_2))\sin(\theta_1-\theta_2)}{l_2(\mu - \cos^2(\theta_1-\theta_2))}. \end{aligned} \quad \text{(C.6)}$$

Finally, we define two new variables, ω_1 and ω_2, to create four equations that can be solved numerically,

$$\begin{aligned} \dot{\omega}_1 &= \frac{g(\sin\theta_2\cos(\theta_1-\theta_2) - \mu\sin\theta_1) - (l_2\omega_2^2 + l_1\omega_1^2\cos(\theta_1-\theta_2))\sin(\theta_1-\theta_2)}{l_1(\mu - \cos^2(\theta_1-\theta_2))} \\ \dot{\omega}_2 &= \frac{g\mu(\sin\theta_1\cos(\theta_1-\theta_2) - \sin\theta_2) + (\mu l_1\omega_1^2 + l_2\omega_2^2\cos(\theta_1-\theta_2))\sin(\theta_1-\theta_2)}{l_2(\mu - \cos^2(\theta_1-\theta_2))} \\ \dot{\theta}_1 &= \omega_1 \\ \dot{\theta}_2 &= \omega_2. \end{aligned}$$

Perhaps not surprisingly, it is easy to find parameter values that will lead to chaotic behavior. Figure C.3 shows a chaotic time series for the angular velocity of the lower mass m_2. Similar time series were obtained by Stump (1986).

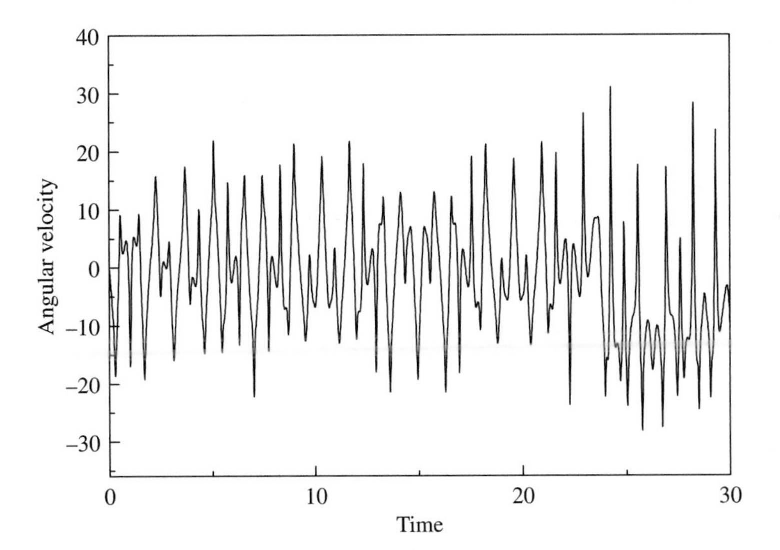

Fig. C.3
Time series of the angular velocity of the lower mass of the double pendulum. The irregular motion indicates the presence of chaotic dynamics. Initial conditions $\theta_1 = 0.5$ rad, $\theta_2 = 2.0$ rad, $\dot{\theta}_1 = 0$, $\dot{\theta}_2 = 0$.

Appendix D: The cradle pendulum

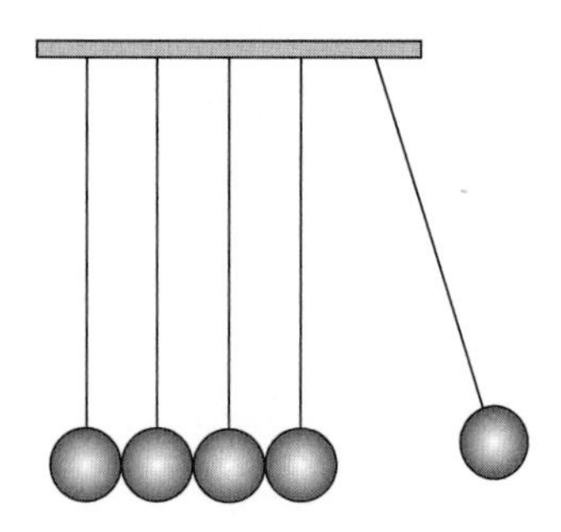

Fig. D.1
Typical arrangement of Newton's cradle.

Newton's cradle is an interesting example of coupled pendulums and is also an amusing toy. It can provide its owner with many moments of fascinating study. In brief, Newton's cradle consists of several, usually five, equal pendulums each suspended by a V-shaped fiber placed perpendicularly to the line of the pendulum bobs. Thus the motion of the bobs is one-dimensional along their line of suspension. The bobs are typically hardened steel balls and the pendulums are arranged so that the balls are just touching each other. The configuration confines the trajectories of the balls to one dimension. See Fig. D.1.

The motion is quite interesting. If, for example, the leftmost ball is displaced and then released to collide with the line of remaining balls, one observes that the collision, for the most part, causes the rightmost ball to move away from the group with a speed similar to the original speed of the leftmost ball. Similarly, if the two leftmost balls are initially displaced and released, the two rightmost balls will be displaced by the collisions of the leftmost balls with the line of the remaining balls. Again the speeds seem to have the same magnitude. The motion just before the collision is approximately mirrored by the motion after the collision provided that the "mirror" is located vertically through the middle pendulum.

The most primitive explanation of these phenomena (Gavenda and Edgington 1997) is that the observations are the result of conservation of momentum and conservation of kinetic energy. Consider the situation where just two balls move. Suppose that the initial momentum and energy of the leftmost ball are

$$p_i = mv_0 \text{ and } E_i = \frac{1}{2}mv_0^2.$$

After the collision the final momentum and energy of the rightmost ball are

$$p_f = mv_0 \text{ and } E_f = \frac{1}{2}mv_0^2.$$

But even assuming that the collisions are perfectly elastic, a unique solution to the problem with this model is only possible if there are just two balls. For three balls, A, B, and C, suppose that initially ball A has a velocity v_0 whereas the velocities of balls B and C are zero. After the collision, the simple explanation suggests that the velocities of balls A and B are zero, whereas $v_C = v_0$. Certainly this scenario is approximately observed and the conservation laws are satisfied. But, for three balls, we may also construct solutions that, while they obey these two conservation laws, are not observed in a physical experiment. Suppose that the leftmost ball is again given a velocity v_0. The following two solutions for the final velocities also obey the conservation laws, but are not typically observed: (Herrmann and Schmälzle 1981)

$$(a)\ v_A = -\frac{1}{3}v_0,\ v_B = v_C = +\frac{2}{3}v_0$$

and (Chapman 1960)

$$(b)\ v_A = -\frac{1}{6}v_0,\ v_B = \frac{[7-\sqrt{21}]}{12}v_0,\ v_C = \frac{[7+\sqrt{21}]}{12}v_0.$$

In both cases, significant motion is predicted for the center ball—an unobserved result.

Clearly some further mechanism must be involved. It has been postulated that one can model the balls as barely touching or not quite touching. (Kline 1960; Piquette and Wu 1984). If such were the case, then the momentum and energy could possibly be transferred along the line of balls, one ball at a time. This would result in the observed motion for only the outside balls. Yet the nontouching model does not seem appropriate. For one thing the balls *do* touch, and for another, the time of the collision is much longer than the time it takes for a sound wave to cross the ball, further suggesting that there could be momentary connections among more than just a single pair of balls. Another interesting effect, not accounted for by the simple model, is the observation that small motions of the inner balls do occur. Sometimes attempts are made to explain away these motions as being the result of less than perfect conservation of momentum and energy. But the actual motions, while small, seem to be due to more than these spurious effects.

A different approach is required; one for which explicit account is taken of the elastic connections between the balls. In 1881 Heinrich Hertz developed a theory of the elastic deformation of colliding bodies (Love 1892). According to this theory one can think of the balls as connected by springs whose force law is not the typical linear Hooke's law, but rather is of the form

$$F = -kx^{3/2}$$

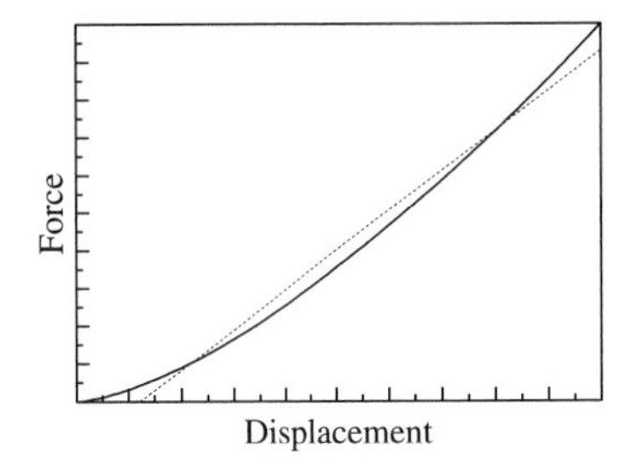

Fig. D.2
Hertz's law for the interaction of colliding spheres as expressed by connection of equivalent springs. The straight line is an approximately equivalent Hooke's law spring (k = 1).

as is shown graphically in Fig. D.2. The straight line gives an approximately equivalent Hooke's law behavior and this will provide us with one way to think about the collisions. If one imagines that the balls interact through equivalent Hooke's law springs, then the graph makes it apparent that the balls act, in a small region where $x < 0.2$ (in arbitrary units), as if they are *not* connected. There is no force between the balls until $x = 0.2$. And if they are not connected, then the relatively simple earlier model of the balls not being connected leads to only the n rightmost outer balls acquiring motion as a result of the collision from the n leftmost balls. By making this comparison of the Hertz spring to some sort of disconnected equivalent Hooke spring, we obtain some intuitive idea as to how the Hertzian spring provides a better model. However, a more correct procedure is to actually do the mathematics required to model the balls as oscillators coupled by the Hertzian springs. Because the Hertz spring is nonlinear, it is necessary to use a computer simulation. In a comparison of such a model with computer simulation of a set offully connected, coupled Hooke's law oscillators, it was found that the interaction of the collisions traveled through the set of balls somewhat differently in each case (Herrmann and Seitz 1982). With the fully connected Hooke's law springs, the interaction is dispersed among several of the balls at a given moment and computer simulations did not provide results that adequately modeled the observations. But with the Hertz law springs, the interaction is more, but not totally, confined to single pairs of balls. This sort of interaction is said to be approximately *dispersion free*. This behavior is especially true after the first collision. We noted above that the interior balls did exhibit small motions. After the first collision these small motions do provide a little space between balls for the subsequent collisions. In this way the interaction is not further dispersed after the first collision, and again the observed motion of the outer balls is predicted by computer simulation. In some way the intuitive picture of the role of the Hertz spring is validated.

Newton's cradle is therefore an example of a system that appears to be deceptively simple but, on closer examination, is found to be, like many pendulum phenomena, complex and fascinating. (Note added in proof. For a recent experimental and simulation study of Newton's cradle, see (Hutzler et al, 2004).)

Appendix E: The Longnow clock

In the April 2, 2000 edition of the New York Times, an article appeared describing an extraordinary torsion pendulum clock, designed by Danny Hillis and funded by The Long Now Foundation (www.longnow.org). The accompanying photograph, reproduced here (Fig. E.1) depicted the prototype of this startling design.

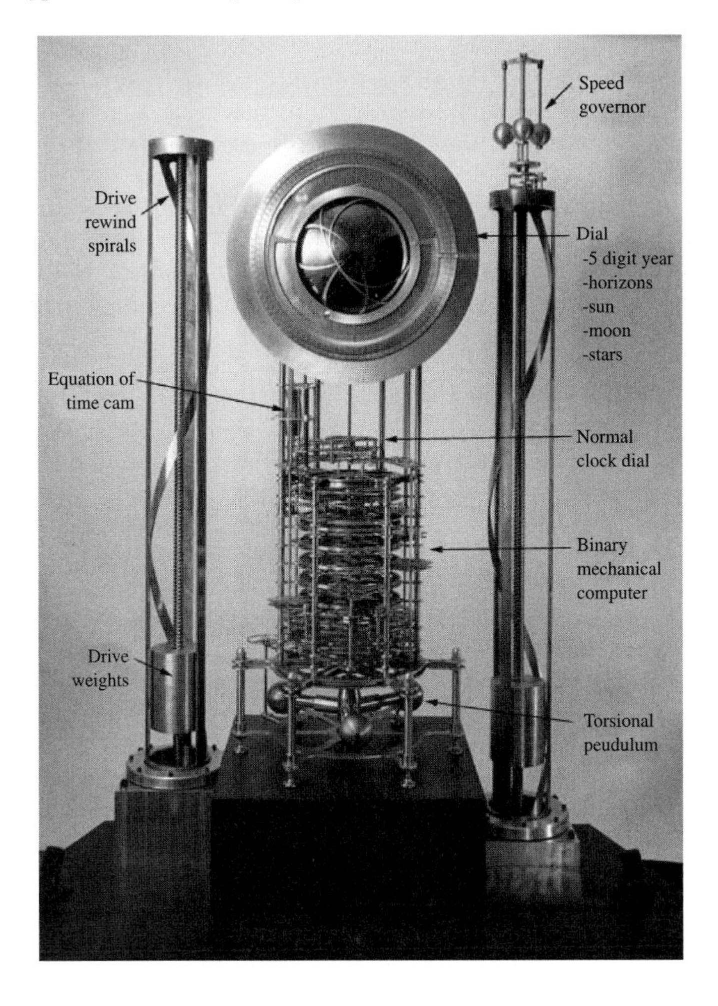

Fig. E.1
Overall view of the Longnow clock. (Photo credit Rolfe Horn, courtesy of the Long Now Foundation.)

Fig. E.2
Close up view of the triple mass torsion pendulum that is the heart of the Longnow clock. (Photo credit Rolfe Horn, courtesy of the Long Now Foundation.)

It is nine feet tall, but the plan is to build a full scale version ten times larger and locate it in a remote desert environment in Nevada. It is intended to keep time for 10,000 years, a symbolic span comparable to the time elapsed since the end of the last ice age.

The clock is based on a torsion pendulum (Fig. E.2) oscillating once every 30 s.

As in any clock, energy loss in the oscillator is made up through the action of an escapement which is powered, in this design, by falling weights.

In the photo, above the 22 pound triple mass tungsten pendulum, can be seen the first few layers in a stack of interacting gears. The full stack is visible in Fig. E.1. This assembly is in fact an elaborate mechanical computer, designed to keep track of time through the millennia, accounting for such factors as the 26,000 year precession of the equinoxes. The evolving appearance of the night sky is calculated and displayed on the black sphere at the top of the clock.

The conception of this clock is not so much about timekeeping, but about the intersection of art and technology. Nowadays very precise clocks are relatively inexpensive and very precise. Computers are equally cheap and powerful. Therefore, a mechanical clock and a mechanical computer are seemingly a throwback to earlier epochs when individual invention and high craftsmanship were hallmarks of human ingenuity and creativity. Harrison's clocks and marine chronometer of the eighteenth century come to mind. The Long Now clock reasserts the value of mankind in technology and looks to the future in a wonderfully anachronistic way.

Appendix F: The Blackburn pendulum

Finally, given the subject of this book and the name of one of the authors, we add this brief historical note on the Blackburn pendulum.[1] The somewhat obscure Blackburn pendulum, not to be confused with the Blackwood ballistic pendulum, is discussed in several papers by Whitaker (1991, 2001). Hugh Blackburn was born near Glasgow, Scotland, in 1823 and died in 1909. He was a lifelong friend from schooldays at Trinity College, Cambridge, of William Thomson (later, Lord Kelvin). In 1849 Blackburn was appointed Professor of Mathematics at the University of Glasgow and, in that same year, he married Jemima Wedderburn who was a first cousin of James Clerk Maxwell.

The Blackburn pendulum utilizes a form of double suspension where the hanging strings are in a "Y" shape, as illustrated in Fig. F.1.

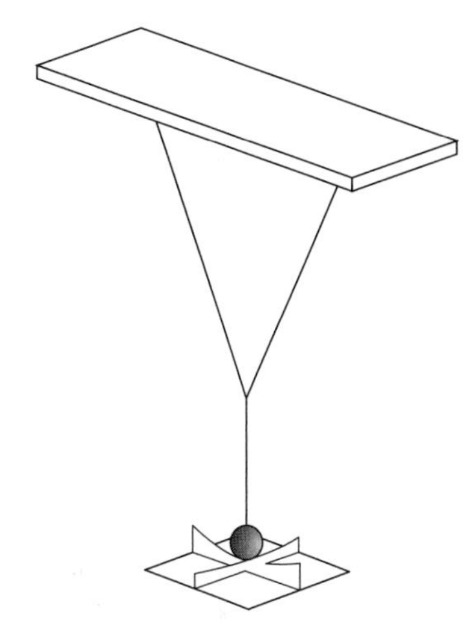

Fig. F.1
A pendulum with a "Y" suspension. The crossed elements beneath the bob are meant to suggest the two planes of oscillation, each involving a different effective pendulum length.

The bob is attached to the bottom of the "Y" while each of the upper two forks is attached to a rigid support. Two orthogonal swinging modes are possible. In the first, the pendulum oscillates in and out of the plane defined by the Y-shape (i.e. in and out of the page), pivoting on both attachment points at the top of the forks. The effective length of the pendulum in this oscillation mode is the vertical height of the "Y". The second mode consists of motion in the plane of the "Y" (in the plane of the page). For this oscillation, the pendulum is shorter—its length is the distance from the bob to the bottom of the V formed by the supporting strings (i.e. the vertical stroke in the "Y").

With two lengths, the Blackburn pendulum can oscillate in modes that are complex superpositions of two pure frequencies and so can be used to give fascinating visual representations of harmonic combinations similar to Lissajous figures; it belongs to the family of mechanical devices known as "harmonographs" (Whitaker 2001).

[1] So far as is known, there is no family connection between Hugh Blackburn and James Blackburn.

Bibliography

H. D. Abarbenal, R. Brown, and J. B. Kadtke, "Prediction in chaotic nonlinear systems: Methods for time series with broadband Fourier spectra," *Physical Review* **A41**, 1782–1807 (1990).

M. Abramowitz and I. A. Stegun, Eds. *Handbook of Mathematical Functions*, (Dover reprint, New York, 1972).

R. H. Abraham and C. D. Shaw, *Dynamics, the Geometry of Behavior*, (Ariel Press, Santa Cruz, CA, 1984).

R. Adams, "Where Europe's cultural soup bubbled over," *The Globe and Mail (Toronto)* (July 24), p. F3 (1999).

R. Aldrovandi and P. Leal Ferreira, "Quantum pendulum," *American Journal of Physics* **48**, 660–664 (1980).

Catherine Aird, *His Burial too*, (Doubleday and Co., New York, 1973.) (Bantam Books, 1981).

Charles J. Allen, *Journal of the Franklin Institute* **52**(whole series), 38–42, 103–104 (1851).

E. Altshuler and R. Garcia, "Josephson junctions in a magnetic field: insights from coupled pendula," *American Journal of Physics* **71**, 405–408 (2003).

E. E. Anderson, *Modern Physics and Quantum Mechanics*, (W. B. Saunders, Philadelphia, 1971).

R. L. Armstrong, G. L. Baker, and K. R. Jeffrey, "Motional averaging of the electric field gradient at chlorine nuclear sites in K_2PtCl_6 and K_2PdCl_6 by lattice vibrations," *Physical Review* **B1**, 2847–2851 (1970).

H. Atmanspacher and H. Scheingraber, "A fundamental link between system theory and statistical mechanics," *Foundations of Physics* **17**, 939–963 (1987).

Charles H. Bagley and Gabriel G. Luther, "Preliminary results of a determination of the Newtonian constant of gravitation: A test of the Kuroda hypothesis," *Physical Review Letters* **78**, 3047–3050 (1997).

G. L. Baker, "Control of the chaotic driven pendulum," *American Journal of Physics* **63**, 832–838 (1995).

G. L. Baker, J. A. Blackburn, and H. J. T. Smith, "Intermittent synchronization in a pair of coupled pendula," *Physical Review Letters* **81**, 554–557 (1998); "A stochastic model of synchronization for chaotic pendulums," *Physics Letters* **A252**, 191–197 (1999).

G. L. Baker, J. A. Blackburn, and H. J. T. Smith, "The quantum pendulum: small and large," *American Journal of Physics* **70**(5), 525–531 (2002).

G. L. Baker and J. P. Gollub, *Chaotic dynamics: an introduction*, 2nd edition (Cambridge University Press, Cambridge, 1996).

G. L. Baker, J. P. Gollub, and J. A. Blackburn, "Inverting chaos: extracting system parameters from experimental data," *CHAOS* **6**, 528–533 (1996).

R. A. Baker, J. Bird, and M. Town, "South pole Foucault pendulum," http:/www.phys-astro.sonoma.edu/people/baker/SouthPoleFoucault.html (2001).

M. K. Bantel and R. D. Newman, "A cryogenic pendulum: progress report," *Classical and Quantum Gravity* **17**, 2313–2318 (2000).

J. Bardeen, "Tunneling into superconductors," *Physical Review Letters* **9**, 147–149 (1962).

J. Bardeen, L. N. Cooper, and J. R. Schrieffer, "Theory of superconductivity," *Physical Review* **108**, 1175–1204 (1957).

A. Barone and G. Paterno, *Physics and Applications of the Josephson Effect*, (Wiley, New York 1982).
J. Barrow-Green, *Poincare and the Three-Body Problem*, (Am. Math. Soc., Providence, RI, 1997).
Silvio A. Bedini, *The Pulse of Time*, (Biblioteca di Nuncius, 1991).
M. Bennett, M. F. Schatz, H. Rockwood, and K. Wiesenfeld, "Huygens's clocks," *Proc. R. Soc. Lond.* **A458**, 563–579 (2002).
J. P. Berdahl and K. V. Lugt, "Magnetically driven chaotic pendulum," *American Journal of Physics* **69**(9), 1016–1019 (2001).
J. A. Blackburn, G. L. Baker, and H. J. T. Smith, "Intermittent synchronization of resistively coupled chaotic Josephson junctions," *Physical Review* **B62**, 5931–5935 (2000).
J. A. Blackburn, Binruo Wu, and H. J. T. Smith, "Analog simulation of superconducting loops containing one or two Josephson junctions," *Journal of Applied Physics* **64**, 3112–3118 (1988).
J. A. Blackburn and H. J. T. Smith, "Dynamics of double-Josephson-junction interferometers," *Journal of Applied Physics* **49**, 2452–2455 (1978).
J. A. Blackburn and H. J. T. Smith, "Single Josephson junction superconducting memory cell," *Electronics Letters* **14**, 597–599 (1978).
J. A. Blackburn, S. Vik, Wu Binruo, and H. J. T. Smith, "Driven pendulum for studying chaos," *Review of Scientific Instruments* **60**, 422–426 (1989).
J. A. Blackburn, S. Vik, Yang Zhou-Jing, H. J. T. Smith, and M. A. H. Nerenberg, "Chaos in a Josephson junction" in *Structure, Coherence and Chaos in Dynamical Systems*, edited by Peter L. Christiansen and Robert D. Permentier, (Manchester University Press, Manchester, 1989) 563–567.
R. J. Blackwell, translation of Christiaan Huygens' *Horologium Oscillatorium*, (The Iowa State University Press, Ames, Iowa 1986).
P. Bouguer, Section vii of "La Figure de la Terre," Paris, 1749, in *The Laws of Gravitation*, edited by A. Stanley Mackenzie, (American Book Co., NY, 1900).
P. Bouguer *Encyclopaedia Britannica Online* (2000).
C. B. Boyer and U. C. Merzbach, *A History of Mathematics*, (John Wiley and Sons, New York, 1991).
C. V. Boys, "On the Newtonian constant of gravitation," *Nature* **50**, 330 (1894).
V. B. Braginsky, A. G. Polnarev, and K. S. Thorne, "Foucault pendulum at the south pole: Proposal for an experiment to detect the earth's general relativistic gravitomagnetic field," *Physical Review Letters* **53**, 863–866 (1984).
Y. Braiman, J. F. Linder, and W. L. Ditto, "Taming spatiotemporal chaos with disorder," *Nature* **378**, 465–467 (1995).
J. S. Brown, *Journal of the Franklin Institute* **52**(whole series), 352–5, 426–7, (1851).
R. Brown, N. F. Rulkov, and E. R. Tracy, "Modeling and synchronizing chaotic systems from experimental data," *Physics Letters* **A194**, 71 (1994).
P. J. Bryant and J. W. Miles, "On a periodically forced, weakly damped pendulum, part 2: horizontal forcing," *Journal of the Australian Mathematical Society*, Series **B32**, 23–41 (1990).
K. E. Bullen, *The Earth's Density* (Chapman and Hall, London, 1975).
David M. Burton, *The History of Mathematics: An Introduction*, (McGraw Hill, New York, 1999).
Eugene I. Butikov, "On the dynamic stabilization of an inverted pendulum," *American Journal of Physics* **69**, 755–768 (2001).
G. Cagnoli, L. Gammaitoni, J. Hough, J. Kovalik, S. McIntosh, M. Punturo, and S. Rowan, "Very high Q measurements on a fused silica monolithic pendulum for use in enhanced gravity wave detectors," *Physical Review Letters* **85**, 2442–2445 (2000).
W. B. Case and M. A. Swanson, "The pumping of a swing from the seated position," *American Journal of Physics* **58**, 463–467 (1990).

T. K. Caughey, "Hula-Hoop: an example of heteroparametric excitation," *American Journal of Phsyics* **28**(2), 104–109 (1960).
H. Cavendish, *Philosophical Transactions of the Royal Society of London* **17**, 469–526 (1798).
A. Cenys, A. N. Anagnostopoulos, and G. L. Bleris, "Symmetry between laminar and burst phases for on–off intermittency," *Physical Review* **E56**, 2592–2596 (1997).
S. Chapman, "Misconception concerning the dynamics of the impact ball apparatus," *American Journal of Physics* **28**, 705–711 (1960).
F. Charron, *Bulletin de la Société Astronomique de France* **45**, 457–462 (1931).
Tai L. Chow, *Classical Mechanics*, (Wiley, New York, 1995).
John Clarke, "The Josephson effect and e/h," *American Journal of Physics* **38**, 1071–1095 (1970).
B. E. Clotfelter, "The Cavendish experiment as Cavendish knew it," *American Journal of Physics* **55**, 210–213 (1987).
E. U. Condon, "The physical pendulum in quantum mechanics," *Physical Review* **31**, 891–894 (1928).
G. P. Cook and C. S. Zaidins, "The quantum point-mass pendulum," *American Journal of Physics* **54**, 259–267 (1986).
H. C. Corben and Philip Stehle, *Classical Mechanics*, 2nd edition, (Dover, New York, 1977).
G. G. Coriolis, "Sur le principe des forces vives dans les mouvements relatifs des machines," *Journal de L'Ecole Polytechnique* **13**, 265–302 (1832).
H. Richard Crane, "Short Foucault pendulum: a way to eliminate the precession due to ellipticity," *American Journal of Physics* **49**, 1004–1006 (1981).
H. Richard Crane, "The Foucault pendulum as murder weapon and a physicist's delight," *The Physics Teacher* (May), 249–269 (1990).
H. Richard Crane, "Foucault pendulum wall clock," *American Journal of Physics* **63**, 33–39 (1995).
K. M. Cuomo and A. V. Oppenheim, "Circuit implementation of synchronized chaos with applications to communications," *Physical Review Letters*, **71**, 65–68 (1993).
Daedalon Corp., P.O. Box 2028, Salem, MA., Instruction Manual for Reversible Pendulum, p.1 (2000).
Daedelon Corp., Salem, MA., Model EM-50 Chaotic Pendulum.
G. D'Anna, "Mechanical properties of granular media, including snow, investigated by a low-frequency forced torsion pendulum," *Physical Review* **E62**, 982–992 (2000); "Dissipative dynamics of granular media investigated by a forced pendulum," *Europhysics Letters* **51**, 293–299 (2000).
M. Debeau and H. Poulet, "Analyse vibrationnelle de complexes hexahalogenes a l'etat cristallise," *Spectrochimica Acta* **25A**, 1553–1562 (1969).
S. Deligeorges, *The pendulum under the eye of God*, Centre des monuments nationaux, (Editions du patrimonie, Paris, French edition 1995, English edition 2000).
R. DeSerio, "Chaotic pendulum: the complete attractor," *American Journal of Physics* **71**(3), 250–257 (2003).
R. H. Dicke, "The Eötvös experiment," *Scientific American* **205**, 84–94 (1961).
Dictionary of Scientific Biography.
S. Drake, *Galileo at Work: His Scientific Biography*, (University of Chicago Press, Chicago 1978).
S. Drake, *Galileo: A Very Short Introduction*, (Oxford University Press, 1980).
J. P. Eckmann, S. O. Kamphorst, D. Ruelle, and S. Ciliberto, "Liapunov exponents from time series," *Physical Review* **A34**, 4971 (1986).
Umberto Eco, *Foucault's Pendulum*, (Harcourt Brace Jovanovich, Inc., Orlando, Fl, 1989).
A. Einstein, "Zur Elektrodynamik bewegter Körper," *Annalen der Physik*, **17**, 891–921 (1905).

A. Einstein, "Über das Relativitätsprinzip und die aus demselben gezogenen Folgerungen," *Jahrb. Radioakt. Eleckt r.* **4**, 411–462 (1908).
Encyclopedia Brittanica, 11th edition, vol. 10, p. 734 (Cambridge, 1910).
R. V. Eötvös, D. Pekár, and E. Fekete, "Beiträge zum Gesetze der Proportionalität von Trägheit und Gravität," *Annalen der Physik* (*Leipzeg*) **68**, 11–66 (1922).
H. Erlichson, "Galileo's pendulum," *The Physics Teacher* **37**, 478–479 (1999).
F. M. Fedchenko, "Astronomical clock AChF-1 with isochronous pendulum," *Astronomical Journal of the Academy of Sciences of the USSR* **34**, 652–663 (1957).
F. M. Fedchenko, "Astronomical clock with an electromagnetically driven pendulum," *Research in the Field of Time Measurement* **58**, 92–99 (1962).
F. M. Fedchenko and A. G. Fleer, "Spring Suspension for a Pendulum," *Measurement Techniques* **8**, 27–30 (August, 1966).
Richard P. Feynman, Robert B. Leighton, and Matthew Sands, *The Feynman Lectures*, vol. I, II, III, (Addison Wesley, Reading MA, 1965).
E. Flores, "The tautochrone under arbitrary potentials using fractional calculus," *American Journal of Physics*, **67**, 718–722 (1999).
M. L. Foucault, "Demonstration physique du mouvement de rotation de la terre au moyen du pendule," *Comptes Rendus, Acad. Sci.* (*Paris*) **32** (Feb 3), 135–138, (1851); In translation, *J. Franklin Insti.*, 3rd Ser., **21**, 350–353 (1851).
S. T. Fryska and M. A. Zohdy, "Computer dynamics and shadowing of chaotic orbits," *Physics Letters* **A166**, 340–346 (1992).
T. A. Fulton, L. N. Dunkleburger, and R. C. Dynes, "Quantum interference properties of double Josephson junctions," *Physical Review* **B6**, 855–875 (1972).
J. D. Gavenda and J. R. Edgington, "Newton's cradle and scientific explanation," *The Physics Teacher*, **35**, 411–417 (1997).
G. T. Gillies and R. C. Ritter, "Torsion balances, torsion pendulums, and related devices," *Review of Scientific Instruments* **64**, 283–309 (1993).
C. Stewart Gillmor, *Coulomb and the Evolution of Physics and Engineering in Eighteenth Century France*, (Princeton University Press, Princeton, NJ 1971).
Herbert Goldstein, *Classical Mechanics*, (Addison-Wesley, Reading MA, 1950).
B. F. Gore, "The child's swing," *American Journal of Physics* **38**, 378–379 (1970).
B. F. Gore, "Starting a swing from rest," *American Journal of Physics* **39**, 347 (1971).
P. Grassberger, Comment on "Intermittent synchronization in a pair of coupled chaotic pendula," *Physical Review Letters* **82**, 4146 (1999).
P. Grassberger and I. Procaccia, "Measuring the strangeness of strange attractors," *Physica* **9D**, 189–208 (1983).
C. Grebogi, S. M. Hammel, J. A. Yorke, and T. Sauer, "Shadowing of physical trajectories in chaotic dynamics: containment and refinement," *Physical Review Letters* **65**, 1527–1530 (1990).
T. B. Greenslade, (Description and photo of Foucault's disk), *American Journal of Physics* **70**, 1135 (2002).
J. H. Gundlach and S. M. Merkowitz, "Measurement of Newton's constant using a torsion balance with angular acceleration feedback," *Physical Review Letters* **85**, 2869–2872 (2000).
E. G. Gwinn and R. M. Westervelt, "Intermittent chaos and low-frequency noise in the driven damped pendulum," *Physical Review Letters*, **54**, 1613–1619 (1985).
E. G. Gwinn and R. M. Westervelt, "Fractal basin boundaries and intermittency in the driven damped pendulum," *Physical Review* **A33**, 4143–4155 (1986).
P. K. Hansma and G. I. Rochlin, "Josephson weak links: shunted junction and mechanical-model results," *Journal of Applied Physics* **43**, 4721–4727 (1972).
J. F. Heagy, N. Platt, and S. M. Hammel, "Characterization of on-off intermittency," *Physical Review* **E49**, 1140–1150 (1994).

K. T. Hecht, "The Crane Foucault pendulum: an exercise in action-angle variable perturbation theory," *American Journal of Physics* **51**, 110–114 (1983).

J. L. Heilbron, "The politics of the meter stick," *American Journal of Physics* **57**, 988–992 (1989).

A. Heidmann, P. F. Cohadon, and M. Pinard, "Thermal noise of a plano-convex mirror," *Physics Letters* **A263**, 27–32 (1999).

A. Hellemans and B. Bunch, *The Timetables of Science*, (Simon and Schuster, New York, 1991).

W. A. Heiskanen and F. A. Vening Meinesz, *The Earth and Its Gravity Field*, (McGraw Hill, NY, 1958).

F. Herrmann and P. Schmälzle, "Simple explanation of a well-known collision experiment," *American Journal of Physics* **49**(8), 761–764 (1981).

F. Herrmann and M. Seitz, "How does the ball-chain work," *American Journal of Physics* **50**(11), 977–981 (1982).

G. Herzberg, *Infrared and Raman Spectra*, (Van Nostrand, Princeton, 1945).

S. Hildebrandt and A. Tromba, *The Parsimonious Universe*, (Copernicus, Springer-Verlag, New York, 1996).

Zhong-Kun Hu and Juu Luo, "Amplitude dependence of quality factor of the torsion pendulum" *Physics Letters* **A 268**, 255–259 (2000).

Joan R. Hundhausen, "Zeroing in on the Delta Function," *The College Mathematics Journal* **29**, 27–32 (1998).

S. Hutzler, G. Delaney, D. Weaire and F. MacLeod, "Rocking Newton's Cradle," *American Journal of Physics* **72**(12), 1508–1516 (2004).

Christiaan Huygens *Horologium Oscillatorium* (1673), Facsimile Reprint, (Dawsons of Pall Mall, London, 1966).

Christiaan Huygens, *Horologium Oscillatorium*, translated by Richard J. Blackwell, *The pendulum clock or geometrical demonstrations concerning the motion of pendula as applied to clocks*, (Iowa State University Press, Ames, Iowa, 1986).

E. Attlee Jackson, *Perspectives of nonlinear dyanmics*, Vol. 1. (Cambridge University Press, Cambridge, 1989).

G. Johnson, "Here they are, science's 10 most beautiful experiments," New York Times, (September 24) (2002).

P. Jordan, *Zeitschrift für Physik* **44**, 1 (1927).

B. D. Josephson, "Possible new effects in superconductive tunnelling," *Physics Letters*, **1**, 251–253 (1962).

H. P. Kalmus, "The inverted pendulum,"*American Journal of Physics* **38**(7), 874–878 (1970).

H. Kamerlingh Onnes, "Further experiments with liquid helium. C. On the change of electric resistance of pure metals at very low temperatures etc. IV. The resistance of pure mercury at helium temperatures," *Communications of the Univ. of Leiden* **120b**, 1479–1481 (1911*a*).

H. Kamerlingh Onnes, "Further experiments with liquid helium. D. On the change of electric resistance of pure metals at very low temperatures etc. V. The disappearance of the resistance of pure mercury," *Communications of the Univ. of Leiden* **122b**, 81–83 (1911*b*).

H. Kamerlingh Onnes, "Further experiments with liquid helium. G. On the electrical resistance of pure metals, etc. VI . On the sudden change in the rate at which the resistance of Mercury disappears," *Communications of the Univ. of Leiden*, **124c**, 799–802 (1911*c*).

J. L. Kaplan and J. A. Yorke, Chaotic behavior of multi-dimensional difference equations, in *Functional differential equations and the approximations of fixed points*, eds. H. O. Peitgen and H. O. Walther, Lecture notes in mathematics, vol. 730, 204–227 (Springer-Verlag, Berlin, 1979).

M. Kesteven, "On the mathematical theory of clock escapements," *American Journal of Physics* **46**, 125–129 (1978).

R. B. Kidd and S. L. Fogg, "A simple formula for the large-angle pendulum period," *The Physics Teacher* **40** (February), 81–83 (2002).

E. Kim and M. H. W. Chan, "Probable observation of a supersolid helium phase," *Nature* **427**, 225–227 (2004).

Charles Kittel, *Introduction to Solid State Physics*, 7th edition, (Wiley, New York, 1996).

C. Kittel, W. D. Knight, and M. A. Ruderman, *Mechanics*, (Berkeley Physics Series, vol. 1), (McGraw Hill, New York, 1962).

J. V. Kline, "The case of the counting balls," *American Journal of Physics*, **28**, 102–103 (1960).

Morris Kline, *Mathematical Thought from Ancient to Modern Times*, (Oxford University Press, New York, 1972).

A. N. Kolmogorov, "Entropy per unit time as a metric invariant of automorphisms," *Dokl. Akad. Nauk. SSSR*, **124**, 724 (1959). English summary in *Math. Rev.* **21**, 2035 (1959).

D. N. Langeberg, D. J. Scalapino, and B. N. Taylor, *Sci. Am.* (May), pp. 30–39, (1996).

Henry Charles Lea, *A History of the Spanish Inquisition, vol. 3.*, (MacMillan Co., New York, 1907.)

T. Leiber and H. Risken, "Stability of parametrically excited dissipative systems," *Physics Letters* **A129**, 214–218 (1988).

R. W. Levin, B. Pompe, C. Wilke, and B. P. Koch, "Experiments on periodic and chaotic motions of a parametrically forced pendulum," *Physica* **16D**, 371–384 (1985).

R. D. Levine and M. Tribus, *The Maximum Entropy Formalism*, (MIT Press, Cambridge, MA, 1979).

T. Li and J. A. Yorke, "Period three implies Chaos," *American Mathematical Monthly* **82**, 927–930 (1975).

R. L. Liboff, *Introductory Quantum Mechanics*, (Holden-Day, San Francisco, 1980).

Juan Antonio LLorente, *A Critical History of the Inquisition of Spain*, English ed., (John Lilburne Co., Williamstown, MA, 1823).

F. London, *Superfluids, Volume I: Macroscopic Theory of Superconductivity*, (Dover Publications, New York, 1961).

F. London and H. London, *Proceedings of the Royal Society (London)* **1**, 1175 (1935).

William H. Louisell, *Coupled Mode and Parametric Electronics*, (John Wiley and Sons, Inc., New York, 1960).

E. N. Lorenz, "Deterministic non-periodic flow," *Journal of Atmospheric Science* **20**, 130–140 (1963).

A. E. H. Love, *A treatise on the mathematical theory of elasticity* (1892) (Dover edition, New York, 1944).

Jun Luo, Zhong-Kun Hu, Xiang-Hui Fu, Shu-Hua Fan, and Meng-Xi Tang, "Determination of the Newtonian gravitational constant G with a nonlinear fitting method," *Physical Review* **D59** 042001–1 to 042001–6 (1998).

W. D. MacMillan, *Theoretical Mechanics: Statics and the Dynamics of a Particle*, (McGraw Hill, New York, 1927).

B. B. Mandelbrot, *Fractals—form, chance and dimension*, (1977, revised as *The fractal geometry of nature* W. H. Freeman, San Francisco, 1982).

E. Mathieu, "Memoire sur le mouvement vibratoire d'une membrane de forme elliptique," *Journal de Mathématiques pures et appliquees* **13**, 137–203 (1868).

J. Matisoo, "The superconducting computer," *Scientific American* **242** (May), 50–65 (1980).

J. Matricon and G. Waysand, *The Cold Wars: a history of superconductivity*, (Rutgers Univ. Press, New Brunswick, NJ, 2003).

M. R. Matthews, *Time for Science Education: How Teaching the History and Philosophy of Pendulum Motion can contribute to Science Literacy* (Kluwer Academic/Plenum Publishers, New York, 2000).

A. Matthiessen, *Reports of the British Association* **32**, 144 (1862).

D. E. McCumber, "Effect of ac impedance on dc voltage—current characteristics of superconductor weak-link junctions," *Journal of Applied Physics* **39**, 3113–3118 (1968).

D. G. McDonald, "Superconducting electronics," *Physics Today* **34** (February) 36–47 (1981).

D. G. McDonald, "The Nobel laureate versus the graduate student," *Physics Today* **54** (July), 46–51 (2001).

J. M. McKinley, "Brachistochrones, tautochrones, evolutes, and tessalations," *American Journal of Physics* **47**, 81–86 (1979).

N. W. McLachlan, *Theory and applications of Mathieu functions*, (Clarendon Press, Oxford, England, 1947).

J. T. McMullan, "On Initiating Motion in a Swing," *American Journal of Physics* **40**, 764–767 (1972).

J. Miles, "Directly forced oscillations of an inverted pendulum," *Physics Letters* **A133**, 295–297 (1988*a*).

J. Miles, "Resonance and symmetry breaking for the pendulum," *Physica* **D31**, 252–268 (1988*b*).

Z. Neda, E. Ravasz, T. Vicsek, Y. Brechet, and A. L. Barabasi, "Physics of the rhythmic applause," *Physical Review* **E61**, 6987–6992 (2000).

Robert A. Nelson and M. G. Olsson, "The pendulum—rich physics from a simple system," *American Journal of Physics* **54**, 112–121 (1986).

F. H. Newman and V. H. L. Searle, *The General Properties of Matter*, p. 138 (Edward Arnold, London, 1961).

Geoffrey. I. Opat, "The precession of a Foucault pendulum viewed as a beat phenonmenon of a conical pendulum subject to a Coriolis force,"*American Journal of Physics* **59**, 822–3 (1991).

E. Ott, C. Grebogi, and J. A. Yorke, "Controlling chaos," *Physical Review Letters* **64**, 1196–1199 (1990).

E. Ott and M. Spano, "Controlling chaos," *Physics Today* **48** (May), 34–40 (1995).

R. J. Ouellette and J. D. Rawn, *Organic Chemistry*, (Prentice Hall, Upper Saddle River, NJ, 1996).

U. Parlitz, "Estimating model parameters from time series by auto-synchronization," *Physical Review Letters* **76**, 1232–1235 (1996).

P. Parmananda, P. Sherard, R. W. Rollins, and H. D. Dewald, "Control of chaos in an electrochemical cell," *Physical Review* **E47**, R3003–R3006 (1993).

L. M. Pecora and T. L. Carroll, "Synchronization in chaotic systems," *Physical Review Letters* **64**, 821–824 (1990).

H. L. Pecseli, *Fluctuations in Physical Systems*, (Cambridge University Press, Cambridge, 2000).

R. D. Peters, "Student-Friendly Precision Pendulum," *The Physics Teacher* **37**, 390–393 (1999).

I. Peterson, *Newton's Clock: Chaos in the Solar System*, (Freeman, New York, 1993).

John Philips, *Proceedings of the Royal Society* **VI**(80), 78–82 (1851).

A. Pikovsky, M. Rosenblum, and J. Kurths, *Synchronization: A Universal Concept of Nonlinear Sciences*, (Cambridge University Press, Cambridge, 2001).

A. B. Pippard, "The inverted pendulum," *European Journal of Physics* **8**, 203–206 (1987).

A. B. Pippard, "The parametrically maintained Foucault pendulum and its perturbations," *Proceedings of the Royal Society (London)* **A420**, 81–91 (1988).

A. B. Pippard, "Foucault's Pendulum," *Proceedings of the Royal Institute* **63**, 87–100 (1991).
J. C. Piquette and Mu-Shiang Wu, "Comment on simple explanation of a well-known collision experiment," *American Journal of Physics* **52**, 83 (1984).
Edgar Allen Poe, *Complete Stories and Poems of Edgar Allen Poe*, 196–207 (Doubleday and Co., Inc., Garden City, NY, 1996).
H. Poincaré, *The Foundation of Science: Science and Method*, (1913), English translation, (The Science Press, Lancaster, PA, 1946).
S. D. Poisson, *Comptes Rendus, Académie des Sciences (Paris)* **5**, 660 (1837).
S. D. Poisson, "Sur le mouvement des Projectiles dans l'air, en ayant egard a la rotation de la Terre," *Journal de L'Ecole Polytechnique*, **26**, 1–176 (1838) and **27**, 1–50 (1839).
T. Pradhan and A. V. Khare, "Plane pendulum in quantum mechanics," *American Journal of Physics* **41**, 59–66 (1973).
W. H. Press, B. P. Flannery, S. A. Teukolsky, and W. T. Vetterling, *Numerical Recipes: The Art of Scientific Computing*, (Cambridge University Press, Cambridge, 1986).
I. Prigogine, *From being to becoming*, (Freeman, New York, 1980).
A. L. Rawlings, *The Science of Clocks and Watches*, 2nd edition, (Pitman and Sons, London, 1948).
R. H. Romer, "A double pendulum 'art machine', "*American Journal of Physics* **38**(9), 1116–1121 (1970).
R. Roy, T. W. Murphy, Jr., T. D. Maier, Z. Gills, and E. R. Hunt, "Dynamical control of a chaotic laser: experimental stabilization of a globally coupled system," *Physical Review Letters* **68**, 1259–1262 (1992).
R. Roy, Z. Gills, and K. S. Thornburg, "Controlling chaotic lasers," *Optics and Photonic News*, **5**, 8–15 (1994).
L. Ruby, "Applications of the Mathieu equation," *American Journal of Physics* **64**, 39–44 (1996).
J. R. Sanmartin, "O Botafumeiro: Parametric pumping in the Middle Ages," *American Journal of Physics* **52**, 937–945 (1984).
D. S. Saxon, *Elementary Quantum Mechanics*, (Holden Day, San Francisco, 1968).
E. O. Schulz-DuBois, "Foucault pendulum experiment by Kamerlingh Onnes and Degenerate Perturbation theory," *American Journal of Physics* **38**, 173–188 (1970).
B. B. Schwartz and S. Foner, *Superconductor Applications: SQUIDSs and Machines*, edited by Brian B. Schwartz and Simon Foner, proceedings of the NATO Advanced Study Institute on Small Scale Applications of Superconductivity, Gardone Riviera (Italy 1976), (Plenum Press, 1977).
D. Sciama, *The Unity of the Universe*, (Doubleday, New York, 1961).
A. S. Sedra and K. C. Smith, *Microelectronic Circuits*, 5th edition (Oxford University Press, New York, 2004).
Raymond A. Serway, Clement J. Moses, and Curt A. Moyer, *Modern Physics*, 77–93 (Saunders College Publishing, Philadelphia, 1989).
T. Shachtman, *Absolute Zero and the Conquest of Cold* (Houghton, Mifflin Co. New York, 1999).
D. Sher, "Foucault's Pendulum," *Journal of the Royal Astronomical Society of Canada* (*R. A. S. C. Jour.*) **63**, 227–228 (1969).
W. L. Shew, H. A. Coy, and J. F. Lindner, "Taming chaos with disorder in a pendulum array," *American Journal of Physics* **67**, 703–708 (1999).
T. Shinbrot, C. Grebogi, J. Wisdom, and J. A. Yorke, "Chaos in a double pendulum," *American Journal of Physics* **60**, 491–499 (1992).
A. E. Siegman, "Comments on pumping on a swing," *American Journal of Physics* **37**, 843–844 (1969).
A. D. C. Simpson, "The pendulum as the British length standard: a nineteenth-century legal aberration" in *Making Instruments Count: Essays on historical*

scientific instruments presented to Gerard L' Estrange Turner, edited by R. G. W. Anderson, J. A. Bennett, W. F. Ryan, (VARIORIUM, Ashgate Publishing, Aldershot, Hampshire, UK, 1993).

A. Smith, *Clocks and Watches* (The Connoisseur, London, 1975).

H. J. T. Smith and J. A. Blackburn, "Multiple quantum flux penetration in superconducting loops," *Physical Review* **B19**, 940–942 (1975).

H. J. T. Smith and J. A. Blackburn, "Experimental study of an inverted pendulum," *American Journal of Physics* **60**, 909–911 (1992).

H. J. T. Smith, J. A. Blackburn, and G. L. Baker, "Experimental observation of intermittency in coupled chaotic pendulums," *International Journal of Bifurcation and Chaos* **9**, 1907–1916 (1999).

H. J. T. Smith, J. A. Blackburn, and G. L. Baker, "When two coupled pendulums are equal to one: a synchronization machine," *International Journal of Bifurcation and Chaos* **13**, 7–18 (2003).

D. Sobel, *Longitude: the True Story of a Lone Genius who Solved the Greatest Scientific Problem of His Time* (Penguin, New York, 1996).

D. Sobel and W. J. H. Andrews, *The Illustrated Longitude* (Penguin Books, New York, 1998).

J. Starrett and R. Tagg, "Control of a chaotic parametrically driven pendulum," *Physical Review Letters* **74**, 1974–1977 (1995).

S. H. Strogatz, *Nonlinear dynamics and chaos* (Addison Wesley, Reading MA, 1994).

S. H. Strogatz, *Sync: The Emerging Science of Spontaneous Order*, (Hyperion, New York, 2003).

D. R. Stump, "Solving classical mechanics problems by numerical integration of Hamilton' equations," *American Journal of Physics* **60**, 1096–1100 (1986).

Y. Su, B. R. Heckel, E. G. Adelberger, J. H. Gundlach, M. Harris, G. L. Smith, and H. E. Swanson, "New tests of the universality of free fall"*Physical Review* **D50**, 3614–3636 (1994).

D. B. Sullivan and J. E. Zimmerman,"Mechanical analogs of time dependent Josephson phenomena," *American Journal of Physics* **39**, 1504–1517 (1971).

John L. Synge and Bryon A. Griffith, *Principles of Mechanics*, 3rd edition, chapters 12 and 13 (McGraw-Hill, New York, 1959).

H. Takayasu, *Fractals in the Physical Sciences*, (Manchester University Press, Manchester, 1990).

Jim Tattersall and Shawnee McMurran, "An interview with Dame Mary L. Cartwright, D.B.E., F.R.S.," *The College Mathematics Journal* **32**, 242–254 (2001).

L. W. Taylor, *Physics, the Pioneer Science*, (Houghton and Mifflin, Boston, 1941. Reprinted Dover, New York, 1959).

P. L. Tea and H. Falk, "Pumping on a Swing," *American Journal of Physics* **36**, 1165–1166 (1968).

J. J. Thomsen and D. M. Tcherniak, "Chelomei's pendulum explained," *Proceedings of the Royal Society* (*London*) **A457**, 1889–1913 (2001).

K. S. Thornburg, Jr., M. Moller, R. Rajarshi Roy, T. W. Carr, R.-D. Li, and T. Erneux, "Chaos and coherence in coupled lasers," *Physical Review* **E55**, 3865–3869 (1997).

W. Tobin and B. Pippard, "Foucault, his pendulum and the rotation of the earth," *Interdisciplinary Science Reviews* **19**(4), 326–337 (1994).

William J. Tobin, *The Life and Science of Leon Foucault*, (Cambridge University Press, Cambridge, 2003).

Y. Ueda, "Strange attractors and the origin of chaos," in *Chaos Avant-Garde: Memories of the Early Days of Chaos Theory* (World Scientific Publishing Co., Singapore, 2000).

T. Van Duzer and C. W. Turner, *Principles of Superconductive Devices and Circuits* (Elsevier, New York, 1981).

G. D. VanWiggeren and R. Roy, "Communication with chaotic lasers," *Science* **279**, 1198–1200 (1998).
F. A. Vening Meinesz, *Theory and Practice of Pendulum Observations at Sea* (Publication of the Netherlands Geodesic Committee, Waltman, Delft, 1929).
Product of Visual Numerics, Inc. Houston, Texas.
J. Walker, "The Amateur Scientist," *Scientific American* **260** (March), 106–109 (1989).
D.-H. Wang, X.-L. Wang, L. Zhao and L.-C. Tu, "Eddy current loss testing in the torsion pendulum," *Physics Letters* **A290** 41–48 (2001).
F. A. B. Ward, *Time Measurement: Historical Review* (Science Museum, London, 1970).
A. S. Weigend and N. A. Gershenfeld, (Eds.) *Time Series Prediction: Forecasting the Future and Understanding the Past* (Addison-Wesley, Reading MA, 1994).
C. Wheatstone, "Note relating to M. Foucault's new mechanical proof of the rotations of the earth," *Proceedings of the Royal Society* **6**(80), 65–68 (1851).
Robert J. Whitaker, "A note on the Blackburn pendulum," *American Journal of Physics* **59**(4), 330–333 (1991).
Robert J. Whitaker, "Harmonographs I. Pendulum design," *American Journal of Physics* **69**, 162–173 (2001); "Harmonographs II. Circular design," *American Journal of Physics* **69**, 174–183 (2001).
L. Pierce Williams, *Album of Science*, vol. **3** (Charles Scribner's Sons, New York, 1978).
S. Wirkus, R. Rand, and A. Ruina, "How to pump a swing," *College Mathematics Journal* **29**, 266–75 (1998).
W. Wolf, J. B. Swift, H. L. Swinney, and J. A. Vastano, "Determining Lyapunov exponents from a time series," *Physica* **16D**, 285–317 (1985).
Philip Woodward, *My Own Right Time* (Oxford University Press, Oxford, 1995).
A. M. Zhabotinskii, "A History of chemical oscillations and waves," *Chaos* **1**, 379–86 (1991).
D. G. Zill and M. R. Cullen, *Differential equations with boundary value problems* (PWS-Kent, Boston, 1993).

Index